Homological Algebra
In Strongly Non-Abelian Settings

Homological Algebra

In Strongly Non-Abelian Settings

Marco Grandis

Università di Genova, Italy

World Scientific

NEW JERSEY · LONDON · SINGAPORE · BEIJING · SHANGHAI · HONG KONG · TAIPEI · CHENNAI

Published by

World Scientific Publishing Co. Pte. Ltd.

5 Toh Tuck Link, Singapore 596224

USA office: 27 Warren Street, Suite 401-402, Hackensack, NJ 07601

UK office: 57 Shelton Street, Covent Garden, London WC2H 9HE

British Library Cataloguing-in-Publication Data
A catalogue record for this book is available from the British Library.

ISBN 978-981-4425-91-9

Printed in Singapore by World Scientific Printers.

To my granddaughters

Alessandra and Elena

Preface

We propose here a study of 'semiexact' and 'homological' categories as a basis for a generalised homological algebra. Our aim is to extend the homological notions to deeply non-abelian situations.

This is a sequel of a book [G20], referred to as Part I, and achieves its goals but can be read independently of the latter.

Part I develops homological algebra in p-exact categories, i.e. exact categories in the sense of Puppe and Mitchell [P2, Mt, HeS, AHS, FS] - a moderate generalisation of abelian categories that is nevertheless crucial for a theory of 'coherence' and 'universal models' of (even abelian) homological algebra. The main motivation of the present, much wider extension is that the exact sequences or spectral sequences produced by unstable homotopy theory cannot be dealt with in the previous framework.

As in Part I, our research follows a sort of 'projective' approach, where lattices (and sublattices) of normal subobjects play an important role. The properties of modularity or distributivity of these lattices are crucial, in contrast with the abelian case where modularity is automatically granted and distributivity is impossible at all.

The role of additivity is more important here than in Part I. Indeed, while the *additive* (or even semiadditive) p-exact case reduces to the abelian one, and was not developed in Part I, additive homological categories need not be abelian. Homotopies of chain complexes on these categories are considered here, if in a marginal way; in fact, this study is mainly related to such constructions as mapping cylinder and suspension, derived and triangulated categories, that should rather be viewed as a part of *homotopical algebra*, and need an appropriate investigation.

According to the present definitions, a semiexact category is a category equipped with an ideal of 'null' morphisms and provided with kernels and cokernels *with respect to this ideal*. A homological category satisfies some further conditions that allow the construction of subquotients and induced morphisms, in particular the homology of a chain complex or the spectral sequence of an exact couple. Both settings are self-dual.

These notions have been introduced in [G10, G11, G13, G14]. Extending abelian categories, and also the p-exact ones, they include the usual *domains* of homology and homotopy theories, e.g. the category of 'pairs' of topological spaces or groups; they also include their *codomains*, since the sequences of homotopy 'objects' for a pair of pointed spaces or a fibration can be viewed as exact sequences in a homological category, whose objects are actions of groups on pointed sets.

Our view is thus quite distinct from the more usual *affine* generalisations of the abelian framework, that are based on finite limits: Barr-exact categories, their extensions and variations. There seems to be a sort of opposition between a projective and an affine approach, that will be further analysed in the Introduction.

Contents

Introduction

0.1 Categorical settings for homological algebra

The core of homological algebra can be described as the study of exact sequences and of their preservation properties by functors. It was established in *categories of modules* (see Cartan and Eilenberg [CE]) and extended to *abelian categories* (Grothendieck [Gt]), with formal advantages (e.g. duality) and a concrete extension of its domain (to sheaves, Serre's quotients [Ga], etc.).

Many homological procedures can be freed from additivity and extended to a *p-exact category*, i.e. an exact category in the sense of Puppe and Mitchell [P2, Mt, BP, GrV1, GrV2, HeS, AHS, FS]. This is a category with zero object, where every map has a kernel and a cokernel, and factorises as a normal epi followed by a normal mono.

Such extensions are dealt with in a previous book [G20], that will be cited here as Part I. Developing an approach of the present author (see [G1]-[G9]), Part I shows that the setting of p-exact categories permits a notion of *distributive homological algebra*, which cannot be formulated in the abelian framework. On the one hand, distributivity is at the core of *coherence of induced morphisms*, as we prove in a Coherence Theorem for homological algebra (Theorem 2.7.6 of Part I); on the other, it yields useful universal models for 'distributive' theories, namely for spectral sequences; these models - a formalisation of Zeeman diagrams [Ze, HiW] - live in the distributive p-exact category $\mathcal{I}$ of sets and partial bijections.

Yet, this extension is far from covering all the situations in which exact sequences are considered; the main exception is probably given by the homotopy sequences of a pair of pointed spaces or of a Serre fibration, which are not even confined to the category Gp of groups but degenerate in low degree into pointed sets and actions of groups on the latter. Therefore, their 'exactness' is usually described and studied step by step (cf. [HaK],

1.1), even in complicated situations as the homotopy spectral sequence of a tower of fibrations ([BouK], IX.4) or of cofibrations ([Bau], III.2).

From a concrete viewpoint the present framework - semiexact and homological categories - includes, besides all p-exact categories, various categories of interest in algebraic topology, generally lacking a zero object but provided with a natural ideal of 'null morphisms', and kernels and cokernels with respect to the latter. From a formal viewpoint, it suffices to define and study the basic homological notions.

This setting has a similarity with that studied by Lavendhomme and Van Den Bossche [La, LaV, Vb], mostly developed for 'categories of pairs'. C. Ehresmann also considered kernels with respect to an ideal, in connection with the cohomology of categories ([Eh2], p. 546-547; see also the Comments, p. 845-847).

The hierarchy of categorical settings we are using here:

(a) *semiexact, homological, generalised exact, Puppe-exact category,*

should not be confused with a different, better known system based on the existence of finite limits:

(b) *Barr-exact* [Ba], *protomodular* (Bourn [Bou1]), *Borceux-Bourn homological* [BoB, BoC] and *semiabelian category* in the sense of Janelidze, Márki and Tholen [JMT, Bo2, Bou2] or in the sense of Raĭkov [Ra]. (The definition of Barr-exact category is recalled in the Appendix, in A2.6.)

Such settings are appropriate to investigate other non-abelian properties, where the existence of finite limits is important, and for the study of algebraic varieties.

It would be good to have a clearer understanding of the cleavage between these two approaches. As already discussed in Part I, the present approach can be called 'projective', at the light:

- of the crucial use of lattices of (normal) subobjects and their direct and inverse images (in analogy with the fact that a projectivity between two groups is an isomorphism between their lattices of subgroups);

- of the construction of the projective category associated to a p-exact category (see Part I, Section 2.3 or [G4]), of its extension to homological categories given here (in 2.1.4), and of the characterisation of the projective categories associated to the abelian ones established by Carboni and the author in [CaG].

Line (b) might perhaps be called 'cartesian', because of the weight it gives to finite limits. Or also 'affine', by opposition with 'projective' and because of Carboni's characterisation of 'categories of affine spaces' [Ca1].

We refer to [BoG] for a different setting in line (a), based on Burgin's

γ-categories. There is also a (categorical) notion of 'homological monoid', introduced by Hilton and Ledermann [HiL] (a reference that the author owes to G. Janelidze). Finally, in the domain of homotopical algebra, there is a notion of *Quillen-exact* category (see A2.6), introduced by Quillen [Qu] as a foundation of higher K-theory.

One can find in Part I, 1.3.1, a digest of the various meanings that the term 'exact category' has assumed in category theory, homological algebra and algebraic topology. Literature on homological algebra is referred to in Section 8 of the Introduction of Part I.

0.2 Semiexact, homological and generalised exact categories

The first two chapters introduce the categorical settings used here.

Our basic notion, a *semiexact* category, or *ex1*-category, is a category *equipped* with an ideal of 'null morphisms' and provided with kernels and cokernels *with respect to this ideal* (Section 1.3). A semiexact category is said to be *pointed*, or *p-semiexact*, if it has a zero object (both initial and terminal) and its null morphisms coincide with the zero morphisms; then, kernels and cokernels have the usual (categorical) meaning.

For instance, let us consider the standard domain of Eilenberg-Steenrod homology theories, namely the category Top_2 of *pairs* (X, A) of topological spaces, in the usual sense: X is a space and A is a subspace of X; a morphism $f\colon (X, A) \to (Y, B)$ is a *map of pairs*, i.e. a continuous mapping $f\colon X \to Y$ that takes A into B. Let us recall that the pair (X, A) is often read, and thought of, as X *modulo* A, while a space X is identified with the pair $(X, \emptyset)$.

Top_2 is not pointed, but has a natural ideal of *null* morphisms, the maps $f\colon (X, A) \to (Y, B)$ such that $f(X) \subset B$ (weak inclusion), i.e. the maps which factorise through the *null* pairs (X, X). Top_2 is semiexact with respect to this ideal: kernels and cokernels exist and are easily calculated (cf. 1.4.2). Every short exact sequence is of the following type (up to isomorphism), for a 'triple' $X \supset A \supset B$ of spaces:

$$(A, B) \rightarrowtail (X, B) \twoheadrightarrow (X, A). \tag{0.1}$$

Now, the same sequence, for $B = \emptyset$, shows that the pair (X, A) *is indeed the quotient* X/A *in the present semiexact structure*: namely, the cokernel of the embedding of $A = (A, \emptyset)$ in $X = (X, \emptyset)$.

The categories Set_2 and Gp_2 of pairs of sets and groups respectively, have a similar semiexact structure. Abstract categories of 'pairs', over a category with assigned distinguished subobjects, are constructed in Section 1.6.

The category Set_2 is regular and not Barr-exact (see 1.4.6). But it is more important to notice that the short exact sequence (0.1), *determined by the assigned ideal of null morphisms*, has nothing to do with the short exact sequences of Set_2 *in the sense of regular categories* - characterised in 1.4.6(c). Top_2 is not even regular, essentially because Top is not.

In a semiexact category E, every morphism has a *normal factorisation* through its normal coimage and its normal image; exact sequences and exact functors can be introduced; normal subobjects and normal quotients form anti-isomorphic lattices. Direct and inverse images of normal subobjects can be organised as a *transfer functor* Nsb: $\mathsf{E} \to \mathsf{Ltc}$, with values in the category of lattices and Galois connections (Section 1.5).

A semiexact category is *ex2* if its normal monos and normal epis are closed under composition; it is *homological* (or *ex3*) if, moreover, the construction of subquotients is possible: given two normal subobjects $N \leqslant M$ of A, the composed map $M \rightarrowtail A \twoheadrightarrow A/N$ is an *exact* morphism (i.e. factorises normal epi - normal mono), which allows us to define the subquotient M/N of A as the central object of this factorisation. All the previous examples, in this section, are homological categories. Ltc is pointed homological; we shall see that it is semiadditive (see 1.2.7), with an idempotent sum (therefore not additive). The category Gp of groups is just semiexact: but *homological facts concerning groups are here studied in larger, homological categories*, like Gp_2 (cf. 1.6.1).

More particularly, a *g-exact*, or *generalised exact* category, or *ex4*-category (Section 2.2), is here a semiexact category whose maps are all exact; then the null morphisms are identified by the mere categorical structure (as the maps which factorise through the quasi-zero object of their connected component, see 2.2.6). A p-exact category is thus the same as a pointed g-exact category; the gap between these two notions is discussed in 2.2.7.

A semiexact category E is said to be *modular* (or *distributive*) if its transfer functor actually takes values in the p-exact category Mlc (resp. Dlc) of modular (resp. distributive) lattices and modular connections (Section 2.3), which has been used in Part I as the codomain of the transfer functor of p-exact (resp. distributive p-exact) categories. In the modular case, *elementary* diagram lemmas can be proved, by diagram chasing of normal subobjects (2.3.5-2.3.7). Every g-exact category is modular (cf. 2.3.3); an abelian one (more generally, an additive p-semiexact category) cannot be distributive unless it is trivial (reduced to zero-objects): see 2.4.4.

Chapter 2 ends with a study of the epi-mono completion $\mathsf{Fr}\,\mathsf{C}$ of a category C, in Sections 2.6-2.8. It is a generalisation of the well-known Freyd embedding of a *stable* homotopy category into an *abelian* category [F1]-[F3]: *without assuming stability*, a homotopy category C produces a *homological* category, as we shall prove in Theorem 2.7.6. But the interest of this construction is diminished by the fact that Fr C is not well-powered, in general, as briefly discussed in Section 2.8.

0.3 Subquotients and homology

Chapter 3 shows that homological categories form a natural framework in order to study subquotients and their (regularly) induced morphisms.

Induction is the key for 'non-elementary' diagram chasing in homological categories, replacing to a certain extent the use of the categories of relations for g-exact categories. For instance, this tool allows one to construct the connecting morphism (3.3.3-3.3.4) and to get a homology sequence for any short exact sequence of chain complexes over any homological category E (3.3.5).

This homology sequence is always *of order two.* It is *exact*, for every short exact sequence of complexes, if and only if E is *modular* (Theorem 3.3.6). However, for a particular short exact sequence of complexes, specific modularity conditions can be given in order to get an exact homology sequence (Theorem 3.3.5); further, a very simple condition on the middle complex (the exactness of its differential) gives a partial exactness property for the homology sequence (3.3.5(b)), sufficient for studying the universality of chain homology (Section 4.5).

Finally, Massey's exact couples [Mas1] can be considered in any homological category (Section 3.5); their derived couples are still exact and produce the associated spectral sequence.

0.4 Satellites

Connected sequences of functors and satellites are introduced for semiexact categories, together with their limit-construction as pointwise Kan extensions (Chapter 4).

Thus, the classical construction of the left satellite of an additive functor $F \colon \mathsf{E} \to \mathsf{E}'$ between categories of modules [CE] has various extensions, including the non-additive case of categories of pairs (Section 3.5).

Satellites can be *detected* by means of effacements, as in the classical abelian case treated by Grothendieck [Gt]. The extension is easier for *exact* sequences of functors with values in a *g-exact* category (4.3.1-4.3.3), but can also be done in the general case of a *connected* sequence of functors with values in a *semiexact* category (Sections 4.3.5-4.3.6).

As an application, the sequence of chain (or cochain) homology functors is proved to be universal for all abelian categories and also for some interesting homological, non-abelian ones (Section 4.5); these include: all the additive homological categories, the p-exact category $\mathcal{I}$ of sets and partial bijections, the distributive homological category Inj_2 of pairs of sets and injective mappings of pairs, the p-exact category Mlc of modular lattices and modular connections.

Notice that the additive homological categories Ban and Hlb (of Banach and Hilbert spaces) are not modular, so that their connected sequence of homology functors is not exact. A deeper study of the homological categories whose sequence of chain homology functors is universal could be of interest.

0.5 Exact centres, expansions, fractions and relations

In chapter 5 we deal with some universal constructions which can be performed in the present settings.

Thus, the *exact centre* $\mathrm{Exc}\,\mathsf{E}$ of an ex2-category is formed by its exact objects and morphisms; for instance, $\mathrm{Exc}\,\mathsf{Ltc} = \mathsf{Mlc}$. This procedure satisfies a universal property: technically, it is the 'coreflector' (or right adjoint) of the embedding of g-exact categories into ex2-categories (for exact functors which are full with respect to normal subobjects). It is dealt with in Section 5.1 with other 'coreflective' constructions (right adjoints of inclusions), such as the modular centre, or the modular expansion, or the exact expansion of an ex2-category.

Then Section 5.2 studies the categories of *ex1-fractions*, solving the universal problem of inverting assigned morphisms *within* semiexact categories and exact functors. These fractions lack a binary calculus and even a bounded one (excepting the g-exact case), which makes them rather awkward to handle. However, various 'reflective' constructions (left adjoints of inclusions of 2-categories) can be obtained by these procedures, like the g-exact quotient of a semiexact category (Section 5.3).

Relations for p-exact categories have been studied in Section 1.6 of Part I; their construction is easily extended to g-exact categories, in Section 5.4. They yield a three-arrow calculus for fractions, already studied in detail in [G12] for p-exact categories and extended here to the g-exact case (Section 5.5).

In particular, this provides *a homological structure on the whole category* EX_4 *of g-exact categories*, and on the category AB of abelian categories (Section 5.6).

0.6 Applications

Chapter 6 deals with applications to Algebraic Topology.

Taking into account the aforementioned characterisation of short exact sequences in the homological category Top_2 (Section 2 of this Introduction), the first four axioms of Eilenberg-Steenrod [ES] for a homology theory $H = (H_n, \partial_n)$ defined on all topological pairs amount to saying that H is an *exact sequence* of functors from Top_2 to some category of modules.

More generally, homology theories can be defined over a *homological basis* - a semiexact category equipped with a 'homotopy' relation and 'excision' maps (Section 6.1). Transformations of these bases, that allow one to 'pullback' the theories, are considered; in this way, the obvious adjunction between the homological categories Top_2 and $\mathsf{Top}_\bullet$ (of pointed spaces) produces an equivalence between relative theories satisfying the relative-homeomorphism axiom (over Top_2) on the one hand, and single-space theories (over $\mathsf{Top}_\bullet$) on the other; the argument restricts to compact Hausdorff spaces, encompassing single space homology theories for locally compact spaces (Section 6.1).

A singular homology theory for *fibrations* is proposed in Section 6.2. Restricting it to *covering maps*, we do obtain an exact sequence of functors over a homological category, but the theory works for *weak* fibrations (in a more general sense than Serre's) which just form a 'category with distinguished short exact sequences', a (non-additive) notion related to Quillen-exact categories (see A2.6) which might deserve to be studied.

In Section 6.3, we show that a tower of fibrations between *path connected* spaces defines a homotopy exact couple in the homological category Ngp of *normalised groups* - a category of fractions of Gp_2 obtained 'by excision of the invariant subgroups' and similar to the category of 'homogeneous spaces' of Lavendhomme [La]. The interest of using Ngp is to make *exact* all the morphisms which intervene in the exact couple, since they fail to be exact in Gp_2. Thus, the theory developed here (Section 3.5) applies, producing the spectral sequence of the tower of fibrations.

The next two sections remove the restriction to path connected spaces. Relative homotopy can be seen as an exact sequence of functors from the homological category $\mathsf{Top}_{\bullet 2}$ of pairs of pointed spaces to the homological category Act of *actions* of groups on pointed sets. In this category an object (X, S) consists of a pointed set X with a group S acting on it; a morphism $f = (f', f'') : (X, S) \to (Y, T)$ is formed by a map $f' : X \to Y$ of pointed sets together with a group-homomorphism $f'' : S \to T$ consistent with the former; f is assumed to be null whenever f' is a zero-morphism (Section 6.4).

Finally, in Section 6.5, we introduce a homological category of fractions $\mathsf{Nac} = \Sigma^{-1}\mathsf{Act}$, *the category of normalised actions*, which is adequate for homotopy sequences and exact couples of topological spaces that are not assumed to be path-connected.

Another example, dealt with in [G13], is not recalled here: the Massey relative cohomology groups for pairs of groups $H^n(X, Y)$ [Mas2, Ta, Ri] produce an exact sequence of functors $H^n \colon \mathsf{Gp}_2 \to \mathsf{Ab}$ defined on the homological category Gp_2 of pairs of groups.

0.7 Homological theories and biuniversal models

In Section 7.2 of the last chapter we define homological theories and their biuniversal models, similar to the *proper* EX-theories of Part I.

Then, in the next three sections of Chapter 7, we study three homological theories:

- the *modular bifiltration*,

- the *modular sequence of morphisms*,

- the *modular exact couple*,

and construct their biuniversal model in the distributive homological category Inj_2 of pairs of sets and injective mappings of pairs.

As in Part I, the well-known Birkhoff theorem on free modular lattices, recalled in 7.3.1, plays a crucial role. We give here, in Theorem 7.3.5, a generalised version that can be applied - for instance - to groups equipped with two filtrations, one of them made of invariant subgroups.

0.8 Modularity and additivity

The importance of distributivity for p-exact categories is already stressed in Part I (and has been since [G5]-[G7]). The importance of modularity for homological categories now shows; it is relevant for diagram chasing and characterises the homological categories whose chain-homology functors form an exact sequence (see Section 3 of this Introduction).

These lattice-theoretic properties seem to have had little acknowledgement in homological algebra: a consequence of the prevalent abelian setting where the first property is automatically granted, so that its effects cannot be clearly distinguished, while the second is completely prevented, since a non-trivial abelian category cannot be distributive (Proposition 2.4.4).

On the other hand, additivity has here a secondary interest (cf. Sections 2.4, 3.4, 4.2, 4.5), but not a negligible one. It should be noted that, while the *semiadditive* (or additive) *p-exact* case reduces to the classical abelian setting (see Part I, Theorem 2.1.5), and was not developed in Part I, a semiadditive (or additive) *homological* category need not be abelian, nor even p-exact: for instance, Ltc is semiadditive homological, with an idempotent sum of maps (see 1.2.7), while Ban, the category of Banach spaces and continuous linear maps, is additive homological.

Chain complexes over a semiadditive homological category and their homotopies are briefly studied in Section 3.4. This gives rise to a homotopical structure and - in our opinion - should be viewed as a part of homotopical algebra.

According to this point of view, already developed in [G14], the usual study of chain complexes over abelian categories can be split in two different extensions: on the one hand, chain complexes on a homological category, which form a *homological* category and can be studied via kernels, cokernels and subquotients (as we are doing here); on the other hand, chain complexes on a merely additive category, which form a *homotopical* category and can be investigated by homotopical means, as mapping cone, suspension, Puppe sequence... (see [G18] and references therein). Chain complexes over additive homological categories, including the abelian ones, form the intersection of the two theories.

Analogously, homology theories in the sense of Eilenberg-Steenrod form an exact sequence of functors on the homological category of pairs of spaces, which is essentially a homological notion (Section 6.1), while their 'absolute' presentation based on the suspension endofunctor [DT, Ke1] is an entirely homotopical topic.

0.9 A list of examples

The categories Gp of groups and Rng of rings are p-semiexact (see 1.6.1), not homological: their normal monomorphisms are not stable under composition.

Similarly, the category Gpd of small groupoids is semiexact and not homological, with respect to the ideal of functors whose image is discrete, i.e. consists of identities (1.6.1).

The following categories are p-homological and 'are not' p-exact (or 'need not be', when depending on a parameter):

(a) Ltc: lattices and Galois connections (see 1.4.1),

(b) Abm: abelian monoids (or $\mathbb{N}$-semimodules) (1.6.2),

(c) pAb: preordered abelian groups (1.6.2),

(d) R Smd: semimodules over a unital semiring R (1.6.2),

(e) topological (or Hausdorff) K-vector spaces, topological (or Hausdorff) abelian groups (1.6.3),

(f) Ban, Hlb: Banach or Hilbert spaces and continuous linear mappings (1.6.3),

(g) Ban_1: Banach spaces and linear weak contractions (1.7.4(e), 2.2.9),

(h) Set., Top., Cph.: pointed sets, pointed spaces, pointed compact Hausdorff spaces (1.6.4),

(i) $\mathcal{S}$: sets and partial mappings (equivalent to Set.) (1.6.5),

(j) $\mathcal{C}$: spaces and continuous partial maps (1.6.5),

(k) $\mathcal{T}$: spaces and continuous partial maps, defined on open subspaces (1.6.5),

(l) $\mathcal{T}_{LC}$: locally compact Hausdorff spaces and partial proper maps, defined on open subspaces (equivalent to Cph. via one-point compactification; see 1.7.4(d), 2.2.9),

(m) Fr C: the 'epi-mono completion' of a category C with weak kernels and weak cokernels (Section 2.7).

The categories in (a) - (d) are *semiadditive*, i.e. provided with finite biproducts (that yield a sum of parallel maps, in the usual way); those in (e) - (f) are even additive. This fact will be of interest in Chapter 4, namely in order to prove that the homology functors H_n of chain complexes are left satellites of H_0, in cases (e) - (f).

The following categories are homological, not g-exact (in general), with respect to a specified set of null objects (i.e. with respect to the ideal of morphisms which factorise through them):

(n) Mod: modules (M, R) over arbitrary unital rings, with respect to the *null* modules $(0, R)$ (1.6.6),

(o) Set_2 Top_2, Cph_2, $Top._2$ and Gp_2, with respect to the null pairs (X, X) (1.4.2, 1.6.7, 1.7.4),

(p) C_2: *pairs* of a category C with distinguished subobjects, w.r.t. the identities (Section 2.5),

(q) Cvm: the category of *covering maps*, with respect to homeomorphisms (Section 6.2),

(r) Act: actions of groups on pointed sets, with respect to actions on the singleton (Section 6.4),

(s) the category Ch.E of chain complexes on the homological category E, with respect to the morphisms whose components are null in E (Section 3.3),

(t) the category E^S of functors $S \to E$, where S is a small category and E is a homological category, with respect to the functors which annihilate all the objects of S. (Moreover, E^S has the same exactness properties as E, see 1.7.7).

In particular, every variety of algebras $C = Alg(\Omega, \Sigma)$ gives a homological category of pairs C_2 (see 2.5.1). Augmented rings, supplemented algebras and Lie algebras can be dealt with in convenient semiexact categories of pairs (see 2.5.7).

The category GrAb of graded abelian groups $(A_n)_{n\in\mathbb{N}}$ and morphisms of *any* degree $r \geqslant 0$, is *generalised exact* with respect to the ideal of morphisms which factorise through $(0)_{n\in\mathbb{N}}$ (see 1.4.4); notice that the latter is *not* a zero object (it is quasi-zero, in the sense of 1.3.2) and GrAb is not exact in the sense of Puppe.

Finally, the category EX of p-exact $\mathcal{U}$-categories (for some universe $\mathcal{U}$) and exact functors is homological with respect to the ideal of exact functors which annihilate every object. The normal subobjects coincide with the thick subcategories, the normal quotients with the p-exact categories of fractions. The same holds for g-exact and abelian $\mathcal{U}$-categories (5.6.3, 5.6.4).

0.10 Terminology and notation

The categorical terminology used here is standard, as in Part I and Mac Lane's text [M5]; the main points are briefly reviewed in an appendix, Chapter A.

A universe $\mathcal{U}$, whose elements are called *small* sets, is fixed throughout. A $\mathcal{U}$-*category* has objects and arrows belonging to this universe. The concrete categories we consider are generally large $\mathcal{U}$-categories, like the category Set of small sets (and mappings), or Top, of small topological spaces (and continuous mappings), or Ab of small abelian groups (and homomorphisms), or Gp of small groups. CAT denotes the 2-category of all $\mathcal{U}$-categories, with their functors and their natural transformations; it is a $\mathcal{V}$-category, for any universe $\mathcal{V}$ to which $\mathcal{U}$ belongs.

As in Part I, we do not assume that a $\mathcal{U}$-category has small hom-sets. In fact, the hom-functors do not play a relevant role here, while the subobject-functors do. Hence we are more interested in other conditions of 'local smallness', for instance the fact of being *normally* well-powered for semiexact and homological categories (1.5.1); this amounts to being well-powered in the case of a g-exact category, where all subobjects are normal.

ObC and MorC denote the set of objects and morphisms of a category C. ⊤ and ⊥ denote respectively the terminal and initial object; the zero object, that is both initial and terminal, is usually written as 0. C^{op} stands for the opposite category, with reversed arrows. $F \dashv G$ means that the functor F is left adjoint to the functor G. A natural transformation φ between the functors $F, G\colon \mathsf{C} \to \mathsf{D}$ is also written as $\varphi\colon F \to G\colon \mathsf{C} \to \mathsf{D}$.

Our conventions about subobjects (as *selected* monomorphisms) and quotients (as *selected* epimorphisms) can be found in A1.6. An arrow $\rightarrowtail$ always denotes a *normal* mono (i.e. a kernel of some morphism, with respect to an assigned ideal of null morphisms), while $\twoheadrightarrow$ stands for a normal epimorphism.

General monomorphisms and epimorphisms have here a marginal interest, except of course in the case of g-exact categories where they coincide with the normal ones.

An ordered set is identified with the corresponding category, and an increasing mapping $f\colon X \to Y$ with the corresponding functor. The symbols **1** and **2** denote the categories associated to the corresponding ordinals. A lattice is assumed to have a least and a greatest element: see Section 1.1.

If X is an ordered set, X^{op} is the opposite one (with reversed order). If $a \in X$, the symbols $\downarrow a$ and $\uparrow a$ respectively denote the downward and upward closed subsets of X determined by a:

$$\downarrow a = \{x \in X \mid x \leqslant a\}, \qquad \uparrow a = \{x \in X \mid x \geqslant a\}. \tag{0.2}$$

The symbols $\subset, \supset$ denote *weak* inclusion.

Subsections whose reference number is marked with a star, like 1.4.5*, are addressed to readers with specific interests in category theory, and are not used elsewhere. The reference I.1.2 applies to Part I, Chapter 1, Section 2, while I.1.2.3 applies to Subsection 3 therein.

0.11 Acknowledgements

Part I of this work is dedicated to the memory of Gabriele Darbo, my professor and dear friend, who introduced me to algebraic topology, homological algebra and category theory in the 1960's.

I would like to thank the colleagues with which I had occasions of discussing parts of this work, like F. Borceux, George and Zurab Janelidze (for the first chapters), F.W. Lawvere and P. Freyd (for weak subobjects, in Chapter 2), S. Mardešić and T. Porter (for parts of Chapter 6), F. de Giovanni (for parts of Chapter 7). Their interest, questions and remarks have been of great help.

All 'crossword diagrams', in Chapter 7, have been typeset with a specific - quite effective - package 'zeeman.tex', prepared by F. Borceux when we were writing the paper [BoG]; the other diagrams are composed with 'xy-pic', by K.H. Rose and R. Moore. Both packages are free.

This work was supported by PRIN Research Projects and by research grants or contracts of Università di Genova.

1

Semiexact categories

After a description of the category Ltc of lattices and Galois connections, and an elementary study of its exactness properties (Section 1.2), we introduce in Section 1.3 our main definitions: semiexact, homological and generalised exact categories. All these notions are self-dual.

Various examples of semiexact and homological categories are presented in Sections 1.4 and 1.6. Semiexact categories are studied in Section 1.5, and exact functors between them in Section 1.7; the study of homological categories is deferred to the next chapter.

In every semiexact category E, normal subobjects, with their direct and inverse images, produce a 'transfer functor' Nsb: E → Ltc (see Theorem 1.5.8) with values in the aforementioned category of lattices (that is homological).

This transfer functor is a prominent tool of the present analysis; it is also the natural extension to semiexact categories of a tool developed in Part I: the transfer functor of subobjects Sub: E → Mlc of a p-exact category (where all subobjects are normal), that takes values in the category of modular lattices and modular connections, a p-exact subcategory of Ltc.

1.1 Some basic notions

This elementary review of basic topics, like lattices, monomorphisms and epimorphisms, is meant to form a common basis for readers with different backgrounds. Most of these things can also be found in Part I [G20], but we want to make the present book essentially independent of the former.

1.1.1 Lattices

Lattices of substructures and lattices of quotients will play an important role. We recall here some basic facts about the theory of lattices; the interested reader is referred to the classical texts by Birkhoff and Grätzer [Bi, Grz].

Classically, a lattice is defined as a (partially) ordered set X such that every pair x, x' of elements has a *join* $x \vee x' = \sup\{x, x'\}$ (the least upper bound) and a *meet* $x \wedge x' = \inf\{x, x'\}$ (the greatest lower bound).

But in this book (as in Part I) it is more convenient to adopt a slightly stronger definition: *lattice* will always mean an ordered set with *finite* joins and meets. This is equivalent to requiring the existence of binary joins and meets *together* with the least element $0 = \vee \emptyset$ (the empty join) and the greatest element $1 = \wedge \emptyset$ (the empty meet). These bounds are the unit of the join and meet operations, respectively; they coincide in the one-point lattice, and only there. A lattice is said to be *complete* if every subset has a supremum, or - equivalently - every subset has an infimum.

In the lattice $L(A) = \mathrm{Sub}A$ of subgroups of an abelian group A, the meet of two subgroups H, K is their intersection $H \cap K$, and the join is their 'sum'

$$H + K = \{h + k \mid h \in H, \, k \in K\}.$$

But we always use the notation $H \wedge K$ and $H \vee K$, for the sake of uniformity with lattices of subobjects in an abelian (or p-exact) category. For the same reason we prefer to write this lattice as $\mathrm{Sub}A$, even though the notation $L(A)$ is more frequently used.

Consistently with our choice of terminology, a *lattice homomorphism* has to preserve all *finite* joins and meets; a *sublattice* of a lattice X is closed under such operations, and thus has the same least and greatest element as X.

Occasionally, we speak of a *quasi lattice* when we only assume the existence of *binary* joins and meets; a homomorphism of quasi lattices preserves them. A *quasi sublattice* Y of a quasi lattice X is closed under binary joins and meets in X; therefore, if the latter is a lattice, Y may have different minimum or maximum, or none.

For instance, if X is a lattice and $a \in X$, the downward and upward closed subsets of X generated by a

$$\downarrow a = \{x \in X \mid x \leqslant a\}, \qquad \uparrow a = \{x \in X \mid x \geqslant a\}, \tag{1.1}$$

are quasi sublattices of X (and lattices in their own right).

The following remark, even though of little importance, should not be forgotten. The *free quasi lattice* generated by an element x is obviously the one-point lattice $\{x\}$, but the *free lattice* L generated by x has three elements: $0 < x < 1$, so that every mapping $\{x\} \to X$ with values in a lattice has a unique extension to a lattice homomorphism $L \to X$.

In the same way, the free lattice generated by a set S can be obtained from the corresponding free quasi lattice M by adding a new minimum and a new maximum, even when M *already* has a least and a greatest element.

1.1.2 Distributive and modular lattices

A lattice is said to be *distributive* if the meet operation distributes over the join operation, or equivalently if the join distributes over the meet. (The proof of the equivalence is easy, and can be found in [Bi], I.6, Theorem 9.)

A *boolean algebra* is a distributive lattice where every element x has a (necessarily unique) *complement* x', defined by the obvious properties: $x \wedge x' = 0$, $x \vee x' = 1$.

The subsets of a set X form the classical boolean algebra $\mathcal{P}X$. The lattice Sub$A$ of subgroups of an abelian group (or submodules of a module) is not distributive, generally; but one can easily check that it always satisfies a weaker, restricted form of distributivity, called modularity.

Namely, a lattice is said to be *modular* if it satisfies the following selfdual property (for all elements x, y, z)

(a) if $x \leqslant z$, then $(x \vee y) \wedge z = x \vee (y \wedge z)$.

By Birkhoff's representation theorem ([Bi], III.5, Theorem 5), the free distributive lattice on n generators is finite and isomorphic to a lattice of subsets. The reader may also be interested to know that the free modular lattice on three elements is finite and (obviously!) not distributive, while four generators already give an infinite free modular lattice (see [Bi], III.6).

But we are more concerned with the free modular lattice generated by two chains, that - because of a well-known theorem of Garrett Birkhoff - is always distributive (see Section 7.3).

1.1.3 Galois connections

This is an important issue, that plays a relevant role here (as in Part I), in order to express direct and inverse images of subobjects.

Formally, a Galois connection is an adjunction between (partially) ordered sets, viewed as categories; but one can easily define directly this notion, which is much simpler than a general adjunction (cf. A3.1, in Appendix A) and can serve as an introduction to the latter.

Given a pair X, Y of ordered sets, a (covariant) *Galois connection* between them can be presented in the following equivalent ways.

(i) We assign two increasing mappings $f: X \to Y$ and $g: Y \to X$ such that:

$$f(x) \leqslant y \text{ in } Y \quad \Leftrightarrow \quad x \leqslant g(y) \text{ in } X.$$

(ii) We assign an increasing mapping $g\colon Y \to X$ such that, for every $x \in X$, there exists:

$$f(x) = \min\{y \in Y \mid x \leqslant g(y)\}.$$

(ii*) We assign an increasing mapping $f\colon X \to Y$ such that, for every $y \in Y$, there exists:

$$g(y) = \max\{x \in X \mid f(x) \leqslant y\}.$$

(iii) We assign two increasing mappings $f\colon X \to Y$ and $g\colon Y \to X$ such that $\mathrm{id}X \leqslant gf$ and $fg \leqslant \mathrm{id}Y$.

By these formulas, g determines f (called its *left adjoint*) and f determines g (its *right adjoint*). One writes $f \dashv g$ (borrowing the notation from category theory).

Of course an isomorphism of ordered sets is, at the same time, left and right adjoint to its inverse. But an increasing mapping between ordered sets *may* have a right (resp. left) adjoint, which should be viewed as a 'lower (resp. upper) approximation' to an inverse which may not exist.

For instance, the embedding of ordered sets $i\colon \mathbb{Z} \to \mathbb{R}$ has a well-known right adjoint, the integral-part function, or *floor* function

$$[\ ]\colon \mathbb{R} \to \mathbb{Z}, \qquad [x] = \max\{k \in \mathbb{Z} \mid k \leqslant x\}. \tag{1.2}$$

The left adjoint also exists: it is the *ceiling* function

$$-[-x] = \min\{k \in \mathbb{Z} \mid k \geqslant x\}$$

(that is here linked to the right adjoint by the anti-isomorphism $x \mapsto (-x)$ of the real and integral lines).

Let us come back to a general Galois connection $f \dashv g$. Then f *preserves all the existing joins* (also infinite), while g *preserves all the existing meets*.

In fact, if $x = \vee x_i$ in X, then $f(x_i) \leqslant f(x)$ (for all indices i). Supposing that $f(x_i) \leqslant y$ in Y (for all i), it follows that $x_i \leqslant g(y)$ (for all i); but then $x \leqslant g(y)$ and $f(x) \leqslant y$.

Furthermore, the relations $\mathrm{id}X \leqslant gf$ and $fg \leqslant \mathrm{id}Y$ imply that:

$$f = fgf \quad \text{and} \quad g = gfg.$$

As a consequence, the connection restricts to an isomorphism (of ordered sets) between the sets of *closed elements* of X and Y:

$$\mathrm{cl}(X) = g(Y) = \{x \in X \mid x = gf(x)\},$$
$$\mathrm{cl}(Y) = f(X) = \{y \in Y \mid y = fg(y)\}. \tag{1.3}$$

1.1.4 Contravariant Galois connections

We shall occasionally use *contravariant* Galois connections between ordered sets, a symmetric notion equivalent to the covariant one (and called a 'Galois connection' in some texts, like Mac Lane's [M5]).

A *contravariant Galois connection* is formed by a pair (f, g), where

(i) $f \colon X \to Y$ and $g \colon Y \to X$ are *decreasing* mappings,

(ii) $gf(x) \geqslant x, \qquad fg(y) \geqslant y$.

It can be transformed into a covariant connection by reversing the order in Y (so that $f \dashv g$) *or* in X (so that $g \dashv f$). Even though the two notions are equivalent, the covariant connections can be composed (and *must* be used to form a category, as we shall do in the next section) while the contravariant form has the advantage of being symmetric.

Here, both f and g turn the existing joins into meets. Again $fgf = f$ and $gfg = g$, and our connection restricts to a bijection $\mathrm{cl}X \to \mathrm{cl}Y$ between the ordered subsets of closed elements (defined in formulas (1.3)), that is now an *anti*-isomorphism.

The following properties will be of interest.

(a) If the closed elements of X have (finite) joins *in X*, then $\mathrm{cl}Y$ has (finite) meets that subsist in Y, whereas $\mathrm{cl}X$ has (finite) joins that generally do not subsist in X.

(b) If the closed elements of X have (finite) meets *in X*, these are closed and $\mathrm{cl}Y$ has (finite) joins that generally do not subsist in Y.

(c) If X is a lattice, or a complete lattice, so is $\mathrm{cl}X$ (with the same meets as in X) and also $\mathrm{cl}Y$.

Property (a) is straightforward, since f transforms joins into meets and $\mathrm{cl}X$ is anti-isomorphic to $\mathrm{cl}Y$. As to (b), let (y_i) be a (finite) family in $\mathrm{cl}Y$ and $x = \wedge g(y_i)$ in X. We show first that $f(x) = \vee y_i$ *in* $\mathrm{cl}Y$; indeed, $x \leqslant g(y_i)$ and $f(x) \geqslant y_i$, for all indices i; further, if $y \geqslant y_i$ in $\mathrm{cl}Y$ (for all i), then $g(y) \leqslant g(y_i)$, i.e. $g(y) \leqslant x$ and $y = fg(y) \geqslant f(x)$. The element x is closed, since the restriction $g' \colon \mathrm{cl}Y \to X$ is again (part of) a contravariant connection, hence turns joins into meets and $gf(x) = \wedge g(y_i) = x$. Last, (c) follows from (a) and (b).

A classical example of a contravariant connection (for a commutative unital ring R) concerns the subsets of the affine space R^n on the one hand, and the subsets of the polynomial ring $R[X_1, ...X_n]$ on the other, linked by the equation $f(x_1, ...x_n) = 0$. If R is an algebraically closed field, Hilbert's Nullstellensatz says that the closed elements in $\mathcal{P}(R[X_1, ...X_n])$ coincide with the radical ideals; on both sides the union of closed elements is smaller than their 'closed join'.

1.1.5 Isomorphisms, monomorphisms and epimorphisms

We now review some elementary points of category theory; other issues can be found in Appendix A.

In a category, a morphism $f\colon A \to B$ is called an *isomorphism*, or an *iso*, if it is *invertible* (inside the category), i.e. if it has an *inverse*: a (uniquely determined) morphism $f^{-1}\colon B \to A$ such that $f^{-1}.f = \mathrm{id}A$ and $f.f^{-1} = \mathrm{id}B$.

More generally, monomorphisms and epimorphisms are defined by cancellation properties.

Namely, a morphism $f\colon A \to B$ is a *monomorphism*, or *mono*, if $fu = fv$ implies $u = v$ whenever the compositions make sense, as in the left diagram below

$$\bullet \underset{v}{\overset{u}{\rightrightarrows}} A \xrightarrow{f} B \qquad A \xrightarrow{f} B \underset{v}{\overset{u}{\rightrightarrows}} \bullet \qquad (1.4)$$

while f is an *epimorphism*, or *epi*, if $uf = vf$ implies $u = v$, as in the right diagram above. (The bullets stand for arbitrary objects.)

In the category **Set** of sets (and mappings), or **Ab** of abelian groups (and their homomorphisms), or **Top** of topological spaces (and continuous mappings), all this is quite simple and easy to determine:

- a morphism is mono (resp. epi) if and only if it is an injective (resp. surjective) mapping.

But we shall soon need to analyse these issues in other categories, where they are less obvious.

The reader may notice that **Set** and **Ab** are *balanced* categories: a morphism that is epi and mono is always an isomorphism. Plainly, this is not the case of **Top**.

The derived notions of subobjects (as selected monomorphisms) and quotients (as selected epimorphisms) can be found in A1.6.

1.1.6 Pointed categories

The kernel and cokernel of a morphism are usually defined in a pointed category (but we shall extend them to a more general situation, in Section 1.3).

Now, a category E is said to be *pointed* if there is a *zero object* 0: for every object A there is precisely one morphism $0 \to A$ (which means that 0 is *initial* in E) and precisely one morphism $A \to 0$ (i.e. 0 is *terminal*). The zero object is determined up to isomorphism.

Obviously, Ab is pointed, with zero object given by the null group; Set and Top are not, since their initial and terminal objects are distinct.

In a pointed category, given two objects A, B, the composite $A \to 0 \to B$ is called the *zero morphism* from A to B, and written as $0_{AB} \colon A \to B$, or also as 0.

Plainly, the morphism $0 \to A$ is necessarily a monomorphism, also written as $0_A \colon 0 \to A$ and called the *zero subobject* of A; dually the morphism $A \to 0$ is necessarily an epimorphism, written as $0^A \colon A \to 0$ and called the *zero quotient* of A.

1.1.7 Kernels and cokernels

In the pointed category E, the *kernel* of a morphism $f \colon A \to B$ is defined by a universal property, as an object $\operatorname{Ker} f$ equipped with a morphism $\ker f \colon \operatorname{Ker} f \to A$, such that:

(i) $f . \ker f = 0$,
   every map h such that $fh = 0$ factorises uniquely through $\ker f$.

More explicitly, there exists a unique morphism u such that $h = (\ker f)u$, as in the diagram below (again, the bullet stands for an arbitrary object)

$$
\operatorname{Ker} f \xrightarrow{\ker f} A \xrightarrow{f} B \qquad (1.5)
$$

Notice that $\ker f$ is necessarily a monomorphism of the category (because of the uniqueness part of the universal property).

The solution, if it exists, is determined up to isomorphism; but $\ker f$ usually refers to *the subobject* strictly determined as above (cf. A1.6).

Thus, in Ab, the natural 'kernel-object' is the usual subgroup $\operatorname{Ker} f = f^{-1}\{0\}$, and $\ker f$ is its embedding in the domain of f; however, if $u \colon K \to \operatorname{Ker} f$ is an isomorphism, also the composite $(\ker f)u \colon K \to A$ is *a kernel-morphism* of f.

In a pointed category, a monomorphism is said to be *normal* if it is a kernel of some arrow. It is easy to see that a normal monomorphism that is epi is an isomorphism: if a kernel k of f is epi, then $f = 0$ and the thesis follows.

Dually, the *cokernel* of $f\colon A \to B$ is an object $\operatorname{Cok} f$ equipped with a morphism $\operatorname{cok} f\colon B \to \operatorname{Cok} f$, such that:

(i*) $(\operatorname{cok} f).f = 0$,

every map h such that $hf = 0$ factorises uniquely through $\operatorname{cok} f$.

In other words, there exists a unique morphism u such that $h = u(\operatorname{cok} f)$

$$
A \xrightarrow{\ f\ } B \xrightarrow{\ \operatorname{cok} f\ } \operatorname{Cok} f
$$

$$(1.6)$$

Again, the solution is an epimorphism determined up to isomorphism, and we always choose *the* cokernel as *the quotient* determined by the universal property (cf. A1.6). A *normal epimorphism* is any cokernel of a morphism; a normal epimorphism that is mono is an isomorphism.

In Ab, the natural cokernel-object is the quotient $B/f(A)$, and $\operatorname{cok} f$ is the canonical projection $B \to B/f(A)$.

1.2 Lattices and Galois connections

We now introduce the category Ltc of (small) lattices and Galois connections, that will be the codomain of the functor Nsb: E → Ltc of normal subobjects, for every semiexact category E, and a 'homological category' in its own right. The present study of exactness properties in Ltc can also be seen as an introduction to the general study of exactness in semiexact and homological categories.

General conventions on lattices have been stated in the previous section. Ord denotes the category of (small) ordered sets and increasing mappings.

1.2.1 Definition

An object of the category Ltc of *lattices and connections* is a small lattice (with least element 0 and greatest element 1).

A morphism $f = (f_\bullet, f^\bullet)\colon X \to Y$ is a *covariant Galois connection* between the lattices X and Y, as defined in 1.1.3:

(i) $f_\bullet\colon X \to Y$ and $f^\bullet\colon Y \to X$ are increasing mappings,

(ii) $f^\bullet f_\bullet \geqslant 1_X$, $\qquad f_\bullet f^\bullet \leqslant 1_Y$.

We already know that $f_\bullet$ preserves all the existing joins (including $0 = \vee\emptyset$), $f^\bullet$ preserves all the existing meets (including $1 = \wedge\emptyset$) and

$$f_\bullet f^\bullet f_\bullet = f_\bullet, \qquad f^\bullet f_\bullet f^\bullet = f^\bullet. \qquad (1.7)$$

The mapping $f_\bullet$ (strictly) determines $f^\bullet$, and conversely

$$f^\bullet(y) = \max\{x \in X \mid f_\bullet(x) \leqslant y\},$$
$$f_\bullet(x) = \min\{y \in Y \mid f^\bullet(y) \geqslant x\}. \qquad (1.8)$$

Composition in Ltc is the (obvious) composition of adjunctions:

$$(g_\bullet, g^\bullet).(f_\bullet, f^\bullet) = (g_\bullet f_\bullet, f^\bullet g^\bullet).$$

The category Ltc is *selfdual*, under the contravariant endofunctor

$$X \mapsto X^{\mathrm{op}}, \quad ((f_\bullet, f^\bullet)\colon X \to Y) \mapsto ((f^\bullet, f_\bullet)\colon Y^{\mathrm{op}} \to X^{\mathrm{op}}). \qquad (1.9)$$

The morphism f is an isomorphism in Ltc if and only if the following equivalent conditions hold (use (1.7)):

(a) $f^\bullet f_\bullet = 1_X, \quad f_\bullet f^\bullet = 1_Y,$

(b) $f_\bullet$ is a bijective mapping,

(b*) $f^\bullet$ is a bijective mapping,

(c) $f_\bullet$ is an isomorphism of ordered sets (hence of lattices, in the usual sense),

(c*) $f^\bullet$ is an isomorphism of ordered sets (hence of lattices).

Therefore, no distinction is needed between the isomorphisms of Ltc and the usual isomorphisms of lattices, i.e. the bijective homomorphisms.

In particular, both of the following faithful functors reflect the isomorphisms ($|X|$ denotes the underlying set of the lattice X)

$$U\colon \mathsf{Ltc} \to \mathsf{Set} \qquad \text{(the *forgetful* functor)},$$
$$X \mapsto |X|, \qquad\qquad (f_\bullet, f^\bullet) \mapsto f_\bullet, \qquad (1.10)$$

$$V\colon \mathsf{Ltc} \to \mathsf{Set}^{\mathrm{op}}, \qquad X \mapsto |X|, \quad (f_\bullet, f^\bullet) \mapsto f^\bullet. \qquad (1.11)$$

The category Ltc has a natural order relation (consistent with composition): given two parallel connections $f, g\colon X \to Y$, we say that $f \leqslant g$ if the following two equivalent conditions hold

$$f_\bullet \leqslant g_\bullet, \qquad f^\bullet \geqslant g^\bullet. \qquad (1.12)$$

In fact, $f_\bullet \leqslant g_\bullet$ implies $g^\bullet \leqslant f^\bullet f_\bullet g^\bullet \leqslant f^\bullet g_\bullet g^\bullet \leqslant f^\bullet$. Each ordered set $\mathsf{Ltc}(X, Y)$ is actually a join-semilattice, as we shall see in 1.2.7.

1.2.2 Monos and epis

With regard to the forgetful functor U, the two-point lattice $\Omega = \{0, 1\}$ is the free object on one generator (the element 1). Indeed, for every $x_0 \in X$ there exists a unique connection $h: \Omega \to X$ such that $h_\bullet(1) = x_0$

$$
\begin{aligned}
h_\bullet(0) &= 0, & h_\bullet(1) &= x_0, \\
h^\bullet(x) &= 1 \text{ if } x \geqslant x_0, & h^\bullet(x) &= 0 \text{ otherwise.}
\end{aligned}
\tag{1.13}
$$

In other words, the functor $U: \mathsf{Ltc} \to \mathsf{Set}$ is representable as $U \cong \mathsf{Ltc}(\Omega, -)$ (cf. A1.8; we shall see, in 1.2.3, that the object Ω also represents the normal subobjects).

The functor U reflects the monomorphisms (since it is faithful), and also preserves them. In fact, given a monomorphism $f = (f_\bullet, f^\bullet): X \to Y$ and two elements $x_1, x_2 \in X$ such that $f_\bullet(x_1) = f_\bullet(x_2)$, let (for $i = 1, 2$) the connection $h_i: W \to X$ be defined by $h_{i\bullet}(1) = x_i$, as above; then $f_\bullet h_{1\bullet} = f_\bullet h_{2\bullet}$, whence $fh_1 = fh_2$; cancelling f we get $h_1 = h_2$, and finally $x_1 = x_2$.

(More generally, one shows in the same way that each representable faithful functor with values in Set preserves and reflects monomorphisms.)

It follows easily, by (1.7) and duality, that the monos and epis in Ltc are respectively characterised by the following two lists of equivalent conditions:

(a) f is mono in Ltc,	(a*) f is epi in Ltc,
(b) $f_\bullet$ is an injective mapping,	(b*) $f_\bullet$ is a surjective mapping,
(c) $f^\bullet f_\bullet = 1_X$,	(c*) $f_\bullet f^\bullet = 1_Y$,
(d) $f_\bullet$ is a section in Ord,	(d*) $f_\bullet$ is a retraction in Ord,
(e) $f^\bullet$ is a surjective mapping,	(e*) $f^\bullet$ is an injective mapping,
(f) $f^\bullet$ is a retraction in Ord,	(f*) $f^\bullet$ is a section in Ord.

In particular, Ltc is a balanced category (cf. 1.1.5): a morphism that is both mono and epi is an isomorphism.

Moreover every map $f: X \to Y$ has an epi-mono factorisation, unique up to isomorphism, that will be called the *canonical factorisation* of f

$$
\begin{array}{ccc}
X & \xrightarrow{\ f\ } & Y \\
{\scriptstyle p}\downarrow & & \uparrow{\scriptstyle m} \\
\mathrm{cl}X & \xrightarrow[\ u\ ]{} & \mathrm{cl}Y
\end{array}
\tag{1.14}
$$

Here $\mathrm{cl}X$ and $\mathrm{cl}Y$ respectively denote the set of *closed* elements of X or

Y, with respect to f:

$$clX = f^\bullet(Y) = \{x \in X \mid x = f^\bullet f_\bullet x\},$$
$$clY = f_\bullet(X) = \{y \in Y \mid y = f_\bullet f^\bullet y\}),$$

while the connections p, m, u are defined as follows:

$$p_\bullet(x) = f^\bullet f_\bullet(x), \qquad p^\bullet(x') = x',$$
$$m_\bullet(y') = y', \qquad m^\bullet(y) = f_\bullet f^\bullet(y),$$
$$u_\bullet(x') = f_\bullet(x'), \qquad u^\bullet(y') = f^\bullet(y').$$

We only have to show that the ordered sets clX and clY are indeed lattices. (This has already been proved in 1.1.4(c), but we prefer to repeat the argument in a shorter way, adapted to the present situation.)

First, clX and clY are isomorphic ordered sets, via $u_\bullet$. Further, since X has finite joins preserved by $f_\bullet$, clY has finite joins (consistent with those of Y); therefore clX has finite joins (that need not be consistent with those of X). Dually, as Y has finite meets preserved by $f^\bullet$, clX and clY also have them.

Classical examples show that clX and clY need not be *sublattices* of X and Y, even forgetting about least and greatest elements: see 1.1.4.

1.2.3 Kernels and cokernels

The category Ltc is *pointed* (cf. 1.1.6): the zero object is the one-point lattice $0 = \{*\}$. Indeed each lattice X has just one morphism $m\colon 0 \to X$, and one morphism $p\colon X \to 0$, defined as follows (for all $x \in X$)

$$m_\bullet(*) = 0, \qquad m^\bullet(x) = *,$$
$$p_\bullet(x) = *, \qquad p^\bullet(*) = 1. \tag{1.15}$$

The zero-morphism $f\colon X \to Y$ is described by $f_\bullet(x) = 0$, $f^\bullet(y) = 1$; it is the least element of the ordered set $\mathsf{Ltc}(X, Y)$. It is characterised by each of the following four equivalent conditions:

(a) $f_\bullet(x) = 0$ $(x \in X)$,   (a*) $f^\bullet(y) = 1$ $(y \in Y)$,
(b) $f_\bullet(1) = 0$,   (b*) $f^\bullet(0) = 1$,

Every morphism f has a kernel and a cokernel (as defined in 1.1.7)

$$m = \ker f\colon \downarrow f^\bullet 0 \to X, \qquad m_\bullet(x') = x', \ m^\bullet(x) = x \wedge f^\bullet 0, \tag{1.16}$$

$$p = \operatorname{cok} f\colon Y \to \uparrow f_\bullet 1, \qquad p_\bullet(y) = y \vee f_\bullet 1, \ p^\bullet(y') = y'. \tag{1.17}$$

On the other hand, every element $a \in X$ determines a *normal subobject* and a *normal quotient* of X (cf. 1.1.7)

$$m \colon \downarrow a \to X, \qquad m_{\bullet}(x') = x', \quad m^{\bullet}(x) = x \wedge a, \tag{1.18}$$

$$p \colon X \to \uparrow a, \qquad p_{\bullet}(x) = x \vee a, \quad p^{\bullet}(x') = x'. \tag{1.19}$$

These correspondences establish an isomorphism between the *lattice* X and the ordered set $\mathsf{Nsb}X$ of *normal subobjects* of X in Ltc, as well as an anti-isomorphism of X with the ordered set of *normal quotients* of X. (Since the underlying set $|X|$ can be identified with $\mathsf{Ltc}(\Omega, X)$, the two-point lattice Ω also 'represents' the set of normal subobjects of X.)

The normal mono m and normal epi p determined by any element $a \in X$ form a *short exact sequence*

$$\downarrow a \to X \to \uparrow a, \qquad m = \ker p, \quad p = \operatorname{cok} m. \tag{1.20}$$

Conversely, every short exact sequence (m, p) in Ltc with central object X is isomorphic to a unique sequence of this type, with $a = m_{\bullet}1 = p^{\bullet}0$.

1.2.4 The normal factorisation

The morphism $f \colon X \to Y$ factorises through a normal quotient, the *normal coimage* $\operatorname{ncm} f$, and a normal subobject, the *normal image* $\operatorname{nim} f$, producing a *normal factorisation* $f = (\operatorname{nim} f)g(\operatorname{ncm} f)$ where g *need not be an isomorphism*

$$q = \operatorname{ncm} f = \operatorname{cok}(\ker f) \colon X \to \uparrow f^{\bullet}0,$$
$$q_{\bullet}(x) = x \vee f^{\bullet}0, \qquad q^{\bullet}(x) = x,$$

$$g \colon \uparrow f^{\bullet}0 \to \downarrow f_{\bullet}1,$$
$$g_{\bullet}(x) = f_{\bullet}(x), \qquad g^{\bullet}(y) = f^{\bullet}(y), \tag{1.21}$$

$$n = \operatorname{nim} f = \ker(\operatorname{cok} f) \colon \downarrow f_{\bullet}1 \to Y,$$
$$n_{\bullet}(y) = y, \qquad n^{\bullet}(y) = y \wedge f_{\bullet}1.$$

The normal factorisation can be combined with the canonical one (cf. (1.14)) forming the following commutative diagram

$$\tag{1.22}$$

$$a_{\bullet}(x) = f^{\bullet} f_{\bullet}(x), \qquad\qquad a^{\bullet}(x) = x,$$
$$b_{\bullet}(y) = y, \qquad\qquad b^{\bullet}(y) = f_{\bullet} f^{\bullet}(y).$$

1.2.5 Exact connections

Consider again the connection $f\colon X \to Y$ with its combined canonical and normal factorisations (1.22).

We say that f is *left exact* (resp. *right exact*) if it satisfies the equivalent conditions (a) - (g) (resp. (a*) - (g*)):

(a) $f^{\bullet} f_{\bullet}(x) = x \vee f^{\bullet} 0$, for $x \in X$, (a*) $f_{\bullet} f^{\bullet}(y) = y \wedge f_{\bullet} 1$, for $y \in Y$,

(b) $f^{\bullet} f_{\bullet}(x) = x$, for $x \geqslant f^{\bullet} 0$, (b*) $f_{\bullet} f^{\bullet}(y) = y$, for $y \leqslant f_{\bullet} 1$,

(c) a is an isomorphism, (c*) b is an isomorphism,

(d) a is mono $(a^{\bullet} a_{\bullet} = 1)$, (d*) b is epi $(b_{\bullet} b^{\bullet} = 1)$,

(e) g is mono $(g^{\bullet} g_{\bullet} = 1)$, (e*) g is epi $(g_{\bullet} g^{\bullet} = 1)$,

(f) $p \sim \operatorname{ncm} f$, (f*) $m \sim \operatorname{nim} f$,

(g) $p^{\bullet} p_{\bullet}(x) = x \vee p^{\bullet} 0$, for $x \in X$, (g*) $m_{\bullet} m^{\bullet}(y) = y \wedge m_{\bullet} 1$, for $y \in Y$.

(The equivalence relation $\sim$ between monos with the same codomain, *or* epis with the same domain, is defined in A1.6.)

Indeed the properties (c), (d), (e), (f) are trivially equivalent. (c) $\Rightarrow$ (a): $f^{\bullet} f_{\bullet}(x) = p^{\bullet} p_{\bullet}(x) = q^{\bullet} q_{\bullet}(x) = x \vee f^{\bullet} 0$ (see (1.15)). (a) $\Rightarrow$ (b) is obvious. (b) $\Rightarrow$ (d): $a^{\bullet} a_{\bullet}(x) = f^{\bullet} f_{\bullet}(x) = x$, for $x \geqslant f^{\bullet} 0$. (a) $\Leftrightarrow$ (g) because $f^{\bullet} 0 = p^{\bullet} 0$.

We say that the morphism f is *exact* if it is both left and right exact. This holds if and only if the associated map g is an isomorphism, if and only if f factorises as a normal epi followed by a normal mono.

For instance, given a mapping of sets $h\colon A \to B$, the associated connection $(h_{*}, h^{*})\colon \mathcal{P}A \to \mathcal{P}B$ between their boolean algebras of parts, produced by direct and inverse images, is always right exact; it is left exact if and only if h is injective.

On the other hand, starting from a homomorphism of abelian groups $h\colon A \to B$, the associated connection $(h_{*}, h^{*})\colon \operatorname{Sub}A \to \operatorname{Sub}B$ between their lattices of subgroups is always exact; this fact will be seen to hold much more generally (see 1.2.8).

1.2.6 Normal monos and epis

With the same notation, the following conditions on a Galois connection $f\colon X \to Y$ are equivalent, in each list:

(a) f is a normal epi, (a*) f is a normal mono,

(b) $f \sim \mathrm{ncm}\, f$, (b*) $f \sim \mathrm{nim}\, f$,

(c) f is epi and left exact, (c*) f is mono and right exact,

(d) f is epi and exact, (d*) f is mono and exact,

(e) $f^\bullet f_\bullet x = x \vee f^\bullet 0,\ f_\bullet f^\bullet y = y$, (e*) $f^\bullet f_\bullet x = x,\ f_\bullet f^\bullet y = y \wedge f_\bullet 1$,

(f) g and $\mathrm{nim}\, f$ are iso, (f*) g and $\mathrm{ncm}\, f$ are iso.

1.2.7 The semiadditive structure

We now show that the pointed category Ltc is *semiadditive*, i.e. it has finite biproducts (see A1.5); this defines a sum of parallel maps, *that lacks opposites*.

First, we want to prove that, if X and Y are lattices, their cartesian product $X \times Y$ (i.e. their categorical product in the category of lattices and homomorphisms) is their *biproduct* in Ltc, by means of the following connections, where the injections i, j are normal monos and the projections p, q are normal epis

$$X \underset{p}{\overset{i}{\rightleftarrows}} X \times Y \underset{q}{\overset{i}{\rightleftarrows}} Y \qquad i_\bullet \dashv i^\bullet = p_\bullet \dashv p^\bullet, \qquad (1.23)$$

$$i_\bullet(x) = (x, 0_Y), \qquad i^\bullet(x,y) = x = p_\bullet(x,y), \qquad p^\bullet(x) = (x, 1_Y),$$

$$i^\bullet i_\bullet(x) = x, \qquad i_\bullet i^\bullet(x,y) = (x, 0_Y) = (x,y) \wedge i_\bullet(1_X),$$

$$p^\bullet p_\bullet(x,y) = (x, 1_Y) = (x,y) \vee p^\bullet(0_X), \qquad p_\bullet p^\bullet(x) = x.$$

To show that $(X \times Y; p, q)$ is the product of X and Y in Ltc (according to the general notion, recalled in A1.4) it suffices to take two connections $f\colon Z \to X$ and $g\colon Z \to Y$ and define the connection $h\colon Z \to X \times Y$ as:

$$h_\bullet(z) = (f_\bullet z, g_\bullet z), \qquad h^\bullet(x,y) = f^\bullet(x) \wedge g^\bullet(y), \qquad (1.24)$$

$$h^\bullet h_\bullet(z) = (f^\bullet f_\bullet z) \wedge (g^\bullet g_\bullet z) \geqslant z,$$

$$h_\bullet h^\bullet(x,y) = (f_\bullet(f^\bullet x \wedge g^\bullet y),\ g_\bullet(f^\bullet x \wedge g^\bullet y))$$
$$\leqslant (f_\bullet f^\bullet x,\ g_\bullet g^\bullet y) \leqslant (x,y).$$

In fact, h satisfies the conditions $ph = f$, $qh = g$ and is determined by them, since all this trivially holds on the covariant parts. The second fact, namely that $(X \times Y; i, j)$ is the categorical sum of X and Y in Ltc (cf. A1.5), follows from the first, according to the duality (1.9).

The sum $f + g$ of two maps $f, g \colon X \to Y$ can now be introduced by means of the diagonal d of X (defined by $pd = qd = \mathrm{id}X$) and of the codiagonal ∂ of Y (defined by $\partial i = \partial j = \mathrm{id}Y$)

$$X \xrightarrow{\;d\;} X \times X \xrightarrow{\;f \times g\;} Y \times Y \xrightarrow{\;\partial\;} Y \qquad (1.25)$$

$$d_{\bullet}(x) = (x, x), \qquad\qquad d^{\bullet}(x, x') = x \wedge x',$$
$$\partial_{\bullet}(y, y') = y \vee y', \qquad\qquad \partial^{\bullet}(y) = (y, y),$$
$$f + g = \partial(f \times g)d = (f_{\bullet} \vee g_{\bullet}, f^{\bullet} \wedge g^{\bullet}).$$

Notice that $f + g$ *is the join* $f \vee g$ with respect to the order relation (1.12). Hence *the sum is idempotent*: $f + f = f$; the cancellation law does not hold and *our structure is not additive*.

1.2.8 Modular connections

A *modular connection* $f = (f_{\bullet}, f^{\bullet}) \colon X \to Y$ (an essential topic of Part I) can now be defined as an *exact* Galois connection (cf. 1.2.5) between *modular* lattices.

Taking into account the modularity of X and Y, the connection f has to satisfy the equivalent conditions (a), (b) (for all $x \in X$, $y \in Y$)

(a) $\quad f^{\bullet} f_{\bullet} x = x \vee f^{\bullet} 0, \qquad\qquad f_{\bullet} f^{\bullet} y = y \wedge f_{\bullet} 1,$

(b) $\quad f^{\bullet}(f_{\bullet} x \vee y) = x \vee f^{\bullet} y, \qquad\quad f_{\bullet}(f^{\bullet} y \wedge x) = y \wedge f_{\bullet} x.$

Trivially, (b) implies (a). Conversely, from (a) and the modularity of lattices we get

$$x \vee f^{\bullet} y = (x \vee f^{\bullet} y) \vee f^{\bullet} 0 = f^{\bullet} f_{\bullet}(x \vee f^{\bullet} y) = f^{\bullet}(f_{\bullet} x \vee f_{\bullet} f^{\bullet} y)$$
$$= f^{\bullet}(f_{\bullet} x \vee (y \wedge f_{\bullet} 1)) = f^{\bullet}((f_{\bullet} x \vee y) \wedge f_{\bullet} 1))$$
$$= f^{\bullet} f_{\bullet} f^{\bullet}(f_{\bullet} x \vee y)) = f^{\bullet}(f_{\bullet} x \vee y)).$$

Now, the connections satisfying the condition (b) are obviously closed under composition. We have thus redefined the category Mlc of *modular lattices and modular connections*, used in Part I as the codomain of the transfer functor of p-exact categories. (Later, in 5.1.4, we shall see that this category comes out of a universal construction, as the *exact centre* of the homological category Ltc.)

We are also interested in the full subcategory Dlc $\subset$ Mlc of *distributive lattices and modular connections*. In a modular connection $f = (f_{\bullet}, f^{\bullet})$:

$X \to Y$ between distributive lattices, the mappings $f_\bullet$ and $f^\bullet$ are always homomorphisms of *quasi* lattices, i.e. preserve binary meets and joins

$$
\begin{aligned}
f_\bullet(x \wedge y) &= f_\bullet f^\bullet f_\bullet(x \wedge y) = f_\bullet((x \wedge y) \vee f^\bullet 0) \\
&= f_\bullet((x \vee f^\bullet 0) \wedge (y \vee f^\bullet 0)) = f_\bullet(f^\bullet f_\bullet x \wedge f^\bullet f_\bullet x) \\
&= f_\bullet f^\bullet(f_\bullet x \wedge f_\bullet y) = (f_\bullet x \wedge f_\bullet y) \wedge f_\bullet 1 = f_\bullet x \wedge f_\bullet y.
\end{aligned}
\tag{1.26}
$$

Mlc and Dlc do *not* inherit the semiadditive structure of Ltc. If X and Y are modular lattices, the maps of the biproduct-decomposition of $X \times Y$ (cf. (1.23)) are indeed modular connections but the universal property for the product (see (1.24)) or sum is not satisfied, even in the distributive case.

(In fact, we have already seen in Part I that the p-exact category Mlc lacks binary products and sums, and cannot even be exactly embedded in an abelian category; see I.2.1.8, I.2.2.8, I.2.3.5. Dlc is distributive p-exact, and therefore cannot have binary products or sums.)

1.3 The main definitions

Loosely speaking, a semiexact category is a category *equipped* with an ideal of 'null morphisms' and provided with kernels and cokernels with respect to this ideal. Some exactness conditions on its morphisms yield the stronger notions of homological and generalised exact category; in the pointed case, the latter notion is equivalent to an exact category in the sense of Puppe and Mitchell.

General conventions about subobjects (as selected monomorphisms) and quotients (as selected epimorphisms) can be found in A1.6, in the Appendix.

As recalled in Section 1 of the Introduction, kernels and cokernels with respect to an ideal can already be found in a paper by C. Ehresmann [Eh2]. Categories 'with α-kernels and α-cokernels', studied in three papers by R. Lavendhomme and G. Van Den Bossche [La, LaV, Vb], are equivalent to semiexact categories in the present sense (see 1.3.4).

1.3.1 Ideals of null morphisms

As in semigroup theory, an *ideal* of a category E is a set of morphisms such that, if f belongs to it, every (legitimate) composite hfg in E does.

In particular, if E is pointed (cf. 1.1.6), its zero morphisms form an ideal (determined by the categorical structure), that will be called the *pointed ideal* of E. But we shall see that a pointed category can have other, more important, ideals.

An *N-category* will be a category E equipped with an ideal $\mathcal{N} = \mathrm{Nul}\,\mathsf{E}$, whose maps are called the *null morphisms* of E; an object is null if so is its identity. An *N-functor* is a functor between N-categories that preserves the null maps, *hence* also the null objects; it is *null* if it annihilates all maps, or equivalently all objects.

This very elementary setting will not have here an autonomous treatment; yet some notions that typically belong to the theory of N-categories will appear now and then - identified by the prefix 'N' - and are not without interest or applications (see 1.4.5; 1.5.2; 1.7.0; 1.7.2; 1.7.5; Chapter 4).

Let E be an N-category. Extending the usual notion (reviewed in 1.1.7), the *kernel* of a morphism $f\colon A \to B$, written as

$$\ker f\colon \mathrm{Ker}\,f \to A, \qquad (1.27)$$

is a morphism satisfying the universal property *with respect to the ideal* $\mathcal{N}$ (and for every morphism h in E):

$f.(\ker f)$ is null,

if fh is null, then h factorises uniquely through $\ker f$. $\qquad (1.28)$

This implies that $\ker f$ is mono, and determined up to the equivalence relation of monomorphisms with values in A (cf. A1.6). We always choose $\ker f$ to be a *subobject* of A, so that it is strictly determined (if it exists).

Analogously, the *cokernel* of f

$$\mathrm{cok}\,f\colon B \to \mathrm{Cok}\,f, \qquad (1.29)$$

will be the *quotient* of B strictly determined by the dual conditions (if it exists).

As in [La], we say that the morphism f is *N-mono* (resp. *N-epi*) if it satisfies the condition (a) (resp. (a*))

(a) if fh is null then h is null,   (a*) if kf is null then k is null.

Both properties are closed under composition. Every split monomorphism is N-mono, and every split epi is N-epi.

If E has kernels and cokernels, condition (a) is equivalent to each of (b) and (c), while (a*) is equivalent to each of (b*) and (c*) (assuming that the composites fh and kf make sense)

(b) $\ker f$ is null,   (b*) $\mathrm{cok}\,f$ is null,

(c) $\ker\,(fh) = \ker h$ (for each h),   (c*) $\mathrm{cok}\,(kf) = \mathrm{cok}\,k$ (for each k).

1.3.2 Closed ideals

In every category E there is a covariant Galois connection $N \dashv O$ between sets of objects $\mathcal{O}$ and ideals of morphisms $\mathcal{N}$ (with regard to the inclusion relations)

$$N(\mathcal{O}) = \{a \in \mathrm{Mor}\mathsf{E} \mid a \text{ factorises through some object in } \mathcal{O}\},$$
$$O(\mathcal{N}) = \{A \in \mathrm{Ob}\mathsf{E} \mid 1_A \in \mathcal{N}\}, \tag{1.30}$$
$$\mathcal{O} \subset ON(\mathcal{O}), \qquad NO(\mathcal{N}) \subset \mathcal{N}.$$

The ideal $\mathcal{N}$ is *closed* in this connection if and only if $\mathcal{N} = N(\mathcal{O})$ for some set of objects $\mathcal{O}$, if and only if $\mathcal{N} = NO(\mathcal{N})$; this means that *every morphism of $\mathcal{N}$ factorises through some identity in $\mathcal{N}$*. Such an ideal will be called a *closed* ideal of E.

Analogously the set of objects $\mathcal{O}$ is *closed* in this connection if and only if $\mathcal{O} = O(\mathcal{N})$ for some ideal of morphisms $\mathcal{N}$, if and only if $\mathcal{O} = ON(\mathcal{O})$. Since $ON(\mathcal{O})$ is easily seen to be the set of retracts of the objects of $\mathcal{O}$, the latter is closed if and only if it is *closed under retracts*.

We shall generally use, as an ideal of null morphisms, a *closed ideal*. The interest of this fact will be clear in the sequel (cf. 1.5.3, 1.5.5, 1.6.9). If E and E$'$ are N-categories and NulE is closed, a functor $F \colon \mathsf{E} \to \mathsf{E}'$ is an N-functor if and only if it preserves the null objects.

If E has a zero object, the pointed ideal defined above is obviously closed.

More generally, to take into account a situation arising for instance in categories of graded objects (that we shall encounter in 1.4.4), an object Z will be said to be *quasi initial* if

(a) for every object A there is *some* morphism $Z \to A$ (i.e. Z is weakly initial),

(b) for every pair of maps $u, v \colon Z \to A$ there is some automorphism $i \colon Z \to Z$ such that $u = vi$.

Plainly, if Z and Z' are both quasi initial, every map $Z \to Z'$ is iso (and such maps exist); every split monomorphism with values in a quasi initial object is an isomorphism.

The category E will be said to be *quasi pointed* if it has a *quasi zero* object (quasi initial and quasi terminal); such objects are all isomorphic (not canonically) and closed under retracts.

Then, the *quasi zero maps* (that factorise through a quasi zero object) form the *quasi pointed* ideal of E; again, this ideal is closed and determined by the categorical structure of E. Given two objects A and B, the quasi zero maps $A \to B$ are in bijective (not canonical) correspondence with the automorphism group $\mathsf{E}(Z, Z)$ of any quasi zero object Z.

1.3.3 Semiexact categories

A *semiexact* category, or *ex1-category*, is a pair $(E, \mathcal{N})$ that satisfies the following self-dual axioms:

(ex0) E is a category and $\mathcal{N}$ is a closed ideal (see 1.3.2) of E,

(ex1) every morphism $f \colon A \to B$ of E has a kernel and cokernel (with respect to $\mathcal{N}$, in the sense of 1.3.1).

The pair $(E, \mathcal{N})$ will usually be written as E. Again, the arrows of $\mathcal{N}$ are called *null morphisms* of E and the ideal $\mathcal{N}$ is also written as NulE. The objects of $O(\mathcal{N})$, i.e. those whose identity is null, are called *null objects* of E; they are closed under retracts and determine NulE (cf. 1.3.2).

A *normal* monomorphism is any mono equivalent to the kernel of some morphism, while a *normal* epimorphism is any epi equivalent to a cokernel. The arrows $\rightarrowtail$, $\twoheadrightarrow$ are reserved for such maps.

As a consequence of (ex1), the morphism $f \colon A \to B$ factorises uniquely through its *normal coimage* ncm f, as well as through its *normal image* nim f

$$f = f'p, \qquad p = \operatorname{ncm} f = \operatorname{cok} \ker f, \tag{1.31}$$

$$f = mf'', \qquad m = \operatorname{nim} f = \ker \operatorname{cok} f, \tag{1.32}$$

$$\begin{array}{c} \operatorname{Ker} f \rightarrowtail A \xrightarrow{\ f\ } B \longrightarrow \operatorname{Cok} f \\[2pt] \quad {\scriptstyle p}\downarrow \;\; \times \;\; \uparrow{\scriptstyle m} \\[2pt] \operatorname{Ncm} f \xrightarrow{\ g\ } \operatorname{Nim} f \end{array} \tag{1.33}$$

and it is easy to see that $\ker \operatorname{ncm} f = \ker f$ and $\operatorname{cok} \operatorname{nim} f = \operatorname{cok} f$.

We shall show that f factorises through *both* p and m, by a unique morphism g (in 1.5.5). But, *as of now*, we define f to be an *exact* morphism if there exists an isomorphism g, obviously unique, that makes the previous diagram commutative.

A semiexact category E will be said to be *pointed* (as a semiexact category), or *p-semiexact*, if its ideal NulE is pointed (cf. 1.3.1), i.e. if E has a zero object *and* its null morphisms coincide with the zero morphisms; kernels and cokernels amount then to the usual notions (recalled in 1.1.7).

Analogously, and more generally, a semiexact category is said to be *quasi pointed* if so is its ideal of null maps (cf. 1.3.2).

1.3.4 Remarks

Every abelian category is pointed semiexact. Every category C has a *trivial* semiexact structure with NulC $= $ C and, for every $f \colon A \to B$, $\ker f = 1_A$

and $\operatorname{cok} f = 1_B$. In the former case all maps are exact whilst in the latter only the isomorphisms are.

These arguments also show that a category can have different semiexact structures; however, we shall see that such structures cannot be *exactly comparable* (in 1.7.2).

(In every category C, the empty set of maps is also a closed ideal; but C has no kernels and cokernels for this ideal - unless it is empty itself.)

We shall prove in Theorem 1.5.4(b) that a semiexact category is the same as a category 'having α-kernels and α-cokernels' with respect to a given ideal α, according to Lavendhomme's terminology [La]; this is not straightforward, since in [La] the ideal α is not assumed to be closed but all α-kernels are assumed to be α-mono (in the same sense of 1.3.1) and all α-cokernels to be α-epi.

1.3.5 Kernel duality and short exact sequences

Let E be a semiexact category (or, more generally, an N-category satisfying (ex1)).

For an object A, we write $\operatorname{Nsb}A$ for the ordered subset of $\operatorname{Sub}A$ consisting of the normal subobjects. We write $\operatorname{Nqt}A$ for the ordered subset of $\operatorname{Quo}A$ formed by the normal quotients.

Cokernels and kernels define two mappings

$$\operatorname{cok} : \operatorname{Sub}A \rightleftarrows \operatorname{Quo}A : \ker, \tag{1.34}$$

between the set of subobjects and the set of quotients of A. These two mappings form a contravariant Galois connection (between possibly non-small ordered sets, cf. 1.1.4) whose closed elements are, respectively, the normal subobjects and the normal quotients of A. Indeed:

- if $x \leqslant y$ in $\operatorname{Sub}A$ then $\operatorname{cok} x \geqslant \operatorname{cok} y$ in $\operatorname{Quo}A$, by the following diagram where $\operatorname{cok} y$ annihilates x, hence factorises through $\operatorname{cok} x$;

$$\tag{1.35}$$

- $\ker \operatorname{cok} x \geqslant x$ in $\operatorname{Sub}A$, since x annihilates $\operatorname{cok} x$, hence factorises through $\ker \operatorname{cok} x$;

- last, every kernel is the kernel of a (normal) epi, by 1.3.3: $\ker f = \ker \operatorname{cok} \ker f$; this proves that the closed elements of $\operatorname{Sub}A$ are precisely the normal subobjects.

We thus have two 'closure operators'

$$x \mapsto \operatorname{nim} x = \ker \operatorname{cok} x$$
$$= \text{the least normal subobject of } A \text{ greater than } x, \tag{1.36}$$

$$p \mapsto \operatorname{ncm} p = \operatorname{cok} \ker p$$
$$= \text{the least normal quotient of } A \text{ greater than } p, \tag{1.37}$$

and an *anti-isomorphism* of ordered sets between normal subobjects and normal quotients of A

$$\operatorname{cok} : \operatorname{Nsb}A \rightleftarrows \operatorname{Nqt}A : \ker, \tag{1.38}$$

that will be called *kernel duality*. These ordered sets will be proved to be lattices (in 1.5.8).

Let us insist on the fact that kernel duality operates *in* E and *reverses* the order relation. It should not be confused with categorical duality, that transforms a normal subobject of E into a normal quotient of the *opposite* semiexact category E^{op}, and *preserves* the order relation.

A *short exact sequence* in E will be a pair of maps (m, p) with

$$\bullet \xrightarrow{\ m\ } A \xrightarrow{\ p\ } \bullet \qquad m \sim \ker p, \qquad p \sim \operatorname{cok} m, \tag{1.39}$$

so that, up to isomorphism, m and p are a normal subobject and a normal quotient of the central object A, and are in kernel duality.

It will be useful to note that, for a morphism $f \colon A \to B$

$$\operatorname{nim} f = \min\{y \in \operatorname{Nsb}B \mid f \text{ factorises through } y\}, \tag{1.40}$$

since if f factorises through y, then $\operatorname{cok} y$ annihilates f, hence $\operatorname{cok} y \leqslant \operatorname{cok} f$ and $y \geqslant \ker \operatorname{cok} f = \operatorname{nim} f$.

1.3.6 Homological and generalised exact categories

These notions are introduced here, in order to give a better treatment of the examples and introduce as of now the global framework that we want to establish. But they will be studied in Section 2.2, after developing the basics of semiexact categories, and will not be used in the theoretic parts until then.

A semiexact category E will be said to be an *ex2-category* if it satisfies the axiom:

(ex2) normal monos and normal epis are closed under composition,

and to be a *homological category*, or *ex3-category*, if, moreover:

(ex3) given a normal mono $m \colon M \rightarrowtail A$ and a normal epi $q \colon A \twoheadrightarrow Q$, with $m \geqslant \ker q$ (or equivalently, $\operatorname{cok} m \leqslant q$), the morphism qm is exact (*subquotient axiom*, or *homology axiom*).

Both these axiom are self-dual. The second allows one to construct subquotients, like the homology of a complex or the terms of a spectral sequence (Chapter 3).

Indeed, it can also be expressed in the following form: given an object A and two subobjects $m\colon M \rightarrowtail A$ (the *numerator*), $n\colon N \rightarrowtail A$ (the *denominator*) with $m \geqslant n$, there is a commutative square determined up to isomorphism, where $q = \operatorname{cok} n$

$$
\begin{array}{ccc}
M & \overset{m}{\rightarrowtail} & A \\
{\scriptstyle h}\downarrow & & \downarrow{\scriptstyle q} \\
S & \underset{k}{\rightarrowtail} & Q
\end{array}
\qquad \textit{(subquotient square)}.
\tag{1.41}
$$

(We shall prove, in 2.2.1, that this square is always bicartesian, i.e. pullback and pushout.)

The subquotient S will be written, as usual, as M/N.

It can be noted that, without (ex3), we should distinguish between a 'left subquotient' (.*i.e.* $\operatorname{Ncm} qm$) and a 'right subquotient' (Nim qm), the first being *a quotient of a subobject* (M) *of A*, the second *a subobject of a quotient* (Q) *of A*.

As a further motivation for (ex3), it will be shown that in the basic example Set_2 (pairs of sets) a composed map qm is exact *if and only if* $m \geqslant \ker q$ (see 1.4.2), so that nothing stronger can be asked in this situation. The axioms (ex2) and (ex3) are independent (cf. 1.6.1; 1.6.9).

Finally, a *g-exact* category (i.e. generalised exact), or *ex4-category*, will be a semiexact category E in which:

(ex4) each morphism is exact.

It is important to note that all these conditions require certain maps to be exact; indeed (ex2) will be shown to be equivalent to asking that the composite of two normal monos or two normal epis is an exact morphism (in 2.1.3). It will follow that every g-exact category is homological.

If the ideal NulE is pointed (produced by the zero object), we speak of a *p-homological* or *p-exact* category. We shall see in 2.2.3 that every g-exact category has *quasi pointed* connected components (cf. 1.3.2), so that its structure is merely categorical.

In particular, a p-exact category is a category with zero-object having kernels and cokernels and such that every map factorises conormal epinormal mono. In other words, it is an exact category in the sense of Puppe (see Part I).

It is also well known that a p-exact category is abelian (with a determined additive structure) if and only if it has finite products (Theorem I.2.1.5; also stated in [HeS, AHS] and proved in [FS]).

1.3.7 Subcategories

We also need now the following notions, that will be better analysed in 1.7.3.

Let E' be a subcategory of the ex1-category E. We say that E' is a *semiexact subcategory*, or *ex1-subcategory*, of E if E', equipped with the ideal $E' \cap \text{Nul}E$, is semiexact and the inclusion $E' \to E$ preserves kernels and cokernels.

In the same way one defines an *ex2-*, or *homological*, or *g-exact*, or *p-exact subcategory* of a category of the same kind.

1.4 Structural examples

We treat here some examples of homological and p-exact categories that have a structural relevance to the theory itself. Other examples are dealt with in Section 1.6, or listed in Section 9 of the Introduction, with a reference to other parts of this book.

1.4.1 Lattices and connections

The category Ltc of lattices and connections, studied in Section 1.2, is p-homological.

Indeed we already know that it has a zero-object $0 = \{*\}$, kernels and cokernels (see 1.2.3), whence it is p-semiexact.

The normal subobjects and quotients of the lattice X are represented by quasi sublattices of type $\downarrow a$ and $\uparrow a$ respectively, for $a \in X$ (cf. (1.18)); a general short exact sequence, as in (1.20), is of the following type

$$\downarrow a \xrightarrow{\ m\ } X \xrightarrow{\ p\ } \uparrow a \tag{1.42}$$

$$m_\bullet(x) = x, \quad m^\bullet(x) = x \wedge a; \qquad p_\bullet(x) = x \vee a, \quad p^\bullet(x) = x.$$

The exact morphisms (cf. 1.3.3) are precisely the exact connections $f = (f_\bullet, f^\bullet)$ considered in 1.2.5, since one of their characterisations is to factorise $f = (\text{nim } f).g.(\text{ncm } f)$ with g an isomorphism; non-exact connections exist (see 1.2.5), whence Ltc is not p-exact.

The characterisation of normal subobjects and quotients shows that the axiom (ex2) is satisfied; it also shows that (ex3) holds, and subquotients can be identified with the closed intervals of X

$$
\begin{array}{ccc}
\downarrow a & \rightarrowtail & X \\
\downarrow & & \downarrow \\
[b,a] & \rightarrowtail & \uparrow b \qquad a \geqslant b \text{ in } X.
\end{array}
\qquad (1.43)
$$

The pointed subcategory Mlc of modular lattices and modular connections (cf. 1.2.8) is closed in Ltc under subobjects and quotients, whence it is a p-homological subcategory of the latter. Furthermore, it is p-exact, since a modular connection is by definition an exact morphism (see 1.2.5).

The full subcategory Dlc of distributive lattices (introduced in 1.2.8) is a p-exact subcategory of the former. We already know, from Part I, that both lack products and are not abelian.

1.4.2 A basic homological category

The category Set_2 of *pairs of sets*, in the sense of algebraic topology [ES], is a non-pointed example that explains even better the difference between semiexact (or homological) categories and p-exact (or abelian) categories. (The categories Top_2 and Gp_2, of pairs of topological spaces or groups, have a similar structure, see 1.6.7.)

An object is a pair (X, A) where X is a small set and A is a *subset* of X; a morphism $f: (X, A) \to (Y, B)$ is a *map of pairs*, i.e. a mapping $f: X \to Y$ that takes A into B. The pair (X, A) is often read as X *modulo* A.

This category has a natural semiexact structure where f is *null* if and only if $f(X) \subset B$. The null objects are the pairs (X, X) and f is null if and only if it factorises through the null pair (B, B).

The kernel, cokernel and normal factorisation of f are displayed in the following diagram, with $A' = f^{-1}(B)$ and $B' = f(X) \cup B$

$$
\begin{array}{ccccccc}
(A', A) & \overset{k}{\rightarrowtail} & (X, A) & \overset{f}{\longrightarrow} & (Y, B) & \overset{c}{\twoheadrightarrow} & (Y, B') \\
& & {\scriptstyle p}\downarrow & & {\scriptstyle m}\uparrow & & \\
& & (X, A') & \underset{g}{\longrightarrow} & (B', B) & &
\end{array}
\qquad (1.44)
$$

Here k, c, p, m are defined by inclusions and g is induced by f (i.e. $g(x) = f(x)$, for $x \in X$). The morphism f is exact if and only if it is injective and $f(X) \supset B$.

Every normal subobject and every normal quotient of (X, A) is determined by a set M with $X \supset M \supset A$

$$(M, A) \xrightarrow{\ m\ } (X, A) \xrightarrow{\ p\ } (X, M) \tag{1.45}$$

and every short exact sequence (cf. (1.39)) in Set_2 is of this type, up to isomorphism.

Axiom (ex2) is obviously satisfied, and (ex3) is proved by the diagram below, that also shows the construction of subquotients

$$
\begin{array}{ccc}
(A, Y) & \xrightarrow{\ m\ } & (X, Y) \\
{\scriptstyle h}\big\downarrow & & \big\downarrow{\scriptstyle q} \\
(A, B) & \xrightarrow[\ k\ ]{} & (X, B)
\end{array}
\tag{1.46}
$$

where (X, Y) is a pair of sets and $X \supset A \supset B \supset Y$.

Notice that the morphisms $(A, Y) \rightarrowtail (X, Y) \twoheadrightarrow (X, B)$ are defined as soon as $X \supset A \supset Y$ and $X \supset B \supset Y$, but their composite is an exact morphism if and only if $A \supset B$, by the above characterisation of exact maps: the hypothesis $m \geqslant \ker q$ of (ex3) cannot be weakened, in Set_2.

Note also the following properties of Set_2, in contrast with the behaviour of p-exact or abelian categories:

(a) the null maps $(X, A) \to (Y, B)$ between two given pairs need neither exist (take $X \neq \emptyset$, $B = \emptyset$) nor be unique;

(b) monomorphisms (i.e. the injective mappings of pairs) need not have a null kernel and are of little interest;

(c) a null morphism need not be exact;

(d) an exact monomorphism need not be a normal mono (e.g. the normal quotient in (1.45) is mono but is not a normal one, if $A \neq M$);

(e) the initial object $(\emptyset, \emptyset)$ and the terminal object $(\{*\}, \{*\})$ are distinct. They are both *null* (a general property of semiexact categories, cf. 1.5.3(h)); but these objects do not cover the null ones.

The normal subobjects of any pair (X, A) plainly form a boolean algebra; yet the behaviour of direct and inverse images of normal subobjects, in Set_2, is not 'modular', as we shall see later (cf. 2.3.2).

Furthermore, products and sums exist and are the obvious ones, set-theoretic products and sums on each component of the pairs; but, of course, the functors $- \times (X, A)$ and $- + (X, A)$ do not preserve the null objects (unless (X, A) is null), a fact that restricts their interest in the present context. The relevant tensor product of Set_2 is similar to the smash product of pointed sets (see 1.4.5, 2.4.6).

We prove below, in 1.4.6, that Set_2 is a regular category, not Barr-exact.

But it is more important to notice that the short exact sequences (1.45), determined by the assigned ideal of null morphisms, have nothing to do with the short exact sequences of Set_2 *in the sense of regular categories -* characterised in 1.4.6(c).

The sequences of type (1.45) are the relevant ones in algebraic topology (in Set_2 and even more in the analogous categories Top_2 and Gp_2).

Finally, it will be of interest to note that Set can be embedded in Set_2 by identifying the set X with the pair $(X, \emptyset)$. In this way each object (X, A) is indeed a quotient X/A, i.e. the cokernel of the embedding of A in X, by means of the short exact sequence

$$A \stackrel{m}{\rightarrowtail} X \stackrel{p}{\twoheadrightarrow} (X, A) . \tag{1.47}$$

1.4.3 A p-exact category

The category $\mathcal{I}$ of (small) *sets and partial bijections* has played an important role in Part I. (Indeed, each 'distributive exact theory' has a classifying p-exact category that can be realised as a p-exact subcategory of $\mathcal{I}$; a precise formulation of this fact can be found in I.5.5.6.)

An object of $\mathcal{I}$ is a small set, a map $f \colon X \to Y$ is a *partial bijection*, i.e. a set-theoretic relation that is injective and single-valued, or equivalently a bijection between a subset $\mathrm{Def}\, f$ of X (the *definition* of f) and a subset $\mathrm{Val}\, f$ of Y (the *values* of f) The composition is obvious (inherited from relations).

The zero object is the empty set, and the zero map $0_{XY} \colon X \to Y$ is the empty bijection. The kernel of $f \colon X \to Y$ is the *inclusion* in X of $\mathrm{Ker}\, f$, the set of points of X where f is not defined, while the cokernel of f is the *coinclusion* of Y with values in the set of points that f does not cover

$$\begin{aligned} \ker f &= (\mathrm{Ker}\, f \rightarrowtail X), & \mathrm{Ker}\, f &= X \setminus \mathrm{Def}\, f, \\ \mathrm{cok}\, f &= (Y \twoheadrightarrow \mathrm{Cok}\, f), & \mathrm{Cok}\, f &= Y \setminus \mathrm{Val}\, f. \end{aligned} \tag{1.48}$$

The normal factorisation of f is the following

$$\begin{array}{ccccc} X \setminus \mathrm{Def}\, f & \rightarrowtail & X & \stackrel{f}{\longrightarrow} Y & \longrightarrow & Y \setminus \mathrm{Val}\, f \\ & & \downarrow & \uparrow & \\ & & \mathrm{Def}\, f & \underset{f_0}{\longrightarrow} \mathrm{Val}\, f & \end{array} \tag{1.49}$$

Plainly, each morphism is exact; the coimage and the image of f are $\mathrm{Def}\, f$ and $\mathrm{Val}\, f$. The (normal) subobjects of the set X can be identified with its subsets: $\mathcal{I}$ is distributive (i.e. its lattices of subobjects are distributive), whence it cannot be abelian (see 2.4.4).

$\mathcal{I}$ is also an *inverse category*, since every morphism $f\colon X \to Y$ has precisely one *generalised inverse* $g\colon Y \to X$ (satisfying $f = fgf$ and $g = gfg$); namely, $g = f^\sharp$ is the reverse relation. As any inverse category, the category $\mathcal{I}$ is self-dual, by means of the (contravariant) involution $f \mapsto f^\sharp$ (see I.2.7).

We have also used the p-exact subcategory $\mathcal{I}_0$ of (small) *sets and common parts*, introduced in I.4.6.2. A morphism $L\colon X \to Y$ is given by any common subset L of X and Y (identified with the corresponding 'partial identity' $X \to Y$ of $\mathcal{I}$); the composition is the intersection. (Recall that all small sets, being elements of the universe $\mathcal{U}$, are also subsets of $\mathcal{U}$.) A sequence

$$X \xrightarrow{\ L\ } Y \xrightarrow{\ L'\ } Z \tag{1.50}$$

is exact in Y if and only if L and L' are complementary subsets of Y.

1.4.4 Graded objects

Consider now the category GrAb of graded abelian groups $A = (A_n)_{n \in \mathbb{Z}}$ with morphisms of *any* degree $k \in \mathbb{Z}$

$$f\colon A \to B, \qquad f = (\!{}^\centerdot f_n\colon A_n \to B_{n+k})_{n \in \mathbb{Z}}. \tag{1.51}$$

More precisely, a morphism should be defined as pair (f, k), so that its degree is determined (even when all the components of A and B are equal).

Its object $(0)_{n \in \mathbb{Z}}$ is *quasi zero* (in the sense of 1.3.2) and produces the quasi pointed ideal of morphisms whose components are zero homomorphisms. Plainly, GrAb has kernels and cokernels with respect to this ideal and every morphism is exact; it is a *generalised* exact category (see 1.3.6), it is not p-exact and lacks products.

More generally, given a semiexact (or homological, or g-exact) category E and an abelian group K, the category $\mathrm{Gr}_K \mathsf{E}$ of K-graded objects over E has a natural structure of the same kind, with respect to the ideal of morphisms whose components are null. If E is pointed, $\mathrm{Gr}_K \mathsf{E}$ is quasi pointed. The study of spectral sequences can be simplified by working in such categories and deferring to the end the tedious arguments on *bidegrees*, that vary in $K = \mathbb{Z}^2$ (see Section 3.5).

Notice that the subcategory E_+ of GrAb with the same objects and the morphisms *of degree* $\geqslant 0$ is a homological subcategory of GrAb (in the sense of 1.3.7) but is not g-exact: an isomorphism of GrAb of positive degree is not invertible in E_+, even though its kernel and cokernel are null. Similarly, $\mathbb{N}$-graded abelian groups, with the obvious null morphisms, only give a homological category.

1.4.5 * The canonical enriched structure

(This section and the following, both marked with a star, are addressed to readers with specific interests in category theory, and are not used elsewhere.)

The category of pairs of sets is also important from a structural point of view, as it plays for N-categories (cf. 1.3.1) - in particular for the semiexact ones - a role similar to that of Set for general categories, or of Ab for preadditive categories.

Firstly, the category Set_2 of pairs of sets has a natural monoidal closed structure (see A4.2), with a tensor product related with the smash product of pointed sets (see 2.4.6)

$$(X, A) \otimes (Y, B) = (X \times Y, (X \times B) \cup (A \times Y)). \tag{1.52}$$

The internal hom is:

$$\mathrm{Hom}((X, A), (Y, B)) = (\mathsf{Set}_2((X, A), (Y, B)), N((X, A), (Y, B)), \tag{1.53}$$

where $N((X, A), (Y, B)) = \mathsf{Set}(X, B)$ is the subset of null morphisms.

The adjunction isomorphism is plain (writing $\hat{X}$ for the pair (X, A), and so on)

$$\mathrm{Hom}(\hat{X} \otimes \hat{Z}, \hat{Y}) \to \mathrm{Hom}(\hat{X}, \mathrm{Hom}(\hat{Z}, \hat{Y})),$$
$$f \mapsto (x \mapsto f(x, -) \colon \hat{Z} \to \hat{Y}). \tag{1.54}$$

Each of the functors (1.52), (1.53) is an N-functor in each variable.

Secondly, an N-category E with small hom-sets is the same as a category enriched over Set_2

$$\mathrm{Hom}(A, B) = (\mathsf{E}(A, B), \mathrm{NulE}(A, B)). \tag{1.55}$$

It can also be noted that Set_2 has a *classifier of normal subobjects*. In fact, the following object Ω equipped with the map t ('true') defined on the terminal object $\top = (\{*\}, \{*\})$

$$\Omega = (\{0, 1\}, \{1\}),$$
$$t \colon \top \rightarrowtail \Omega, \qquad t(*) = 1, \tag{1.56}$$

'classifies' the normal subobjects m of any object (X, A). This means that the mapping that assigns to any morphism $\chi \colon (X, A) \to \Omega$ the pullback m of t along χ is a bijective correspondence between the set $\mathsf{Set}_2((X, A), \Omega)$

and the set Nsb(X, A)

$$
\begin{array}{ccc}
(Y, A) & \longrightarrow & (\{*\}, \{*\}) \\
\llap{m}\downarrow & & \downarrow\rlap{t} \\
(X, A) & \underset{\chi}{\longrightarrow} & (\{0, 1\}, \{1\})
\end{array}
\qquad . \qquad (1.57)
$$

(The map χ is called the *characteristic map* of the associated normal subobject m.)

A reader acquainted with the theory of elementary toposes, introduced by F.W. Lawvere and M. Tierney, will notice that the domain of t is the terminal object and *not* the identity of the monoidal product, $I = (\{*\}, \emptyset)$.

1.4.6 * Proposition

The category Set$_2$ *has the following properties, with respect to limits and colimits (cf. Section A2 of the Appendix).*

(a) It is complete and cocomplete.

(b) It is regular, not Barr-exact (cf. A2.6). A 'short exact sequence' in the sense of regular categories is of the following type, up to isomorphism

$$
(R, R_0) \rightrightarrows (X, A) \xrightarrow{p} (Y, B) \qquad (1.58)
$$

$$
R = \{(x, x') \in X \times X \mid p(x) = p(x')\}, \quad R_0 = R \cap (A \times A),
$$

$$
p(X) = Y, \qquad\qquad p(A) = B.
$$

Note. Top$_2$ is not regular; Gp$_2$ is regular, not Barr-exact.

Proof (a) We have already observed, in 1.4.2, that products and sums in Set$_2$ are the obvious ones, set-theoretic products and sums on each component of the pairs

$$
\Pi_i(X_i, A_i) = (\Pi_i X_i, \Pi_i A_i), \quad \Sigma_i(X_i, A_i) = (\Sigma_i X_i, \Sigma_i A_i). \qquad (1.59)
$$

Equalisers (cf. A1.4), and all limits, are also computed componentwise. On the other hand, the coequaliser p of a pair of maps

$$
(X', A') \rightrightarrows (X, A) \xrightarrow{p} (Y, B) \qquad (1.60)
$$

is computed by letting $p \colon X \to Y$ be the coequaliser of the mappings $X' \rightrightarrows X$ in Set, and then taking $B = p(A)$. (The latter need *not* be the coequaliser of the restrictions $A' \rightrightarrows A$; take $A' = \emptyset$ to construct an easy counterexample.)

(b) The previous point shows that *regular epis* of Set$_2$, i.e. coequalisers, amount to maps $p\colon (X, A) \to (Y, B)$ where $p(X) = Y$ and $p(A) = B$ (while a general epi only has to satisfy the condition $p(X) = Y$). Since pullbacks are computed componentwise on pairs of sets, and epimorphisms are pullback-stable in Set, it follows easily that regular epis are stable under pullbacks, in Set$_2$.

If (1.60) is short exact in the sense of regular categories, (X', A') is a kernel pair of p, and we fall in the situation described in (1.58), up to isomorphism.

Finally, we prove that Set$_2$ is not Barr-exact. It is easy to see that an equivalence relation $(R, R_0) \rightrightarrows (X, A)$ (as defined in the theory of regular categories) amounts to assigning two ordinary equivalence relations $R \subset X \times X$, $R_0 \subset A \times A$ with $R_0 \subset R$. But this is a kernel pair if and only if $R_0 = R \cap (A \times A)$, which need not be true. □

1.5 Semiexact categories and normal subobjects

E is always a semiexact category. We prove that the normal subobjects and the normal quotients of any object A form anti-isomorphic lattices, Nsb$A$ and Nqt$A$, that can be 'transferred' by direct and inverse images (1.5.6 - 1.5.8).

This yields the transfer functor Nsb$\colon$ E $\to$ Ltc of our semiexact category E, with values in the homological category Ltc of lattices and Galois connections. Besides being a useful tool, this functor, that preserves and reflects exact sequences, can be seen as *a description of the logic structure* of E.

1.5.1 Semiexact categories and local smallness

We say that the semiexact category (see 1.3.3) E is a *semiexact $\mathcal{U}$-category* if:

(a) it is a $\mathcal{U}$-category, i.e. all its objects and morphisms belong to the universe $\mathcal{U}$ (see A1.1),

(b) it is *normally well-powered*, i.e. all its sets Nsb$A$ of normal subobjects (cf. 1.3.5) belong to $\mathcal{U}$.

By kernel duality (see 1.3.5), the sets Nqt$A$ of normal quotients are also small (and our condition is self-dual).

Since we shall mainly consider such categories, *semiexact* (or *ex1*) *category will mean, from now on, semiexact $\mathcal{U}$-category*. A similar convention applies to all the derived notions: homological, g-exact, p-exact, and so on. When, as in Sections 2.7 and 5.2, we do not want to assume any hypothesis of 'local' smallness we shall speak of an *unrestricted* semiexact (or homological,...) category.

1.5.2 Exact sequences

In the semiexact category E the sequence

$$A \xrightarrow{f} B \xrightarrow{g} C \tag{1.61}$$

is said to be *of order two* if it satisfies the following three equivalent conditions

$$gf \text{ is null}, \quad \operatorname{nim} f \leqslant \ker g, \quad \operatorname{cok} f \geqslant \operatorname{ncm} g. \tag{1.62}$$

The first equivalence comes from (1.40), the second from kernel duality, (1.38).

The sequence (1.61) is *exact* (in B) if $\operatorname{nim} f = \ker g$, or equivalently $\operatorname{cok} f = \operatorname{ncm} g$. It is *strongly exact* if, moreover, f and g are exact morphisms.

A *short exact* sequence (see 1.3.5), defined by the conditions $f \sim \ker g$ and $g \sim \operatorname{cok} f$, is exact. We shall see that it is the same as 'a *strongly* exact sequence with two null objects and three intermediate objects'; in order to avoid misunderstandings, the reader might take, as of now, a look at this characterisation, in 1.5.5(f).

It is easy to see that the sequence (f, g) is exact if and only if, in the diagram below

$$A \xrightarrow{f} B \xrightarrow{g} C \tag{1.63}$$

(a) gf is null,

(b) whenever gu and vf are null, vu is also.

Actually, as we already know that (a) is equivalent to $\operatorname{nim} f \leqslant \ker g$, it suffices to prove that (b) is equivalent to the opposite inequality. If (b) holds, take $u = \ker g$, $v = \operatorname{cok} f$; then $vu \in \mathcal{N}$ implies that $\ker g = u = \operatorname{nim} u \leqslant \ker v = \ker \operatorname{cok} f = \operatorname{nim} f$. Conversely, if $\ker g \leqslant \operatorname{nim} f$, from the annihilation of gu and vf it follows that $\operatorname{nim} u \leqslant \ker g \leqslant \operatorname{nim} f \leqslant \ker v$, i.e. vu is null.

If our category is p-semiexact (cf. 1.3.3), the conditions (a), (b) trivially amount to saying that the central square below (where D is the zero object) is *semicartesian*, as defined in [G1]. This means that the square is commutative and moreover

$$(gu = f'v' \text{ and } vf = u'g') \Rightarrow (vu = u'v') \tag{1.64}$$

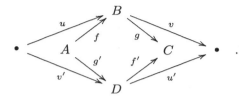

Notice that in any N-category (cf. 1.3.1) one can define the sequence (f, g) to be exact (or N-exact) if it satisfies the previous conditions (a), (b), *without* requiring the existence of kernels or cokernels. This fact has an interest in homotopical situations; for instance, the Puppe sequence [P1] of a continuous map (in Top) is exact with respect to the ideal of contractible maps [G14], but kernels and cokernels with respect to this ideal do not exist, in general.

1.5.3 Lemma (Annihilation properties)

Let E *be a semiexact category.*

(a) Every object A has a least normal subobject and a least normal quotient

$$0_A = \ker 1_A \colon A_0 \rightarrowtail A, \qquad 0^A = \operatorname{cok} 1_A \colon A \twoheadrightarrow A^0, \tag{1.65}$$

that are null *morphisms (they annihilate 1_A) and will both be written as* 0 *when this is not ambiguous. In particular, there exist null morphisms starting from any object A, or ending there. Moreover, 1_A is the greatest normal subobject (the kernel of the null morphism 0^A) and the greatest normal quotient of A.*

For instance, in Set$_2$ *(see 1.4.2) we have*

$$(X, A)_0 = (A, A), \qquad\qquad (X, A)^0 = (X, X)$$

(but the least subobject and quotient of (X, A) are $(\emptyset, \emptyset)$ and $(\{*\}, \{*\})$, respectively).

(b) The morphism $f \colon A \to B$ is null if and only if $\ker f = 1$, if and only if $\operatorname{cok} f = 1$, if and only if $\operatorname{nim} f = 0_B$, if and only if $\operatorname{ncm} f = 0^A$, if and only if f factorises through the least normal subobject 0_B, if and only if f factorises through the least normal quotient 0^A.

(c) Exact sequences determine null maps, N-monos and N-epis. More precisely, $f \colon A \to B$ is null if and only if $(1_A, f)$ is exact, while f is N-mono

if and only if it forms an exact sequence with some null map h with values in A (or equivalently, with every such h). Such maps exist, by (a).

(d) The object A has precisely one null normal subobject, namely 0_A, and precisely one null normal quotient, namely 0^A.

(e) The object A is null if and only if $\mathrm{Nsb}A = 0$ (the one-point lattice, see 1.2.3), if and only if $\mathrm{Nqt}A = 0$, if and only if the sequence of identities $A \to A \to A$ is exact, if and only if the latter is short exact.

(f) The objects A_0 and A^0 introduced in (1.65) are null.

(g) Every normal mono m is N-mono and every normal epi p is N-epi.

(h) The initial and terminal object, if extant, are necessarily null.

Note. The properties (f), (g) depend on the closedness of the ideal NulE; see also criterion 1.5.4(b). Other properties of closed ideals will be stated in 1.7.5 and 1.7.9.

Proof (a) and (b) are obvious; (c) is a trivial consequence.

(d) Any null normal subobject $m \in \mathrm{Nsb}A$ factorises through 0_A by (b), hence it is smaller than the least normal subobject.

(e) If A is null, all its normal subobjects are, and coincide by (c); conversely, if $\mathrm{Nsb}A$ has just one element, then the normal subobjects 1_A and 0_A coincide, i.e. 1_A is null and A too.

(f) We prove that A_0 is null. Since the morphism $0_A \colon A_0 \rightarrowtail A$ is null, it factorises through some null object Z

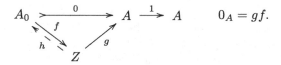

$$A_0 \xrightarrow{\ \ 0\ \ } A \xrightarrow{\ 1\ } A \qquad 0_A = gf.$$

Now g is null (its domain is Z), hence annihilates 1_A and factorises through $0_A = \ker 1_A$ as $g = 0_A h$; therefore $0_A(hf) = gf = 0_A$; but 0_A is mono and $hf = 1$. This means that A_0, as a retract of Z, is a null object.

(g) Take $m \colon M \rightarrowtail A$ and $0_A = mh$ (as $0_A \leqslant m$). Take $a \colon M' \to M$ with ma null; by (b), ma factorises through 0_A, $ma = 0_A b = mhb$; then $a = hb$ factorises through the domain A_0 of h and is null.

(h) If Z is an initial object, consider the normal subobject $0_Z \colon Z_0 \rightarrowtail Z$ and the unique morphism $i \colon Z \to Z_0$; then $0_Z.i \colon Z \to Z$ coincides with 1_Z, that is null. $\qquad\square$

1.5.4 Theorem (Two criteria for semiexact categories)

(a) Criterion A. *An N-category* E *satisfies (ex1), i.e. has kernels and cokernels, if and only if:*

(i) Every morphism has a kernel (existence of kernels);

(ii) every normal mono has a cokernel (existence of normal coimages),

(iii) for every morphism $f \colon A \to B$ *there is a least normal subobject* n *of* B *through which* f *factorises (existence of normal images).*

(b) Criterion B. *Let* E *be an N-category satisfying (ex1). Then the ideal* NulE *is closed (i.e.* E *is unrestricted semiexact) if and only if every normal mono of* E *is N-mono, if and only if every normal epi is N-epi.*

Note. Criterion A gives priority to kernels versus cokernels and will be used when the cokernels of normal monos are more easily constructed than the general ones.

Proof (a) If (ex1) holds, the conditions (i) - (iii) above are satisfied with $n = \operatorname{nim} f$, by (1.40) (which does not depend on the closedness of the ideal NulE).

Conversely, if these three conditions are satisfied, it suffices to prove that $q = \operatorname{cok} n$ is the cokernel of $f = nf'$. Indeed $qf = qnf'$ is null; if gf is null, f factorises through $\ker g$, whence $n \leqslant \ker g$ and $q \geqslant \operatorname{cok} \ker g$, i.e. $\operatorname{cok} \ker g$ factorises through q and g also does.

(b) If the ideal NulE is closed, the thesis follows from 1.5.3(g).

Conversely, if every normal mono is N-mono, we have to prove that any null morphism f factorises through a null identity. Indeed $1_B f = f$ is null, hence f factorises through $m = \ker 1_B \colon M \to B$, that is null; since $m 1_M = m$ is null, 1_M must also be. The second part follows by duality. $\square$

1.5.5 Normal factorisations and exact morphisms

E is always a semiexact category. We can now complete the *normal factorisation* of a morphism $f \colon A \to B$, as suggested in (1.33).

The map f factorises as $f = mf''$, through the normal image $m = \operatorname{nim} f$. Since $m(f''.\ker f) = f.\ker f$ is null, it follows (by 1.5.3(g)) that $f''.\ker f$ is null, i.e. f'' factorises through the normal coimage $p = \operatorname{ncm} f = \operatorname{cok} \ker f$

$$\begin{array}{ccccc}
\operatorname{Ker} f & \rightarrowtail & A & \xrightarrow{f} & B & \twoheadrightarrow & \operatorname{Cok} f \\
& & {\scriptstyle p}\big\downarrow & & \big\uparrow{\scriptstyle m} \\
& & \operatorname{Ncm} f & \xrightarrow[g]{} & \operatorname{Nim} f
\end{array} \qquad (1.66)$$

Since p is epi and m is mono, this factorisation of f is unique, and obviously *natural*.

With the same notation, we have the following consequences.

(a) The morphism f is *exact* if and only if this morphism g is an isomorphism, if and only if f factorises as $f = nq$, with q a normal epi and n a normal mono. Indeed, in this case, ncm $f = $ ncm $q \sim q$ (by 1.5.3(g)) and nim $f \sim n$.

(b) The morphism f is null if and only if it factorises as

$$A \twoheadrightarrow A^0 \to B_0 \rightarrowtail B,$$

through the null quotient of A *and* the null subobject of B. We have already seen that a null morphism need not be exact (in 1.4.2).

(c) The morphism f is a normal mono if and only if it is exact and N-mono. Indeed, if $f \sim \ker a$, then $\ker f = 0$ and $p = 1$; moreover $f \sim \ker \cok \ker a = m$, whence $gp = g$ is an isomorphism; conversely, if $f = mgp$ is exact and N-mono, then $\ker p = \ker f = 0$, p and g are isomorphisms and $f \sim m$.

(d) On the other hand, we have already seen that an exact monomorphism need not be normal (in 1.4.2).

(e) Normal monos and kernels are identified by the exactness of sequences *and* morphisms. It suffices to consider the sequences

$$M_0 \xrightarrow{\;0\;} M \xrightarrow{\;m\;} A \qquad\qquad M_0 \xrightarrow{\;0\;} M \xrightarrow{\;m\;} A \xrightarrow{\;f\;} B \;. \qquad (1.67)$$

By (c), the morphism $m \colon M \to A$ is a normal mono if and only if it is exact and the left sequence above is exact in M; moreover, m is a kernel of f if and only if it is exact and the right sequence above is exact in M and A.

(f) As a consequence, the sequence (m, p) is short exact if and only if the following sequence is exact *and* the morphisms m, p are exact, if and only if the same sequence is strongly exact (cf. 1.5.2)

$$M_0 \xrightarrow{\;0\;} M \xrightarrow{\;m\;} A \xrightarrow{\;p\;} P \xrightarrow{\;0\;} P^0. \qquad (1.68)$$

1.5.6 *Direct and inverse images*

Each morphism $f \colon A \to B$ has associated mappings, of *direct and inverse images for normal subobjects*

$$f_* \colon \mathrm{Nsb}A \to \mathrm{Nsb}B, \qquad f_*(m) = \mathrm{nim}\,(fm) = \ker \cok\,(fm), \qquad (1.69)$$

$$f^* \colon \mathrm{Nsb}B \to \mathrm{Nsb}A, \qquad f^*(n) = \ker\,((\cok n)f), \qquad (1.70)$$

$$
\begin{array}{ccc}
A & \xrightarrow{\ f\ } & B \\
{\scriptstyle m}\big\uparrow & & \big\uparrow{\scriptstyle f_*m} \\
\bullet & \dashrightarrow & \bullet
\end{array}
\qquad\qquad
\begin{array}{ccc}
A & \xrightarrow{\ f\ } & B \\
{\scriptstyle f^*n}\big\uparrow & & \big\uparrow{\scriptstyle n} \\
\bullet & \dashrightarrow & \bullet
\end{array}
\qquad (1.71)
$$

where the left-hand square is commutative while the right-hand square is easily seen to be a pullback.

This proves the existence in E of the *pullback of a normal mono along an arbitrary map.*

Dually, we have *direct and inverse images for normal quotients*

$$ f_\circ \colon \mathrm{Nqt}A \to \mathrm{Nqt}B, \qquad f_\circ(p) = \mathrm{cok}\,(f(\ker p)), \qquad (1.72) $$

$$ f^\circ \colon \mathrm{Nqt}B \to \mathrm{Nqt}A, \qquad f^\circ(q) = \mathrm{ncm}\,(qf), \qquad (1.73) $$

where (1.72) yields the existence and construction of the *pushout of a normal epi along an arbitrary map.*

In this way, acting on normal subobjects and quotients, the morphism f yields four squares of mappings, that are easily seen to commute

$$
\begin{array}{ccc}
\mathrm{Nsb}A & \underset{f^*}{\overset{f_*}{\rightleftarrows}} & \mathrm{Nsb}B \\[4pt]
{\scriptstyle \mathrm{cok}}\big\updownarrow{\scriptstyle \ker} & & {\scriptstyle \mathrm{cok}}\big\updownarrow{\scriptstyle \ker} \\[4pt]
\mathrm{Nqt}A & \underset{f^\circ}{\overset{f_\circ}{\rightleftarrows}} & \mathrm{Nqt}B
\end{array}
\qquad . \qquad (1.74)
$$

The action of f on normal subobjects is thus equivalent to the action on normal quotients

$$
\begin{aligned}
f_*(\ker p) &= \ker f_\circ p, & f^*(\ker q) &= \ker f^\circ q, \\
f_\circ(\mathrm{cok}\,m) &= \mathrm{cok}\,f_*m, & f^\circ(\mathrm{cok}\,n) &= \mathrm{cok}\,f^*n.
\end{aligned}
\qquad (1.75)
$$

In particular:

$$
\begin{aligned}
f_*0 &= 0, & f_*1 &= \mathrm{nim}\,f; & f^*0 &= \ker f, & f^*1 &= 1, \\
f_\circ 0 &= \mathrm{cok}\,f, & f_\circ 1 &= 1; & f^\circ 0 &= 0, & f^\circ 1 &= \mathrm{ncm}\,f.
\end{aligned}
\qquad (1.76)
$$

1.5.7 Lemma (Meets and detection properties)

Let A be an object of the semiexact category E. *In the ordered sets* $\mathrm{Sub}A$ *and* $\mathrm{Quo}A$ *(of general subobjects and quotients of A), the following meets always exist:*

(a) $f \wedge m \sim f(f^*m)$,   for $f \in \mathrm{Sub}A$ and $m \in \mathrm{Nsb}A$,

(a*) $f \wedge p \sim (f_\circ p)f$,   for $f \in \mathrm{Quo}A$ and $p \in \mathrm{Nqt}A$.

Furthermore, the following detection properties *hold (and will be frequently used):*

(b) if $n = fh$ is a normal mono and f is mono, then $h \sim f^*(n)$ is a normal mono;

(b*) if $p = gf$ is a normal epi and f is epi, then $g \sim f_\circ(p)$ is a normal epi;

(c) if mf is exact and m is a normal mono, then f is exact;

(c*) if fp is exact and p is a normal epi, then f is exact.

Finally, the null normal subobjects and the null normal quotients give:

(d) the composite $M_0 \rightarrowtail M \rightarrowtail A$ is always a normal mono, equivalent to $A_0 \rightarrowtail A$;

(d*) the composite $A \twoheadrightarrow P \twoheadrightarrow P^0$ is always a normal epi, equivalent to $A \twoheadrightarrow A^0$.

Proof (a) Follows from the existence of the pullback of the normal mono m along $f\colon X \to A$ (see (1.71)).

(b) Follows from (a), since $fh = n \sim f \wedge n \sim f.f^*(n)$ and f is mono.

(c) Let $mf = np$ be a normal factorisation; then $\ker f = \ker(mf)$ (by 1.3.1) and $\operatorname{ncm} f = \operatorname{ncm}(mf) = \operatorname{ncm}(np) \sim p$; therefore f factorises as $f = gp$ and $mgp = mf = np$, whence $mg = n$; by (a), g is a normal mono and $f = gp$ is exact.

(d) Let $m\colon M \rightarrowtail A$ be a normal mono. The null normal mono $0_A\colon A_0 \rightarrowtail A$ factorises as $0_A = mh$, where $h\colon A_0 \rightarrowtail M$ is a normal mono, by (b); since it is null, $h \sim 0_M$ and $0_A = mh \sim m0_M$. $\qquad\square$

1.5.8 Theorem and Definition (The transfer functor)

In the semiexact $\mathcal{U}$-category E, *all the ordered sets* NsbA *are small lattices. Direct and inverse images produce a functor*

$$\mathrm{Nsb}\colon \mathsf{E} \to \mathsf{Ltc},$$

$$A \mapsto \mathrm{Nsb}A, \tag{1.77}$$

$$f \mapsto \mathrm{Nsb}(f) = (f_*, f^*) \qquad (f_* \dashv f^*),$$

with values in the p-homological category Ltc *of lattices and Galois connections studied in Section 1.2. It will be called the* transfer functor *of* E, *for normal subobjects, or also the* projective functor; *the semiexact category* E *is said to be* projective *if this functor is faithful.*

(Notice that this functor takes values in the category of *small* lattices and connections because of the local smallness hypothesis of 1.5.1. Its exactness properties will be examined in 2.1.2 and 2.1.3, and the associated fibration in 1.5.9(d).)

Dually, there is a transfer functor for normal quotients with values in the category $\mathsf{Ltc}^{\mathrm{op}}$ *of lattices and 'reversed' connections (where the first term is the right adjoint)*

$$\mathrm{Nqt} \colon \mathsf{E} \to \mathsf{Ltc}^{\mathrm{op}},$$

$$A \mapsto \mathrm{Nqt}A, \tag{1.78}$$

$$f \mapsto \mathrm{Nqt}(f) = (f_\circ, f^\circ) \qquad (f^\circ \dashv f_\circ).$$

The lattice operations and the order relation in $\mathrm{Nsb}A$ *and* $\mathrm{Nqt}A$ *can be described as follows, where* $m, n \in \mathrm{Nsb}A$ *and* $p = \mathrm{cok}\, m$, $q = \mathrm{cok}\, n$ *are the corresponding normal quotients:*

$$m \wedge n = m_* m^*(n) = n_* n^*(m), \quad m \vee n = p^* p_*(n) = q^* q_*(m), \tag{1.79}$$

$$p \vee q = m_\circ m^\circ(q) = n_\circ n^\circ(p), \qquad p \wedge q = p^\circ p_\circ(q) = q^\circ q_\circ(p), \tag{1.80}$$

$$n \leqslant m \iff n^*(m) = 1, \qquad q \leqslant p \iff q_\circ(p) = 1. \tag{1.81}$$

Proof First, it is easy to check that the pair (f_*, f^*) is a Galois connection between *ordered sets*, i.e. both mappings are increasing and $f^* f_* m \geqslant m$, $f_* f^* n \leqslant n$.

We now prove that the ordered sets $\mathrm{Nsb}A$ and $\mathrm{Nqt}A$ are lattices. The meet of two normal subobjects m, n of A is already known to exist *in* $\mathrm{Sub}A$: $m \wedge n \sim m(m^*(n))$ (cf. 1.5.7(a)); therefore, *it is a normal mono*, by a general result (in 1.1.4(d)) on the meets of closed elements, with respect to the contravariant connection ker - cok (cf. (1.34)).

$\mathrm{Nsb}A$ is thus a meet-semilattice, with 0 and 1 (1.5.3(a)). By categorical duality (that *preserves* order), $\mathrm{Nqt}A$ is also a meet-semilattice, with 0 and 1. By kernel duality (which *reverses* order, see (1.38)), both are lattices.

The functorial property of $\mathrm{Nsb} \colon \mathsf{E} \to \mathsf{Ltc}$ can be checked on the contravariant part, that determines the covariant one; but this follows from the obvious pasting property of pullbacks.

For the left formula (1.79), we have already seen that the normal mono $m \wedge n$ is equivalent to $m.(m^*(n))$; therefore

$$m \wedge n \sim \mathrm{nim}\,(m(m^*(n))) = m_*(m^*(n)).$$

The other formulas in (1.79), (1.80) follow by categorical and kernel dualities.

Finally, the characterisation (1.81) of the order follows trivially from the fact that $m \wedge n \sim n_* n^*(m)$ is the diagonal of the pullback of m and n.   □

1.5.9 Remarks

(a) *The necessity of* Ltc. The transfer functor of the category Ltc itself is isomorphic to the identity functor of Ltc, via the functorial isomorphism

$$\iota: 1 \to \text{Nsb}: \text{Ltc} \to \text{Ltc}, \qquad \iota_X: X \to \text{Nsb}X,$$
$$(\iota_X)_\bullet: a \mapsto \downarrow a, \qquad (\iota_X)^\bullet: m \mapsto m_\bullet(1). \tag{1.82}$$

This shows that *every lattice X is isomorphic to a lattice of subobjects* of some object (X itself) in some semiexact category (and indeed in a fixed one, Ltc). Similarly, *every lattice-connection f is* (up to isomorphism) *the transfer connection* of some morphism (f itself) in a semiexact category (Ltc).

Therefore, no proper conservative subcategory (see A1.2) of Ltc could suffice to receive the transfer functors coming from all the semiexact $\mathcal{U}$-categories.

The previous result also shows that Ltc is a projective homological category (see 1.5.8). Mlc and Dlc are projective p-exact categories.

(b) In the p-exact category $\mathcal{I}$ of sets and partial bijections (see 1.4.3), direct and inverse images of subobjects along a morphism $f: X \to Y$ (with associated bijection f_0) are computed as follows

$$f_*(A) = f_0(A \cap \text{Def } f) \qquad (A \subset X),$$
$$f^*(B) = f_0^{-1}(B \cap \text{Im } f) \cup \text{Ker } f \qquad (B \subset Y). \tag{1.83}$$

The morphism $f: X \to Y$ is thus determined by its direct images of singletons, and $\mathcal{I}$ too is projective

$$f_*\{x\} = \{f(x)\} \text{ if } x \in \text{Def } f, \qquad f_*\{x\} = \emptyset \text{ if } x \in \text{Ker } f. \tag{1.84}$$

(c) The homological category Set$_2$ of pairs of sets (see 1.4.1) is not projective, since all the morphisms $(X, X) \to (X, X)$, being null, are identified by Nsb (cf. 2.2.4).

(d) *The canonical fibration of a semiexact category.* For a reader interested in category theory, let us note that the category Ltc fully embeds in the category Adj of small categories and adjoint pairs of functors.

Therefore every semiexact category has a canonical functor Nsb$'$: E $\to$ Adj. This can be seen as a 'categorical fibration'

$$P: \text{E}_2 \to \text{E}, \tag{1.85}$$

where the category E_2, obtained by gluing the small categories $NsbA$ over the basis E, is precisely the category 'of pairs' of E with respect to its normal subobjects (Section 2.5) and the forgetful functor P associates to a normal subobject its codomain.

1.6 Other examples of semiexact and homological categories

After Section 1.4, we treat here other examples of semiexact categories.

The first three of them (in 1.6.1) satisfy (ex3) but not (ex2); the last (in 1.6.9) satisfies (ex2) but not (ex3): it is a formal construction, whose only interest is to prove the independence of these two axioms. With a similar goal, we also give an example of a category having kernels and cokernels with respect to a non-closed ideal (in 1.6.9). The other examples are homological categories.

The structured sets we consider are always assumed to be small; in each case, the characterisation we give for the normal subobjects of an object A proves that they form a *small* lattice ($NsbA$), even if this fact is not explicitly remarked. Various of these examples are of interest in algebraic topology.

1.6.1 Groups, rings and groupoids

The categories Gp and Rng, respectively of groups and *general* rings (without unit assumption), are pointed semiexact; they satisfy (ex3) and not (ex2).

As well known, in these categories the normal subobjects 'are' the (embeddings of) invariant subgroups or ideals, and are not closed under composition; every epimorphism is normal; the exact morphisms are those whose image is a normal subobject.

As to (ex3), given two invariant subgroups (or ideals) $M \supset N$ of the object A, it suffices to consider the following commutative diagram (that is bicartesian) and verify that the induced homomorphism $M/N \to A/N$ is indeed a normal mono, the kernel of $A/N \to A/M$

$$
\begin{array}{ccc}
M & \xrightarrow{\ m\ } & A \\
{\scriptstyle h}\downarrow & & \downarrow{\scriptstyle q} \\
M/N & \xrightarrow[\ k\ ]{} & A/N
\end{array}
\qquad . \tag{1.86}
$$

Consider now the N-category Gpd of (small) groupoids and their homomorphisms (i.e. functors), equipped with the ideal $\mathcal{N}$ of homomorphisms taking every map of the domain into some identity; the ideal is closed and the null groupoids coincide with the discrete ones (all of whose maps are identities).

Using Criterion A of 1.5.4, it is easy to see that Gpd is semiexact: kernels, normal subobjects and normal quotients with respect to $\mathcal{N}$ coincide with those usually considered in groupoid theory [Br, Hg]; as above, Gpd satisfies (ex3) but not (ex2).

Actually, the kernel of $f\colon A \to B$ is the subgroupoid of A formed by all the objects, and those maps that f takes to identities; a normal subgroupoid H of A is characterised by being *wide* (same objects as A) and by the following *invariance* property:

$$(h \in H,\ a \in A)\ \Rightarrow\ aha^{-1} \in H, \tag{1.87}$$

(if the composition makes sense, i.e. $\operatorname{Dom} h = \operatorname{Dom} a = \operatorname{Cod} h$).

Then, the quotient A/H has objects and maps obtained from those of A, modulo the equivalence relations below, and the composition described below, in (1.90)

$$a_0 \sim a_1 \text{ in } \mathrm{Ob}A \quad \text{if there is a map } h\colon a_0 \to a_1 \text{ in } H, \tag{1.88}$$

$$a \sim a' \text{ in } \mathrm{Mor}A \quad \text{if } ka = a'h, \text{ for some } h, k \text{ in } H, \tag{1.89}$$

$$[b].[a] = [bha], \quad \text{where } h \in H(\operatorname{Cod} a, \operatorname{Dom} b). \tag{1.90}$$

These semiexact, non-homological categories will play a minor role in the sequel. The present terminology is not well adapted to them, *nor it wants to be.*

In fact, in algebraic topology long exact sequences of homotopy groups have a well-known habit of *getting out of the category*, degenerating into pointed sets and actions of groups on the latter. Therefore we prefer to view homological facts concerning groups in larger, *homological* categories, such as Gp_2 (pairs of groups, see 1.6.7) or the even larger category Act (actions of groups on pointed sets, cf. Section 6.4). We shall see (in 1.6.8 and Chapter 6) that in these categories:

- *all subgroups become normal subobjects,*

- *the exactness of a sequence of group-homomorphisms has its usual meaning.*

For instance the sequences of homotopy 'objects' for a pair of pointed spaces or for a fibration can be viewed as exact sequences in Act (6.4.7, 6.4.8); or also in Gp_2, for path-connected spaces.

On the other hand, if one is interested in studying Gp (and various other categories of 'algebras'), a non-selfdual setting based on factorisations via a normal epi and a general mono, like Barr-exact categories and their variations or [BoG], is probably more adequate.

Indeed, dealing with Gp as a p-semiexact category, one finds the following facts, which only agree in part with the usual analysis. A *short exact sequence*

$$H \rightarrowtail A \twoheadrightarrow A/H$$

is determined by an invariant subgroup H of A, and is consistent with the usual notion. But *a general sequence* (f, g) is *exact in the present sense* (see 1.5.2) if and only if Ker g coincides with Nim f, the normal subobject spanned by the image; it is *strongly exact* (by 1.5.2, again) if and only if Ker $g = $ Im f *and* g is exact. None of these conditions is equivalent to the *usual* one, Ker $g = $ Im f, which should be expressed in a rather strange way, saying that the sequence *and* the morphism f are exact, in the present sense.

1.6.2 Abelian monoids, semimodules, preordered abelian groups

The category Abm of abelian monoids is p-homological. A normal subobject H of A is a submonoid 'closed under difference' in the following sense

$$\text{if } a \in A \text{ and } h, k \in H, \text{ then } a + h = k \text{ implies } a \in H. \tag{1.91}$$

The associated normal quotient A/H is the quotient of the monoid A modulo the congruence $\sim$ spanned by annihilating H, namely:

$$a \sim b \quad \text{if and only if there are } h, k \in H \text{ such that } a + h = b + k. \tag{1.92}$$

If A satisfies the cancellation property, the normal submonoids of A coincide with the traces on A of the subgroups of the associated abelian group. For instance, the normal submonoids of the additive monoid $\mathbb{N}$ are cyclic, of type $a\mathbb{N}$, while the submonoid $\mathbb{N} \setminus \{1\}$, spanned by 2 and 3, is not normal.

The verification of (ex2) by this characterisation of normal monos and epis is straightforward; (ex3) is proved as for groups (in 1.6.1).

More generally, the category R Smd of *semimodules* over a unital semiring R is p-homological, with an analogous characterisation of normal subsemimodules and normal quotients. If R is a ring, this category becomes the abelian category of R-modules (with opposites $-x = (-1_R).x$).

The category pAb of preordered abelian groups and (weakly) increasing homomorphisms is also p-homological. A normal subobject is a subgroup with the restricted preorder, while a normal quotient is a quotient group

with the induced preorder. As Ltc (see 1.2.7), it is a preordered category
and a semiadditive one: the finite biproducts are the usual direct sums
of abelian groups with the componentwise preorder; opposite maps are
missing.

1.6.3 Topological vector spaces

The category K Tvs of topological vector spaces and continuous linear mappings (over a topological field K) is p-homological, as well as the category
K Hvs of Hausdorff vector spaces (over a Hausdorff field K).

The normal subobjects of K Tvs (K Hvs) are the (closed) linear subspaces, the normal quotients are the quotients modulo the former, with the
quotient topology. Axiom (ex2) is trivial.

In order to verify (ex3), consider the bicartesian square (1.86) in the
present situation; we must show that the map k is a topological embedding.
Indeed, if W is open in M/N, $h^{-1}W$ is open in M and $h^{-1}W = m^{-1}U$,
for some U open in A; taking into account the fact that the square is a
pullback, we have

$$W = h(h^{-1}(W)) = h(m^{-1}(U)) = k^{-1}(q(U)), \tag{1.93}$$

so that W is the inverse k-image of $q(U)$. But the latter is open in A/N,
because of a general property of topological groups: the projection on a
quotient is necessarily open.

Of course the same holds for topological modules (over a unital topological ring) and in particular for topological abelian groups. Topological
groups, on the other hand, form a p-semiexact category satisfying (ex3),
as Gp.

The category Ban of Banach spaces (on the real or complex field) and
bounded (i.e. continuous) linear mappings is also p-homological. A normal
subobject of X is a closed linear subspace H, a normal quotient X/H is the
'algebraic' quotient, provided with the norm $||x+H|| = \inf_h ||x+h||$, where
h varies in H. The normal image is the closure of the algebraic image. By
the open-map theorem, the exact morphisms are precisely the bounded linear mappings with closed image (and are not stable under composition);
every monomorphism with closed image is normal, every surjective morphism is a normal epi.

The category Hlb of Hilbert spaces and continuous linear mappings has
a similar p-homological structure. The quotient X/H of a space modulo
a closed linear subspace is now equipped with the scalar product coming
from the canonical isomorphism with the orthogonal subspace of H in X.

While a p-exact category with biproducts is necessarily abelian, it can be noted that the pointed categories we have considered above (namely Abm, R Smd, K Tvs, K Hvs, Ban, Hlb) have biproducts $X \oplus Y$ given by the ordinary cartesian product; this also holds for Ltc (cf. 1.2.7).

1.6.4 Pointed sets and spaces

The category Set. of pointed sets and pointed mappings has zero object, the singleton, and is p-homological.

In fact, kernels and cokernels are obvious; a normal subobject $H \rightarrowtail X$ of a pointed set $X = (|X|, 0_X)$ is the inclusion of a subset $|H| \subset |X|$ containing the base point; a quotient X/H is obtained by identifying all the elements of H with the base point. A morphism is exact precisely when it is injective outside of its kernel; the exact morphisms form a subcategory $\mathcal{I}'$ *equivalent* to the p-exact category $\mathcal{I}$ of sets and partial bijections (see 1.4.3; equivalence of categories is recalled in A1.7).

Notice that a pointed set can be seen as a mapping $\top \to |X|$ of sets, defined on the terminal object $\top = \{*\}$ of Set, while a pointed map $f \colon X \to Y$ can be seen as a mapping $|X| \to |Y|$ between the underlying sets that forms a commutative triangle with $\top \to |X|$ and $\top \to |Y|$. In other words, Set. can be identified with the 'slice category' $\mathsf{Set}^\top$ of 'the objects under $\top$', i.e. the full subcategory of the category of morphisms $\mathsf{Set}^\mathbf{2}$ whose objects are the maps defined on the singleton $\top$.

Similarly, the category Top. (or $\mathsf{Top}^\top$) of pointed topological spaces and pointed continuous maps, is p-homological. Kernels and cokernels are described as in Set., with the subspace or the quotient topology, respectively.

1.6.5 Categories of partial mappings

The category $\mathcal{S}$ of sets and partial mappings, defined on a subset of their domain, is equivalent to this category Set. and therefore p-homological.

The equivalence consists of the functors

$$(\ )^- \colon \mathsf{Set}_\bullet \to \mathcal{S}, \quad X \mapsto X^- = X \setminus \{0_X\}, \quad f \mapsto f^-, \qquad (1.94)$$

$$(\ )^+ \colon \mathcal{S} \to \mathsf{Set}_\bullet, \quad S \mapsto S^+ = (S \cup \{\infty_S\}, \infty_S), \quad g \mapsto g^+. \qquad (1.95)$$

The first deprives a pointed set X of its base point, while the second adds to a set S a base point $\infty_S \notin S$. For a pointed mapping $f \colon X \to Y$, the *partial* mapping $f^- \colon X^- \to Y^-$ is the restriction of f defined on $X \setminus f^{-1}\{0_Y\}$. The action of $(\ )^+$ on partial mappings is obvious.

(A possible canonical choice for the base point ∞_S is S itself, since - by the regularity axiom of set theory - a set cannot belong to itself; but this choice may look unpleasant and misleading.)

The zero object of $\mathcal{S}$ is the empty set. The kernel of a map $f\colon X \to Y$ is (the inclusion of) the complement $X \setminus \mathrm{Def}\,(f)$ of the definition-set, while the cokernel is (the 'co-inclusion' with values in) the complement $Y \setminus f(X)$ of the set-theoretic image (a partial mapping, defined on $Y \setminus f(X)$)

$$X \setminus \mathrm{Def}\, f \rightarrowtail X, \qquad Y \twoheadrightarrow Y \setminus f(X). \qquad (1.96)$$

Therefore, the normal subobjects can be identified with the inclusions of subsets, and the normal quotients with the restrictions to subsets. The exact morphisms are the partial bijections, forming the p-exact category $\mathcal{I}$ (see 1.4.2).

Similarly, the category $\mathcal{C}$ of topological spaces and partial continuous mappings (resp. its subcategory $\mathcal{T}$ formed by the continuous partial mappings defined on an open subset of the domain) is pointed homological. The normal subobjects are the inclusions of subspaces (resp. closed subspaces), the normal quotients are the restrictions to subspaces (resp. open subspaces). The exact morphisms are the homeomorphisms of an (open) subspace of the domain, onto a (closed) subspace of the codomain.

(But notice that $\mathsf{Top}_{\bullet}$ is neither equivalent to $\mathcal{C}$ nor to $\mathcal{T}$, since taking out the base point destroys topological information, namely the knowledge of its neighbourhoods.)

A short exact sequence in the p-homological category $\mathcal{S}$ (resp. in $\mathcal{C}$, in $\mathcal{T}$) is always of the following type, up to isomorphism

$$H \rightarrowtail X \twoheadrightarrow U, \qquad (1.97)$$

where H is a subset (resp. a subspace, a closed subspace) of X, U is the complement $X \setminus H$ and the two morphisms are partial identities, an inclusion and a 'co-inclusion'.

1.6.6 General modules

Consider the category Mod of *general modules* over arbitrary rings. An object is a pair (A, R) where R is a unital ring and A is a (left, unitary) R-module; a morphism $f = (f', f'')\colon (A, R) \to (B, S)$ is given by a (unitary) ring homomorphism $f''\colon R \to S$ together with a group-homomorphism $f'\colon A \to B$ consistent with the previous one: $f'(\lambda.x) = f''(\lambda).f'(x)$, for $\lambda \in R$ and $x \in A$. The morphism f is assumed to be *null* if its 'module-part' f' is zero, i.e. if it factorises through a *null* module $(0, R)$.

Mod is homological, non-pointed (nor quasi pointed). The normal factorisation of f is

$$
\begin{array}{ccccc}
(\operatorname{Ker} f', R) \rightarrowtail (A, R) & \xrightarrow{\;f\;} & (B, S) & \twoheadrightarrow & (B/(fA)^-, S) \\
_{p}\downarrow & & _{m}\uparrow & & \\
(A/\operatorname{Ker} f', R) & \underset{g}{\to} & ((fA)^-, S) & &
\end{array}
\tag{1.98}
$$

where $(fA)^-$ is the sub-S-module of B spanned by $f(A)$.

In particular (and up to isomorphism), normal monos, normal epis, exact morphisms, short exact sequences and subquotients live necessarily in some (abelian) fibre $R\,\mathsf{Mod}$ (with respect to the obvious fibration of Mod over the category of unital rings), where they have the usual meaning.

1.6.7 Categories of pairs

The homological structure of the category Set_2 of pairs of sets, with its natural (non-pointed) ideal of null morphisms, has been considered in 1.4.2.

Analogously one treats the homological category Gp_2 of pairs of groups, or Top_2 of pairs of topological spaces, or $\mathsf{Top}_{\bullet 2}$ of pairs of pointed topological spaces. Of course, subsets and their union are to be replaced with subgroups and their join, or subspaces and their join. The normal factorisation (1.44) shows that a morphism $f\colon (X, A) \to (Y, B)$ in Gp_2 (resp. Top_2, $\mathsf{Top}_{\bullet 2}$) is exact if and only if the mapping f is injective (resp. a topological embedding) and $f(X) \supset B$.

Note that Gp_2 has a zero object, the pair of null groups, that is null; but the null morphisms do not reduce to the zero-morphisms. In all these cases, a short exact sequence has the same form as in Set_2 (see (1.45)), where $X \supset M \supset A$ is now a *triple* of groups, or spaces, or pointed spaces.

A general construction for categories of pairs will be given in Section 2.5. This includes (convenient) categories of augmented rings, supplemented algebras and Lie algebras (see 2.5.7). On the other hand, the categories of pointed sets and pointed spaces do not enter in this pattern (see 2.5.8).

1.6.8 Groups as pairs

The category Gp has an obvious embedding in Gp_2

$$
I\colon \mathsf{Gp} \to \mathsf{Gp}_2, \qquad I(G) = (G, 0),
\tag{1.99}
$$

that has interesting properties. Let G be a group, and $f\colon G \to H$ a homomorphism of groups.

(a) The normal subobjects of G *in* Gp_2 coincide with the subgroups of G. (But the invariant subgroups of G will still play a relevant role for our analysis, as *modular elements* in the lattice of all subgroups, in Sections 7.3-7.5.)

(b) The direct and inverse images along f *in* Gp_2 coincide with the usual ones and form an exact connection (see 1.2.5):

$$f_*(G_0) = f(G_0), \qquad f^*(H_0) = f^{-1}(H_0),$$
$$f^*f_*(G_0) = G_0 \vee \operatorname{Ker} f, \qquad f_*f^*(H_0) = H_0 \wedge \operatorname{Im} f, \qquad (1.100)$$
$$(G_0 \subset G, \ H_0 \subset H).$$

(c) A sequence of homomorphisms of groups is exact in Gp_2 if and only if it is exact in the usual sense.

The homomorphism $f \colon G \to H$ has the following normal factorisation *in* Gp_2

$$f^{-1}(0) \ \overset{k}{\rightarrowtail} \ G \ \overset{f}{\longrightarrow} \ H \ \overset{c}{\twoheadrightarrow} \ (H, f(G))$$

with vertical arrows $p \downarrow$ from G to $(G, f^{-1}(0))$ and $m \uparrow$ into H, and $(G, f^{-1}(0)) \overset{g}{\rightarrow} f(G)$. $\qquad (1.101)$

Notice that I preserves kernels, but not cokernels: $\operatorname{Cok} f = (H, f(G))$ and $\operatorname{Nim} f = f(G)$ do not lose information about the (ordinary) image of f, contrarily to what happens in Gp.

Notice also, as a 'negative' fact, that the homomorphism f is exact in Gp_2 if and only if it is injective. This 'anomaly' will be corrected in the category of normalised groups Ngp (see 6.3.7).

Similar facts hold for the embedding of Rng in the homological category Rng_2 of *pairs of rings*: all subrings of a ring R are normal subobjects of $R = (R, 0)$; their direct and inverse images in Rng_2 are the usual ones; the ideals of R are modular elements in the lattice of all subrings.

1.6.9 *Two examples*

Finally, we give two constructions related with the independence of the axioms of homological categories: a category having kernels and cokernels with respect to a non-closed ideal and a category that satisfies (ex2) but not (ex3).

(a) Let S be an ordered set with at least two elements, viewed as a category. Its non-identity arrows $x < y$ form a (non-empty) ideal $\mathcal{N}$, with $O(\mathcal{N}) =$

$\emptyset$, that is *not* closed. If every element x has a predecessor x^- and a successor x^+, this category S has kernels and cokernels with respect to $\mathcal{N}$

$$\ker(x < y) = 1_x, \qquad \ker 1_x = (x^- < x),$$
$$\mathrm{cok}(x < y) = 1_y, \qquad \mathrm{cok}\, 1_y = (y < y^+), \qquad (1.102)$$

but the factorisation (1.33) of $f = (x < y)$ admits a central morphism g if and only if $x^+ \leqslant y^-$

$$
\begin{array}{ccccccc}
x & \overset{1}{\rightarrowtail} & x & \overset{f}{\longrightarrow} & y & \overset{1}{\twoheadrightarrow} & y \\
 & & \downarrow & \times & \uparrow & & \\
 & & x^+ & \dashrightarrow & y^+ & &
\end{array}
\qquad (1.103)
$$

(b) We now construct an ex2-category E that does not satisfy (ex3) (not even in a restricted form, with $m = \ker q$).

Take the category generated by the following graph, under the relation making the pentagon commutative; by definition, the null objects are M_0, M^0, Q_0, Q^0, with $A_0 = M_0$ and $A^0 = Q^0$

$$
\begin{array}{ccccc}
 & M & \overset{m}{\rightarrowtail} A \overset{q}{\twoheadrightarrow} Q & & \\
\nearrow & \downarrow & & \uparrow & \searrow \\
M_0 & M^0 & \overset{}{\underset{g}{\longrightarrow}} Q_0 & & Q^0
\end{array}
\qquad (1.104)
$$

Here $\ker q = m$ and qm is not exact: it is null, whence $\ker(qm) = 1_M$, $\mathrm{cok}(qm) = 1_Q$ and the central morphism g of the normal factorisation of qm is not invertible.

1.7 Exact functors

We now deal with exact functors between semiexact categories, and with some weaker notions: short exact, long exact functors (cf. 1.7.0), left and right exact functors (cf. 1.7.5). The relationship between closed ideals and full subcategories that are reflective and coreflective (cf. A3.4) is examined in 1.7.5, and more completely in 1.7.9.

E and E' are always ex1-categories.

1.7.0 Basic definitions

An *ex1-functor* $F\colon \mathsf{E} \to \mathsf{E}'$ will be a functor between ex1-categories that preserves kernels and cokernels; it will also be called an *exact* functor between semiexact categories. An *ex1-transformation* $\alpha\colon F \to G\colon \mathsf{E} \to \mathsf{E}'$ is a natural transformation between such functors.

The semiexact $\mathcal{U}$-*categories* (see 1.5.1), with such arrows and cells, form the 2-category EX_1 (see A4.5).

An ex1-functor preserves null morphisms (since f is null if and only if $\ker f = 1$), null objects, normal factorisations, exact morphisms, exact and short exact sequences (cf. 1.5.2). It also preserves normal monos and epis, their direct and inverse images (by (1.69)-(1.73)), and finally their meets and joins (by (1.79)-(1.80)).

We also need weaker notions. A functor $F\colon \mathsf{E} \to \mathsf{E}'$ between semiexact categories will be said to be *short exact* (resp. *long exact* or *N-exact*) if it preserves short exact (resp. exact) sequences.

As we just saw, every exact functor is short exact and long exact. But the functor $\mathsf{Top}_2 \to \mathrm{Ch}_\bullet\mathsf{Ab}$ that associates to a pair of spaces its complex of singular chains is short exact and not long exact, while the transfer functor $\mathsf{Nsb}\colon \mathsf{Gp} \to \mathsf{Ltc}$ will be seen to preserve exact sequences and not the short exact ones (see 2.1.2); both are not exact.

Recalling that the null objects are identified by the exact sequences, and also by the short exact ones (see 1.5.3(e)) it is easy to see that the following properties hold, for functors between semiexact categories

(a) a long exact functor preserves N-monos, N-epis and null maps,

(b) a short exact functor preserves normal monos, normal epis, exact and null maps.

Therefore $F\colon \mathsf{E} \to \mathsf{E}'$ is short exact if and only if it preserves the strongly exact sequences. Furthermore, the following conditions are equivalent:

(i) the functor F is exact,

(ii) F is long exact and preserves exact morphisms (use 1.5.5(e)),

(iii) F is both short and long exact,

(iv) F is short exact and preserves kernels and normal images (use the proof of Criterion A, in 1.5.4).

1.7.1 Exact functors and normal subobjects

Every ex1-functor $F\colon \mathsf{E} \to \mathsf{E}'$ defines a family of mappings indexed over the objects A of E

$$(\mathrm{Nsb}F)_A\colon \mathrm{Nsb}_\mathsf{E}(A) \to \mathrm{Nsb}_{\mathsf{E}'}(FA), \quad x \mapsto \mathrm{nim}\,(F(x)), \qquad (1.105)$$

that are lattice homomorphisms, by a previous remark (in 1.7.0). Note that $F(x)$ is a normal monomorphism with codomain FA, whereas $\mathrm{nim}\,(F(x))$

is *the* normal *subobject* equivalent to the former (often not distinguished from $F(x)$, by abuse of notation).

We say that F is *nsb-faithful* if all these homomorphisms are injective, i.e. satisfy the following conditions (equivalent, by the existence of meets and by the detection property 1.5.7(b)):

(a) if x, x' are normal subobjects of an object of E and $Fx \sim Fx'$ in E', then $x = x'$,

(b) if x, x' are normal subobjects of an object of E and $Fx \prec Fx'$ in E', then $x \leqslant x'$ (*order reflection*),

(c) if x is a normal subobject in E and $Fx \sim 1$ in E', then $x = 1$.

Analogously, we say that F is *nsb-full* if all the homomorphisms (1.105) are surjective: in other words, if y is a normal subobject of FA, there exists some normal subobject x of A such that $Fx \sim y$. Then, if $y \leqslant y'$ in $\mathrm{Nsb}(FA)$, there exist normal subobjects $x \leqslant x'$ in $\mathrm{Nsb}A$ transformed into y and y'. (Indeed if $Fx \sim y$ and $Fx' \sim y'$, we can replace x with $x \wedge x'$ so that $F(x \wedge x') \sim (Fx) \wedge (Fx') \sim y$.)

Formally, the naturality of this family of mappings $(\mathrm{Nsb}F)_A$ (for A in E) can be described as follows. Every morphism $a \colon A \to A'$ in E produces a square diagram

$$
\begin{array}{ccc}
\mathrm{Nsb}(A) & \xrightarrow{(\mathrm{Nsb}F)_A} & \mathrm{Nsb}(FA) \\
{\scriptstyle (a_*,a^*)}\Big\downarrow & & \Big\downarrow{\scriptstyle ((Fa)_*,(Fa)^*)} \\
\mathrm{Nsb}(A') & \xrightarrow[(\mathrm{Nsb}F)_{A'}]{} & \mathrm{Nsb}(FA')
\end{array}
\qquad (1.106)
$$

whose horizontal arrows are lattice homomorphisms, while the vertical arrows are Galois connections. This square is *bicommutative*, in the sense that:

$$
\begin{aligned}
(Fa)_*(\mathrm{nim}\,(Fx)) &= \mathrm{nim}\,(F(a_*x)), \\
(Fa)^*(\mathrm{nim}\,(F(x'))) &= \mathrm{nim}\,(F(a^*x')).
\end{aligned}
\qquad (1.107)
$$

In other words, every ex1-functor $F \colon \mathsf{E} \to \mathsf{E}'$ defines a *horizontal transformation of vertical functors*

$$
\mathrm{Nsb}F \colon \mathrm{Nsb}_{\mathsf{E}} \to \mathrm{Nsb}_{\mathsf{E}'}.F \colon \mathsf{E} \to \mathbb{L}\mathrm{thc}.
\qquad (1.108)
$$

It takes values in the double category $\mathbb{L}$thc of *lattices, lattice homomorphisms* (as horizontal arrows) and *Galois connections* (as vertical arrows), whose double cells are the *bicommutative squares* in the sense described above. This extends a similar result for p-exact categories (see I.2.2.8(e)), based on the double subcategory $\mathbb{M}$lhc of modular lattices, homomorphisms and modular connections (see 1.2.8). (The definition of a double category is sketched in A4.7.)

1.7.2 Conservative exact functors

An *N-reflective* functor will be an N-functor (between N-categories) that reflects the null maps; as a consequence, it also reflects N-monos, N-epis and exact sequences (by their characterisation in 1.5.2).

An N-reflective ex1-functor $F: \mathsf{E} \to \mathsf{E}'$ (between ex1-categories) is necessarily nsb-faithful. In fact, using the condition 1.7.1(c), let us assume that x is a normal mono of E and $Fx \sim 1$; then x is also N-epi, hence an isomorphism.

Let $F: \mathsf{E} \to \mathsf{E}'$ be an ex1-functor that is *conservative*, i.e. reflects the isomorphisms. Then F is also N-reflective: if Ff is null in E' then $\ker(Ff) \sim F(\ker f)$ is an isomorphism, and so is $\ker f$. (The converse is false, see 1.7.4(e).)

Moreover F also reflects, in the appropriate sense, all the exactness notions: kernels and cokernels, normal factorisations, exact morphisms, exact and short exact sequences, direct and inverse images of normal subobjects and quotients, finite meets and joins of both. For instance, F reflects kernels, in the following sense: if h and f are composable morphisms in E and $Fh \sim \ker Ff$ in E', then h is a normal mono, equivalent to $\ker f$ in E; indeed, fh is null in E, hence $h = (\ker f)h'$; since Fh' is an isomorphism, so is h'.

It follows that a category C cannot have two distinct and *exactly comparable* semiexact structures. Namely, if $\mathsf{E}_1 = (\mathsf{C}, \mathcal{N}_1)$ and $\mathsf{E}_2 = (\mathsf{C}, \mathcal{N}_2)$ are semiexact categories and the identity of C - obviously conservative - is an exact functor from E_1 to E_2, then $\mathcal{N}_1 = \mathcal{N}_2$ and $\mathsf{E}_1 = \mathsf{E}_2$.

1.7.3 Proposition and Definition (Semiexact subcategories)

Let E' be a subcategory of the ex1-category E.

(a) E' is a semiexact subcategory, *or* ex1-subcategory, *of E if it satisfies property (i) (already used in 1.3.7) or equivalently the conjunction of properties (i') and (i'')*

(i) E', equipped with the ideal $\mathsf{E}' \cap \mathrm{Nul}\mathsf{E}$, is a semiexact category and the inclusion $U: \mathsf{E}' \to \mathsf{E}$ is exact.

(i') for every f in E' there is some kernel m of f in E that belongs to E' and is such that, if ma belongs to E', so does a,

(i″) for every f in E′ *there is* some *cokernel p of f in* E *that belongs to* E′ *and is such that, if ap belongs to* E′*, so does a.*

Then, the inclusion U : E′ → E *is N-reflective and nsb-faithful (cf. 1.7.2).*

(b) The semiexact subcategory E′ *is said to be* conservative *(in* E*) if the inclusion* E′ ⊂ E *is conservative (as defined above, in 1.7.2). Now, a subcategory* E′ *is a conservative ex1-subcategory if and only if the following two conditions hold:*

(ii) for every f in E′ *there are* some *kernel and* some *cokernel of f in* E *that belong to* E′*,*

(iii) if m is a normal mono of E *that belongs to* E′ *and mf is in* E′*, so is f; dually, if p is a normal epi of* E *that belongs to* E′ *and fp is in* E′*, so is f.*

(c) The semiexact subcategory E′ *is said to be* nsb-full *if the inclusion is (cf. 1.7.1). A subcategory* E′ *is a* conservative nsb-full ex1-subcategory *if and only if it satisfies condition (iii) and:*

(ii′) for every A in E′*, every normal subobject (resp. quotient) of A in* E *is equivalent to some mono (resp. epi) of* E′*.*

Analogously one can characterise ex2-subcategories, ex3-subcategories (or homological subcategories), ex4-subcategories (or g-exact subcategories) and p-exact subcategories (defined in 1.3.7).

Proof (a) if E′ is a semiexact subcategory, it satisfies (i′) and (i″) since, for any f in E′, its kernel m and cokernel p in E′ are preserved by the inclusion.

Conversely, assume (i′) and (i″) and take NulE′ = E′ ∩ NulE. Then, kernels and cokernels of E′ exist and are preserved by U, because of the hypothesis; moreover, NulE′ is a closed ideal, by Criterion B of 1.5.4.

(b) If E′ is a conservative semiexact subcategory, the inclusion reflects kernels and cokernels (see 1.7.2), whence (iii). Conversely, from (iii) it follows that U is conservative: if f is in E′ and has an inverse map g in E, then f is a normal mono in E and fg belongs to E′; therefore g too does.

(c) Is an obvious consequence. □

1.7.4 Examples

(a) The category Set. of pointed sets has a natural full embedding in the category of pairs Set$_2$, via the functor $X \mapsto (X, \{0_X\})$, where 0_X is the base point of X. This embedding is long exact (cf. 1.7.0) but does not preserve cokernels; hence Set. is not a semiexact subcategory of Set$_2$.

(b) The subcategory Inj_2 of Set_2 containing all the objects and just the *injective* mappings is a conservative nsb-full semiexact subcategory of Set_2. In Chapter 7 it will play a role in the construction of the universal models for distributive homological theories, analogous to the role played by $\mathcal{I}$ for distributive p-exact theories, in Part I.

Its subcategory Inc_2 containing all the objects and those mappings that are *inclusions* of subsets is a conservative nsb-full semiexact subcategory of both, again of interest for universal models. Similar semiexact subcategories exist in Top_2 and Gp_2.

(c) The full subcategory Cph_2 (resp. Cpm_2) of Top_2 produced by compact Hausdorff (resp. compact metric) pairs is a conservative nsb-full semiexact subcategory (see Section 6.1 for applications).

(d) Consider the subcategory $\mathcal{T}_{LC}$ of $\mathcal{T}$ (see 1.6.5), consisting of the locally compact Hausdorff spaces and *proper* partial continuous mappings defined on an open subset. It is plainly a conservative nsb-full semiexact subcategory of $\mathcal{T}$; it is equivalent to the category $\mathsf{Cph}_\bullet$ of pointed compact Hausdorff spaces, via Alexandroff one-point compactification. It is of interest in algebraic topology, for *single-space homology theories* (see [ES] X.7, and here, Section 6.1).

(e) The subcategory Ban_1 of Ban consists of Banach spaces and *linear weak contractions*, i.e. bounded linear mappings f with $||f|| \leqslant 1$. The canonical choices for normal subobjects and quotients of Ban (cf. 1.6.3) are in Ban_1, that is thus an nsb-full semiexact subcategory. It is not conservative, as its isomorphisms are the surjective *isometries*, satisfying $||f(x)|| = ||x||$.

In contrast with Ban, this subcategory is complete and cocomplete [Se], with the following products and sums (where $\Pi |X_i|$ denotes the cartesian product of the underlying vector spaces)

$$\Pi_i \, X_i = \{(x_i) \in \Pi \, |X_i| \mid \sup_i ||x_i|| < +\infty\},$$
$$||(x_i)|| = \sup_i ||x_i|| \qquad (l_\infty - norm), \tag{1.109}$$

$$\Sigma_i \, X_i = \{(x_i) \in \Pi \, |X_i| \mid \Sigma_i ||x_i|| < +\infty\},$$
$$||(x_i)|| = \Sigma_i \, ||x_i|| \qquad (l_1 - norm). \tag{1.110}$$

The forgetful functor given by the unit ball

$$B \colon \mathsf{Ban}_1 \to \mathsf{Set},$$
$$B(X) = \{x \in X \mid ||x|| \leqslant 1\} = \mathsf{Ban}_1(K, X), \tag{1.111}$$

is represented by the scalar field K ($\mathbb{R}$ or $\mathbb{C}$). It has a left adjoint, that assigns to a set S the space of l_1-summable S-indexed families of scalars $(\lambda_s)_{s \in S}$

$$l_1(S) = \Sigma_{s \in S} K, \qquad ||(\lambda_s)|| = \Sigma_s |\lambda_s|. \qquad (1.112)$$

(f) **Gp** is a full, conservative and nsb-full semiexact subcategory of **Gpd**, the semiexact category of groupoids (cf. 1.6.1), by the usual identification of a group G with the one-object groupoid whose maps are the elements of G.

1.7.5 Left exact functors and right adjoints

E and **E'** are always semiexact categories. A functor $F \colon \mathsf{E} \to \mathsf{E}'$ is called *left exact* (resp. *right exact*) if it preserves kernels (resp. cokernels).

A left exact functor always preserves the least normal subobject $0_A = \ker 1_A \colon A_0 \rightarrowtail A$ of an object, and any pullback of normal monos. (Indeed, the inverse image $f^*(n)$ in (1.70) can also be obtained as $\ker(vf)$, where v is *any* morphism of which n is a kernel.) Dually, a right exact functor preserves the least normal quotient $0^A = \cok 1_A \colon A \twoheadrightarrow A^0$ of an object and any pushout of normal epis.

In both cases, null morphisms and null objects are preserved.

Such functors can arise in adjoint pairs. Let an adjunction be given between our semiexact categories, with inverse bijections $\varphi_{AA'}, \psi_{AA'}$, unit η and counit ε (a 'variable' object of **E**, **E'** is written as A, A', respectively)

$$F \dashv G, \qquad F \colon \mathsf{E} \to \mathsf{E}', \quad G \colon \mathsf{E}' \to \mathsf{E},$$

$$\varphi_{AA'} \colon \mathsf{E}'(FA, A') \to \mathsf{E}(A, GA'),$$

$$\psi_{AA'} \colon \mathsf{E}(A, GA') \to \mathsf{E}'(FA, A'), \qquad (1.113)$$

$$\eta_A = \varphi(1_{FA}) \colon A \to GFA, \quad \varepsilon_{A'} = \psi(1_{GA'}) \colon FGA' \to A'.$$

We say that the adjunction is *consistent* with the semiexact structure, or an *N-adjunction*, if F and G are N-functors (cf. 1.3.1), i.e. preserve the null morphisms (or objects); or equivalently, if all the bijections $\varphi_{AA'}$ and $\psi_{AA'}$ preserve the null morphisms. (This property is automatically satisfied for p-semiexact categories, since every left adjoint preserves the initial object, and dually.)

Then F is right exact and G is left exact. Indeed, let us take $f \colon A \to B$ and $p = \cok f$ in **E**; then a morphism $g \colon FB \to A'$ in **E'** factorises as $g = g'.Fp$ if and only if $\varphi(g) = \varphi(g').p$ in **E**, if and only if $\varphi(g).f = \varphi(g.Ff)$ is null, if and only if $g.Ff$ is null.

(Notice that, in the present non-pointed case, the right exactness of F is not a straightforward consequence of the usual property of preserving the existing colimits, as stated in A3.3. In facts, $\cok(f \colon A \to B)$ is indeed a pushout of $0^A = \text{cok}\, 1_A \colon A \twoheadrightarrow A^0$, but proving that F preserves the cokernel of an identity - with respect to an ideal of maps! - practically requires the same argument that we have given above for a general cokernel.)

(a) Obvious examples of N-adjunctions come from pointed categories having a monoidal closed structure, like Ab or Set$_\bullet$. A non-pointed example is the monoidal closed structure of Set$_2$, in 1.4.5.

Many other examples come from full reflective subcategories of semiexact categories, like Ab in Gp, Ab in Abm (abelian monoids), Banach spaces in normed vector spaces, pointed Hausdorff spaces in general pointed spaces, Set$_\bullet$ in Set$_2$, etc. (see 5.3.3-5.3.4).

(b) In particular, the (left exact and long exact) embedding $J \colon$ Set$_\bullet \to$ Set$_2$, $J(X, x_0) = (X, \{x_0\})$ (1.7.4(a)) has a left adjoint, that is exact:

$$P \colon \mathsf{Set}_2 \to \mathsf{Set}_\bullet, \qquad P(X, A) = X/A = (|X/A|, x_0). \qquad (1.114)$$

Here $P(X, A) = X/A$ is the pointed set whose underlying set $|X/A|$ is defined by the following pushout

$$
\begin{array}{ccc}
A & \xrightarrow{\ i\ } & X \\
\downarrow & & \downarrow{\scriptstyle p} \\
\top & \xrightarrow[\ x_0\ ]{} & |X/A|
\end{array}
\qquad . \qquad (1.115)
$$

Thus, *if $A \neq \emptyset$, the set X/A is the quotient of X that identifies the points of A* (whose class is the base point x_0); but notice that $X/\emptyset$ is the set X *supplemented with a base-point x_0* (not belonging to X); in this case, p is not surjective.

(c) The following example is structural. The full subcategory E$_0$ of the semiexact category E determined by the null objects is a semiexact subcategory, nsb-full in E. Furthermore, it is reflective and coreflective, i.e. the (exact) inclusion $U \colon$ E$_0 \to$ E has a left adjoint N^0 and a right adjoint N_0

$$
\begin{aligned}
N^0 &\colon \mathsf{E} \to \mathsf{E}_0, & A &\mapsto A^0, & f &\mapsto f^0, & (N^0 &\dashv U), \\
N_0 &\colon \mathsf{E} \to \mathsf{E}_0, & A &\mapsto A_0, & f &\mapsto f_0, & (U &\dashv N_0).
\end{aligned}
\qquad (1.116)
$$

These functors associate to every object A, respectively, its least normal quotient or its least normal subobject (1.5.3(a)); the induced morphisms f_0 and f^0 are determined by the following commutative diagram

$$
\begin{array}{ccccc}
A_0 & \rightarrowtail & A & \twoheadrightarrow & A^0 \\
\downarrow{\scriptstyle f_0} & & \downarrow{\scriptstyle f} & & \downarrow{\scriptstyle f^0} \\
B_0 & \rightarrowtail & B & \twoheadrightarrow & B^0
\end{array}
\qquad (1.117)
$$

(since, for example, the composition $f0_A \colon A_0 \to B$ is null, hence factorises uniquely through 0_B). The two natural transformations

$$UN_0 \to \mathrm{id}\mathsf{E} \to UN^0, \quad 0_A \colon A_0 \rightarrowtail A, \quad 0^A \colon A \twoheadrightarrow A^0, \tag{1.118}$$

are respectively the counit of $U \dashv N_0$ and the unit of $N^0 \dashv U$.

1.7.6 Categories of maps and kernel functors

The category E^2 of morphisms of the semiexact category E has a natural semiexact structure.

Its objects are the maps $a \colon A' \to A''$ of E; an arrow $(f', f'') \colon a \to b$ is a commutative square, as the left diagram below

$$
\begin{array}{ccc}
A' \xrightarrow{\ f'\ } B' & \qquad & A' \longrightarrow B_0' \rightarrowtail B' \\
\ \downarrow a \qquad \downarrow b & & \ \downarrow a \qquad\quad \downarrow b_0 \qquad\quad \downarrow b \\
A'' \xrightarrow[\ f''\ ]{} B'' & & A'' \longrightarrow B_0'' \rightarrowtail B''
\end{array}
\qquad . \tag{1.119}
$$

The arrow (f', f'') is assumed to be *null* if both f' and f'' are null, in E; or equivalently if it factorises as in the right diagram above, through the map $b_0 = N_0(b)$. This proves that our ideal of null morphisms is closed; the construction of kernels and cokernels in E^2 is obvious, and E^2 is semiexact.

All this also proves that the semiexact category E has a *kernel functor*

$$
\begin{aligned}
\mathrm{Ker} \colon \mathsf{E}^2 \to \mathsf{E}, \quad & (a \colon A' \to A'') \mapsto \mathrm{Ker}\, a, \\
& (f', f'') \mapsto (f_K \colon \mathrm{Ker}\, a \to \mathrm{Ker}\, b).
\end{aligned}
\tag{1.120}
$$

The functor Ker is right adjoint to the functor

$$0^- \colon \mathsf{E} \to \mathsf{E}^2, \quad A \mapsto (0^A \colon A \twoheadrightarrow A^0), \quad f \mapsto (f, f^0), \tag{1.121}$$

with unit $\eta_A = 1_A \colon A \to A = \mathrm{Ker}\, 0^A$. Hence $\mathrm{Ker} \colon \mathsf{E}^2 \to \mathsf{E}$ is left exact. Dually, the cokernel functor $\mathrm{Cok} \colon \mathsf{E}^2 \to \mathsf{E}$ is right exact.

1.7.7 Categories of functors

More generally, consider a category of functors E^S (cf. A1.8), where S is a small category and E is a semiexact category. This includes all cartesian powers of E (when S is a discrete category), the category of morphisms of E (when $\mathsf{S} = \mathbf{2}$ is the arrow category $0 \to 1$), the category of commutative squares of E (when $\mathsf{S} = \mathbf{2} \times \mathbf{2}$), the categories of presheaves over E (cf. A1.8), etc.

Then E^S has a natural closed ideal, where a *null morphism* is any natural transformation $\varphi: F \to G: \mathsf{S} \to \mathsf{E}$ whose components are null morphisms, and a *null* object is any functor $F: \mathsf{S} \to \mathsf{E}$ that sends each object of S to a null object of E (and factorises through the inclusion $\mathsf{E}_0 \to \mathsf{E}$ of the null subcategory of E). Indeed, the morphism $\varphi: F \to G$ is null if and only if it factorises through the null functor $N_0 G: \mathsf{S} \to \mathsf{E}_0$, where $N_0: \mathsf{E} \to \mathsf{E}_0$ is the least normal subobject functor (cf. (1.116)).

It is easy to see that E is semiexact: its kernels and cokernels are computed componentwise, on every $X \in \mathrm{Ob}\mathsf{S}$, as:

$$(\mathrm{Ker}\,(\varphi: F \to G))_X = \mathrm{Ker}\,(\varphi X: FX \to GX). \qquad (1.122)$$

Furthermore, if E is homological, or g-exact, or pointed, so is E^{S} (and conversely, provided that S is not empty).

1.7.8* Pseudolimits in EX_1

It is easy to prove that the 2-category EX_1 formed by semiexact $\mathcal{U}$-categories (cf. 1.5.1), ex1-functors and ex1-transformations has all indexed pseudolimits ([Ke3], Section 5), that are created by the forgetful 2-functor $\mathsf{EX}_1 \to \mathsf{CAT}$ (see A4.6).

Indeed, it suffices to examine the following particular cases (see [Ke3], Prop. 16.1(b)); in all of them the ideal of null morphisms is the obvious one, and can be proved to be closed using the fact that a morphism is null if and only if it factorises through the least normal subobject of its codomain (1.5.3(b)).

(a) The *product* $\Pi \mathsf{E}_i$ is the usual cartesian product, provided with the ideal $\mathcal{N} = \Pi \mathrm{Nul}(\mathsf{E}_i)$.

(b) The *isoinserter* of $F, G: \mathsf{E} \to \mathsf{E}'$ is a pair (X, ξ) where $X: \mathsf{X} \to \mathsf{E}$ is an ex1-functor and $\xi: FX \to GX: \mathsf{X} \to \mathsf{E}'$ is a functorial isomorphism, the pair being universal in the obvious sense. Solution: the category X has for objects the pairs $(A, b: FA \to GA)$ where A is in E and b is an isomorphism of E', with obvious morphisms and null morphisms; the functor $X: \mathsf{X} \to \mathsf{E}$ is the first projection and the functorial isomorphism $\xi: FX \to GX$ is defined as $\xi(A, b) = b$.

(c) The *equifier* of two cells $\alpha, \beta: F \to G: \mathsf{E} \to \mathsf{E}'$ is an ex1-functor $X: \mathsf{X} \to \mathsf{E}$ such that $\alpha X = \beta X$, in a universal way. Solution: X is the full, semiexact subcategory of E whose objects A satisfy the condition $\alpha A = \beta A$; the functor $X: \mathsf{X} \to \mathsf{E}$ is the inclusion.

(d) Last, the *cotensor* of E by a small category S is the semiexact category E^{S} considered above (in 1.7.7).

1.7.9* *Proposition (Closed ideals and adjunctions)*

One can consider four levels stronger than N-categories, each of them stronger than the previous one.

(a) A category E *equipped with a* closed *ideal* $\mathcal{N}$, *or equivalently with a full subcategory* E_0 *closed under retracts.*

(b) A category E *equipped with a conservative, full, reflective and coreflective subcategory* E_0. *The full embedding* $U: \mathsf{E}_0 \to \mathsf{E}$ *has adjoints* $N^0 \dashv U \dashv N_0$. *We write the unit of the first adjunction and the counit of the second as follows*

$$0^A: A \to A^0 = N^0(A) \qquad (N^0 \dashv U),$$
$$0_A: N_0(A) = A_0 \to A \qquad (U \dashv N_0),$$
$$(1.123)$$

Furthermore, we assume that the construction of the adjoints has been performed so that, for every Z *in* E_0, *the counit of the first adjunction and the unit of the second are* identities: $N^0(Z) = \mathrm{id}Z = N_0(Z)$; *therefore, the triangular identities reduce to:*

$$N^0(0^A) = \mathrm{id}A^0, \quad 0^Z = \mathrm{id}Z: Z \to N^0 Z,$$
$$N_0(0_A) = \mathrm{id}A_0, \quad 0_Z = \mathrm{id}Z: N_0 Z \to Z \quad (Z \text{ in } \mathsf{E}_0).$$
$$(1.124)$$

(c) A category E *with a closed ideal, where every identity* $\mathrm{id}A$ *has a kernel and a cokernel*

$$0^A: A \twoheadrightarrow A^0 = N^0(A), \quad 0_A: N_0(A) = A_0 \rightarrowtail A, \qquad (1.125)$$

or, equivalently, a category E *equipped with a full reflective and coreflective subcategory* E_0, *where every counit-map* $0_A: A_0 \to A$ *is mono and every unit-map* $0^A: A \to A^0$ *is epi.*

(d) An unrestricted semiexact category, or equivalently a category endowed with a full reflective and coreflective subcategory E_0, *such that every counit-map* $0_A: A_0 \to A$ *is mono and has all pullbacks, while every unit-map* $0^A: A \to A^0$ *is epi and has all pushouts.*

Note. These results were established with George Janelidze, in 2009, while we were working on a common project on radicals. The implication (d) $\Rightarrow$ (b), proved in 1.7.5(c), is already in the preprint [G10].

Proof (b) $\Rightarrow$ (a). It is well known that a full reflective subcategory is closed under retracts, up to isomorphism. Indeed, if Z is in E_0 and $m: A \rightleftarrows Z : p$

is a retraction (i.e. $pm = \mathrm{id}A$), the commutative diagram

$$
\begin{array}{ccccc}
A & \xrightarrow{\;m\;} & Z & \xrightarrow{\;p\;} & A \\
{\scriptstyle\eta}\downarrow & & \| & & \downarrow{\scriptstyle\eta} \\
N^0A & \underset{m'}{\rightarrowtail} & N^0Z & \underset{p'}{\longrightarrow} & N^0A
\end{array}
\tag{1.126}
$$

(where $p'm' = N_0(\mathrm{id}A) = \mathrm{id}N_0A$) shows that the unit $\eta = 0^A$ is a split mono and a split epi, whence an isomorphism: $A \cong N^0A$.

(c) $\Rightarrow$ (b). It is sufficient to check the equivalence of the two properties in (c), and it is easy to see that the first implies the second. Conversely, if a counit-map $0_A\colon A_0 \to A$ is mono, let us prove that it is the kernel of $\mathrm{id}A$; take a null map $f\colon X \to A$ in E, that factors $f = vu$ through an object Z of E_0

$$
\begin{array}{ccccc}
X & \xrightarrow{\;u\;} & Z & \xrightarrow{\;v\;} & A \\
 & & \| & & \uparrow{\scriptstyle\varepsilon} \\
 & & N_0Z & \longrightarrow & N_0A
\end{array}
\quad .
\tag{1.127}
$$

Then f factors through the counit $\varepsilon = 0_A\colon N_0(A) \to A$, uniquely because the latter is a monomorphism.

(d) $\Rightarrow$ (c). Obvious. $\qquad\square$

2

Homological categories

Sections 2.1 and 2.2 deal with characterisations and basic properties of ex2-, homological and g-exact categories. The modular and distributive properties of the lattices of normal subobjects are studied in Section 2.3; in a modular semiexact category 'elementary' diagram chasing is possible, while the non-elementary one will be developed in Chapter 3 for homological categories. Section 2.4 briefly introduces additive semiexact categories.

Extending 'pairs' (X, A) of sets or topological spaces (in the usual sense of algebraic topology, with $A \subset X$), we have in Section 2.5 an abstract construction of the homological category of pairs C_2, for a category C equipped with distinguished subobjects. Every variety of algebras gives thus a homological category of pairs (see 2.5.1).

Finally, Sections 2.6-2.8 study the 'epi-mono completion' $Fr\,C$ of a category C. It is a generalisation of the well-known Freyd embedding of a *stable* homotopy category into an *abelian* category [F1]-[F3], introduced in [G15, G16].

We show that, if C has products and weak equalisers, as HoTop and various other homotopy categories, then $Fr\,C$ is complete; similarly, if C has zero-object, weak kernels and weak cokernels, as the homotopy category of pointed spaces, then $Fr\,C$ is an (unrestricted) *p-homological* category; finally, if C is triangulated, $Fr\,C$ is abelian and the embedding $C \to Fr\,C$ is the universal homological functor on C, as in Freyd's original case.

2.1 The transfer functor and ex2-categories

After studying the exactness properties of the transfer functor Nsb of a semiexact category E, we give various equivalent form of the axiom (ex2) and construct the projective category $Pr\,E$ associated to an ex2-category.

2.1.1 Fully normal monos and epis

First, we need some further notions. Let E be a semiexact category.

Given a normal monomorphism $m \colon A' \to A$, the associated connection $\mathrm{Nsb}(m)$ has a null kernel and is a right exact morphism of Ltc (see 1.2.5)

$$\mathrm{Nsb}(m) = (m_*, m^*) \colon \mathrm{Nsb}A' \to \mathrm{Nsb}A,$$
$$m^*0 = 0, \qquad m_*(m^*x) = m \wedge x = x \wedge m_*(1), \tag{2.1}$$

by 1.5.3(g) and (1.79).

We say that the normal mono m is *fully normal* if it satisfies the following equivalent conditions:

(a) for every $x' \in \mathrm{Nsb}A'$, mx' is a normal mono,

(a') for every $x' \in \mathrm{Nsb}A'$, mx' is an exact morphism,

(b) for every $x' \in \mathrm{Nsb}A'$, $m_*x' \sim mx'$,

(c) the connection $\mathrm{Nsb}(m)$ is mono in Ltc, i.e. $m^*m_*x' = x'$ (for every $x' \in \mathrm{Nsb}A'$),

(c') $\mathrm{Nsb}(m)$ is a normal mono in Ltc,

(c'') $\mathrm{Nsb}(m)$ is an exact connection (i.e. an exact morphism of Ltc, cf. 1.2.5),

(d) $\mathrm{Nqt}(m)$ is mono (or normal mono, or exact) in $\mathrm{Ltc^{op}}$, i.e. $m^{\circ}m_{\circ}y' = y'$ (for every $y' \in \mathrm{Nqt}A'$).

Indeed, the equivalence of (a) and (a') follows plainly from 1.5.5(c), while the equivalence of (c), (c'), (c'') follows from (2.1). (a) $\Rightarrow$ (b): $m_*x' = \ker \operatorname{cok} mx' \sim mx'$. (b) $\Rightarrow$ (c): $m^*m_*x' = m^*(mx') = x'$ (by 1.5.7(b)). (c) $\Rightarrow$ (a): $m_*x' = m \wedge m_*x' \sim m.(m^*(m_*x')) \sim mx'$ (by 1.5.7(a)).

Dually, the connection $\mathrm{Nsb}(p)$ associated to a normal epi $p \colon A \to A''$ has a null cokernel and is left exact (cf. 1.2.5)

$$\mathrm{Nsb}(p) = (p_*, p^*) \colon \mathrm{Nsb}A \to \mathrm{Nsb}A'',$$
$$p_*1 = 1, \qquad p^*(p_*x) = x \vee \ker p. \tag{2.2}$$

We say that p is a *fully normal epimorphism* if it is a normal epi and satisfies the equivalent conditions:

(a*) for every $y \in \mathrm{Nqt}A''$, yp is a normal epi (or equivalently, an exact morphism),

(b*) for every $y \in \mathrm{Nqt}A''$, $p^{\circ}(y) \sim yp$,

(c*) $\mathrm{Nqt}(p)$ is epi (or equivalently a normal epi, or an exact morphism) in $\mathrm{Ltc^{op}}$, i.e. $p_{\circ}p^{\circ}y = y$ (for every $y \in \mathrm{Nqt}A''$),

(d*) $\mathrm{Nsb}(p)$ is epi (or a normal epi, or an exact morphism) in Ltc, i.e. $p_*p^*x = x$ (for every $x \in \mathrm{Nsb}A''$).

Fully normal monos and epis are preserved by all *nsb-full* ex1-functors.

In the category Gp of groups there are normal subobjects that are not fully normal, while all epis are normal, hence fully normal; this also shows that, in a semiexact category, the kernel of a fully normal epi need not be a fully normal mono.

2.1.2 Theorem (Exactness properties of the transfer functor)

(a) The transfer functor Nsb: E → Ltc *of a semiexact category is always long exact: it preserves and reflects the exact sequences.*

(b) Nsb *is left (resp. right) exact if and only if all the normal monos (resp. epis) of* E *are fully normal (see 2.1.1), if and only if normal monos (resp. epis) are closed under composition.*

(c) Nsb *is exact if and only if* E *satisfies both conditions in (b), if and only if (ex2) holds (see 1.3.6).*

Proof (a) The functor Nsb preserves and reflects the null morphisms, by 1.5.3(b) and 1.2.3(b). Further, it preserves and reflects the exact sequences: indeed, given a sequence in E and its image in Ltc

$$A \xrightarrow{f} B \xrightarrow{g} C \qquad \mathrm{Nsb}A \xrightarrow{f'} \mathrm{Nsb}B \xrightarrow{g'} \mathrm{Nsb}C \qquad (2.3)$$

we have $\mathrm{Nim}\, f' = [0, f_* 1]$ and $\mathrm{Ker}\, g' = [0, g^* 0]$.

(b) Let $f \sim \ker g$. Then, by the previous point, $f' \sim \ker g'$ if and only if f' is a normal mono in Ltc, if and only if f is a fully normal mono of E (by 2.1.1). In other words, Nsb preserves kernels if and only if all normal monos are fully normal, if and only if normal monos are closed under composition (see 2.1.1). The rest is an obvious consequence, by duality. □

2.1.3 Ex2-categories

Completing a definition given in 1.3.6, an *ex2-category* E is defined as an ex1-category satisfying the following axioms, equivalent by 2.1.1 - 2.1.2:

(ex2) normal monomorphisms and normal epimorphisms are stable under composition,

(ex2a) the composite of two normal monos or two normal epis is an exact morphism,

(ex2b) every normal mono m and every normal epi p is fully normal (i.e. $m^* m_* = 1$, $p_* p^* = 1$; or, equivalently, (m_*, m^*) and (p_*, p^*) are exact connections; cf. 2.1.1), (ex2c) the transfer functor Nsb: E → Ltc is exact.

Let E be an ex2-category. In the normal factorisation $f = mgp$ (see (1.66)), the morphism g: Ncm f → Nim f has a null kernel and a null cokernel, i.e. it is N-mono and N-epi (see 1.3.1); actually, with the notation of (1.66) and using (ex2b)

$$\ker g = g^*0 = g^*m^*0 = p_*p^*g^*m^*0 = p_*f^*0 = p_*p^*0 = 0. \qquad (2.4)$$

It follows easily that, in an ex2-category, *the normal factorisation* $f = mgp$ *is characterised up to isomorphism* by the following properties: p is a normal epi, m is a normal mono, g is N-mono and N-epi.

The (exact) transfer functor Nsb$_E$: E → Ltc is nsb-faithful and nsb-full, since its associated horizontal transformation (defined in 1.7.1)

$$\varphi\colon \mathrm{Nsb_E} \to \mathrm{Nsb_{Ltc}.Nsb_E},$$
$$\varphi(x) = \mathrm{nim}\,(x_*, x^*) = (\downarrow x \rightarrowtail \mathrm{Nsb}A), \qquad (2.5)$$

coincides with the *iso*morphism $\iota.\mathrm{Nsb_E}$ (see 1.5.9).

2.1.4 *The associated projective category*

Let E be an ex2-category, so that the projective functor Nsb: E → Ltc is exact.

Recall that E is said to be *projective* if the functor Nsb: E → Ltc is faithful (Section 1.5.8). This is the case of Ltc itself, whose transfer functor is isomorphic to the identity.

Every ex2-category has an associated category Pr E, that is the quotient of E (see A1.2) modulo the congruence which identifies two parallel morphisms $f, g\colon A \to B$ of E whenever $\mathrm{Nsb}(f) = \mathrm{Nsb}(g)$, or equivalently $f_* = g_*\colon \mathrm{Nsb}A \to \mathrm{Nsb}B$. A morphism $[f]\colon A \to B$ of Pr E is thus an equivalence class of morphisms of E, in this sense.

Pr E is equipped with the (closed) ideal produced by NulE under the canonical projection E → Pr E. It is easy to see that the latter preserves and reflects null morphisms, kernel and cokernels, normal subobjects, normal quotients and exact sequences; therefore Pr E is a projective ex2-category, called the *associated projective category* of E. Moreover, if E is homological, or g-exact, or p-exact, or modular, so is Pr E.

Also here, Pr E is determined up to isomorphism of categories by an obvious *universal property*: every exact functor $F\colon E \to E'$ with values in a projective ex2-category factorises in a unique way as $F = GP$, where $G\colon \mathrm{Pr}\,E \to E'$ is a functor (necessarily exact).

The p-exact case has already been studied in Part I. If $E = K\,\mathrm{Vct}$ is the abelian category of vector spaces on the commutative field K, it is easy to see that $f_* = g_*$ if and only if there exists a non-zero scalar $\lambda \in K$ such that $f = \lambda g$. One can thus view the p-exact category $\Pr(K\,\mathsf{Vct})$ as (a realisation of) the category of projective spaces over K; notice that this category has no products, unless K is the two-element field (in which case $K\,\mathsf{Vct}$ is already projective). We refer to I.2.3 (or [G4]) for a detailed analysis of these facts, and to [CaG] for a characterisation of the 'projective categories' associated to the abelian ones.

Finally, we can think of $\Pr\mathsf{E}$ as a 'projective' category associated to E, also in the general case of ex2-categories. The present setting, that makes a crucial use of the transfer functor Nsb, can thus be considered as a 'projective view' of non-classical homological algebra, as already said in the Introduction.

The p-homological categories $\mathsf{Set_\bullet}$ and $\mathsf{Top_\bullet}$, of pointed sets and pointed spaces (see 1.6.4) are easily seen to be projective. $\mathsf{Set_2}$ is not, but the exact functor

$$P\colon \mathsf{Set_2} \to \mathsf{Set_\bullet}, \qquad P(X, A) = X/A, \tag{2.6}$$

defined in (1.114) induces an equivalence of categories $\Pr\mathsf{Set_2} \to \mathsf{Set_\bullet}$.

The analogous functor $P\colon \mathsf{Top_2} \to \mathsf{Top_\bullet}$ is also exact, but the induced functor $\Pr\mathsf{Top_2} \to \mathsf{Top_\bullet}$ *is not an equivalence of categories.* As well known in algebraic topology, the procedure $(X, A) \mapsto X/A$ destroys topological information, unless particular hypotheses are assumed on the pairs (X, A).

2.2 Characterisations of homological and g-exact categories

Homological and g-exact categories, introduced in Section 1.3, are now dealt with. Various equivalent forms for the axiom (ex3) are given. E is always a semiexact category.

2.2.1 Lemma (The special 3×3-lemma)

In the semiexact category E*, consider the following commutative square of normal monos and epis*

$$
\begin{array}{ccc}
M & \overset{m}{\rightarrowtail} & A \\
{\scriptstyle h}\downarrow & & \downarrow{\scriptstyle q} \\
S & \underset{k}{\rightarrowtail} & Q
\end{array}
\qquad . \tag{2.7}
$$

The following conditions are equivalent:

(a) $m \geqslant \ker q$,

(a*) $\operatorname{cok} m \leqslant q$,

(b) the square (2.7) is a pullback, i.e. $q^*(k) \sim m$,

(b*) the square (2.7) is a pushout, i.e. $m_\circ(h) \sim q$,

(c) $\ker q = m_*(\ker h)$,

(c*) $\operatorname{cok} k = q_\circ(\operatorname{cok} m)$,

(d) the square (2.7) is bicartesian (i.e. pullback and pushout),

(e) the square (2.7) can be embedded in a commutative diagram (2.8), with central row and central column short exact

$$
\begin{array}{ccccc}
N & = & N & & \\
{\scriptstyle u}\downarrow & & \downarrow{\scriptstyle n} & & \\
M & \xrightarrow{\ m\ } & A & \xrightarrow{\ p\ } & P \\
{\scriptstyle h}\downarrow & & \downarrow{\scriptstyle q} & & \| \\
S & \xrightarrow[\ k\ ]{} & Q & \xrightarrow[\ v\ ]{} & P
\end{array}
\tag{2.8}
$$

(f) the square (2.7) can be embedded in a commutative diagram (2.8), with all rows and columns short exact.

Note. This statement can be seen from different points of view. It characterises bicartesian squares of form (2.7); it also concerns subquotients in semiexact categories, taking for granted that the composition qm, in (2.7), is an exact map (as it always is if (ex3) holds). Finally, the equivalence of (e) and (f) is a special instance of the general 3×3-lemma given below (in 2.3.7), because the (null) map $pn = pmu \colon N \to P$ factorises through a null object that can be added in the right upper corner of diagram (2.8).

Proof In diagram (2.7) we have, by (2.1), (2.2)

$$
q^*(k) = q^*(\operatorname{nim} kh) = q^*(\operatorname{nim} qm) = q^*(q_*(m)) = m \vee \ker q,
$$

$$
m_*(\ker h) = m_*(\ker kh) = m_*(\ker qm) = m_* m^*(\ker q) = m \wedge \ker q,
$$

whence (a), (b) and (c) are equivalent. (a) is trivially equivalent to (a*), and both are equivalent to (e). Finally, (e) and (f) are equivalent because $\ker h = h^* k^* 0 = m^* q^* 0 = m^*(n) \sim u$. $\qquad\square$

2.2.2 Theorem (Homological categories)

Let E *be an ex2-category. Then the following conditions are equivalent, and say that* E *is homological, or an ex3-category (as already defined in 1.3.6)*

(ex3) given a normal mono $m \colon M \rightarrowtail A$ *and a normal epi* $q \colon A \twoheadrightarrow Q$ *with*

$m \geqslant \ker q$ *(cok $m \leqslant q$), the morphism qm is exact (*subquotient axiom, or homology axiom*),*

(ex3a) given a pair of normal monos $n \leqslant m$ (with the same codomain A), one can form the diagram (2.8) commutative, with short exact rows and columns (as in the special 3×3-Lemma above),

(ex3b) the pullback of a pair $\bullet \twoheadrightarrow \bullet \leftarrowtail \bullet$ is of the form $\bullet \leftarrowtail \bullet \twoheadrightarrow \bullet$ (pullback axiom);

(ex3b*) the pushout of a pair $\bullet \leftarrowtail \bullet \twoheadrightarrow \bullet$ is of the form $\bullet \twoheadrightarrow \bullet \leftarrowtail \bullet$ (pushout axiom).

More precisely, (ex3) and (ex3a) are equivalent for all semiexact categories. The equivalence of (ex3) and (ex3b) holds for all semiexact categories where normal epis are closed under composition (e.g. in Gp and Rng). Dually, the equivalence of (ex3) and (ex3b*) holds for all semiexact categories where normal monos are closed under composition.

Note. The significance of this axiom for the construction of subquotients has already been observed (in 1.3.6). Note also that the pullback (resp. pushout) considered above is known to exist, as an inverse image of a normal mono (resp. a direct image of a normal epi).

Proof We have to prove the last statement before the note. The equivalence of (ex3) and (ex3a) has been proved in Lemma 2.2.1, for all semiexact categories.

To prove that (ex3) implies (ex3b), take k and q as below; their pullback is constructed by taking $p = \operatorname{cok} k$ and $m = q^*(k) = \ker pq \geqslant \ker q$

$$
\begin{array}{ccc}
M & \overset{m}{\rightarrowtail} & A \\
{\scriptstyle h}\big\downarrow & & \big\downarrow{\scriptstyle q} \\
S & \underset{k}{\rightarrowtail} & Q \underset{p}{\twoheadrightarrow} \bullet
\end{array}
\qquad . \tag{2.9}
$$

Now we have to show that h is a normal epi; by (ex3), $kh = qm$ is exact and has a normal factorisation $kh = k'h'$, with $k' = \operatorname{nim} qm = q_*(m) = q_*q^*(k) = k$ (by 2.1.1(d*)), provided that normal *epis* are closed under composition); hence $h = h'$ is a normal epi.

Conversely, (ex3b) implies (ex3). Given m and q in the above diagram, we set $k = \operatorname{nim} qm = q_*(m)$ and factorise $qm = kh$; since $q^*(k) = q^*q_*(m) = m \vee \ker q = m$, the square is cartesian and h is a normal epi. $\qquad\square$

2.2.3 Definition and Proposition (Exact ideals)

The closed ideal $\mathcal{N}$ of the semiexact category E *will be said to be* exact *if the following equivalent conditions hold:*

(i) each null morphism is exact,

(ii) each morphism between null objects is an isomorphism,

(iii) the null objects of E *coincide with the quasi zero objects (see 1.3.2) of its connected components (and each component necessarily has some quasi zero object, all of them isomorphic),*

(iv) the connected components of E *are quasi pointed semiexact categories (Section 1.3.3).*

Every p-semiexact category is of this type, as well as the category GrAb *of $\mathbb{Z}$-graded groups that is just quasi pointed (cf. 1.4.4). The homological categories* Set$_2$*,* Top$_2$*,* Gp$_2$ *are not of this type.*

The structure of a semiexact category E *with an exact ideal is determined by the mere categorical data. Furthermore:*

(a) if m is a normal mono and is epi then it is iso; if p is a normal epi and is mono then it is iso,

(b) every monomorphism is N-mono, every exact monomorphism is a normal mono (and dually).

Proof By restriction to a connected component, we can assume that E is connected.

(i) $\Rightarrow$ (ii). If Z and Z' are null objects and $f \colon Z \to Z'$ is a morphism with normal factorisation $f = mgp$, then $p = 1$ (the only normal quotient of Z, cf. 1.5.3) and $m = 1$ (the only normal subobject of Z'); since g is iso by hypothesis, so is f.

(ii) $\Rightarrow$ (iii). First we prove that all the null objects are isomorphic. For each object A, the composed map $A_0 \rightarrowtail A \twoheadrightarrow A^0$ is iso, hence $A_0 \cong A$. For each map $f \colon A \to B$, the composed morphism $A_0 \rightarrowtail A \to B \twoheadrightarrow B^0$ is iso, hence $A_0 \cong B^0 \cong B_0$. By induction, since we have assumed that any two objects A and B are connected in E, it follows that $A_0 \cong B_0$. Last, if Z and W are null objects, then $Z \cong Z_0 \cong W_0 \cong W$.

Now, every null object Z is quasi initial. Indeed, for every A, we have a morphism $Z \cong A_0 \rightarrowtail A$; if $u \colon Z \to A$ is a map, than $u = ni$ where $n = \operatorname{nim} u \colon A_0 \rightarrowtail A$ and $i \colon Z \to A_0$ is invertible, by hypothesis. Conversely, if Z is quasi initial then it is null: consider the null subobject $Z_0 \rightarrowtail Z$ and a map $Z \to Z_0$; their two compositions are isomorphisms (by hypothesis or by a property of quasi initial objects), whence $Z \cong Z_0$.

(iii) $\Rightarrow$ (i). If $f\colon A \to B$ is a null morphism, the morphism $g\colon A^0 \to B_0$ of its normal factorisation connects quasi zero objects, hence is an isomorphism.

(iii) $\Leftrightarrow$ (iv). Obvious.

(a) Let p be a normal epi and a monomorphism; it suffices to show that $\ker p = 0$. If $m = \ker p$, we have a pullback

$$
\begin{array}{ccc}
A & \xrightarrow{\;p\;} & B \\
{\scriptstyle m}\big\uparrow & & \big\uparrow{\scriptstyle 0} \\
M & \underset{g}{\overset{f}{\rightleftarrows}} & B_0
\end{array}
\tag{2.10}
$$

and a morphism $g\colon B_0 \to M$, because of (iii); then $fg\colon B_0 \to B_0$ is invertible and f is a split epi; but f is mono, because p is, therefore f is an isomorphism, M is null and $m = 0_A$.

(b) Follows immediately from (a) and 1.5.5(c). $\qquad\square$

2.2.4 Strict ideals

We say that the semiexact category E is *strict*, or that its ideal $\mathcal{N} = \mathrm{Nul\,E}$ is *strict*, if the following conditions hold (they are equivalent, by 2.2.3):

(i) $\mathcal{N}$ is exact and between two null objects there is at most one morphism,

(ii) if Z, Z' are null objects there is at most one morphism $f\colon Z \to Z'$, which is an isomorphism,

(iii) if Z, Z' are null objects in the same connected component of E, there is precisely one morphism $f\colon Z \to Z'$,

(iv) the null objects of E coincide with the zero objects of its connected components,

(v) the connected components of E are p-semiexact categories.

Therefore, a strict semiexact category will also be called a *componentwise p-semiexact category*. Again, the null morphisms are determined by the mere categorical structure. A pointed semiexact category is strict, and a strict one has an exact ideal of null morphisms; further, pointed semiexact is equivalent to: strictly semiexact, connected and non-empty.

Every projective semiexact category (see 1.5.8) is strict (because its faithful transfer functor takes values in a pointed category).

2.2.5 Definition and Proposition (Identifying the null morphisms)

An ex1-category E *with an* exact *ideal* $\mathcal{N} = \text{Nul}\,\mathsf{E}$ *has an* associated strict ex1-category $\text{Str}\,\mathsf{E} = (\mathsf{E}/R, \mathcal{N}/R)$ *where one identifies all the parallel null morphisms, by the following congruence of categories (cf. A1.2)*

$$f\,R\,g \quad \text{if } f \text{ and } g \text{ are parallel morphisms, equal or both null.} \qquad (2.11)$$

(Universal property.) *The canonical projection* $P\colon \mathsf{E} \to \mathsf{E}/R$ *is an ex1-functor, nsb-faithful and nsb-full. Every ex1-functor* $F\colon \mathsf{E} \to \mathsf{E}'$ *with values in a strict semiexact category factorises uniquely through* P *(in* EX_1*). If* E *is ex2 or homological, so is* $\text{Str}\,\mathsf{E}$ *(i.e. strict ex2 or strictly homological).*

Proof R is plainly a congruence of categories, while $\mathcal{N}' = \mathcal{N}/R$ is a closed ideal of E/R, and is strict.

Given $f\colon A \to B$ in E and $m = \ker f$, we want to show that $[m]$ is a kernel of $[f]$ in E/R. Clearly, it suffices to verify the cancellation property $[m][h] = [m][k] \Rightarrow [h] = [k]$; now, either $mh = mk$ or these morphisms are both null in E, in which case h and k are also null (because m is N-mono).

Hence, E/R is ex1, as well as the functor P. The rest is trivial. $\square$

2.2.6 Generalised exact categories

As already defined in 1.3.6, a g-exact category, or ex4-category, is a semiexact category $\mathsf{E} = (\mathsf{E}, \mathcal{N})$ in which

(ex4) each morphism is exact.

Then E is necessarily homological (use the forms (ex2a) and (ex3) of the axioms, in 2.1.3 and 1.3.6); furthermore, $\mathcal{N}$ is an exact ideal and the null objects are categorically determined, as the quasi zero objects of each connected component (see 2.2.3).

By a remark in 2.1.3, an *ex2*-category is g-exact if and only if it satisfies

(ex4') f is iso if (and only if) $\ker f = 0$ and $\text{cok}\, f = 0$.

Let E be g-exact. By 2.2.3(b) every mono is N-mono and every (exact) N-mono is normal, whence:

(a) monos, N-monos, normal monos and fully normal monos coincide; dually for epis;

(b) every morphism factorises through a (normal) epi, followed by a (normal) mono; this factorisation is unique up to isomorphism;

(c) E is balanced, i.e. a map that is mono and epi is always an isomorphism.

The construction considered in 2.2.5 yields the componentwise p-exact category Str E associated to E.

The $\mathbb{Z}$-graded category GrAb (cf. 1.4.4) is a non-strict g-exact category. Of course, we can replace it with the p-exact category Str GrAb, at the cost of destroying the degree of the null morphisms; but there is no real need for all that.

A 'null groupoid' $\mathbf{G}$ (i.e. a groupoid with $\mathrm{Nul}\,\mathbf{G} = \mathbf{G}$) is again a g-exact category, generally non-strict; it is the same as a g-exact category whose objects are all null.

2.2.7 Componentwise p-exact categories

By 2.2.4 and 2.2.6, the following notions are equivalent:

(i) p-exact category, i.e. a category with zero object, kernels and cokernels, such that every map factorises normal epi - normal mono,

(ii) g-exact category with a strict ideal NulE and a zero object,

(iii) connected, non-empty, g-exact category with a strict ideal NulE.

More generally, we say that a category E is *componentwise p-exact*, or an *ex5-category*, if it is g-exact with a *strict* ideal NulE, or equivalently if *all its connected components are p-exact*.

All this shows that *g-exact* and *componentwise p-exact* are slightly more general than *p-exact*. Most of the theory of p-exact categories extends to g-exact categories (including the construction of the categories of relations, see Section 5.4).

Examples of p-exact, non-abelian categories are recalled in 2.4.5.

2.2.8 Exact functors and transformations

An *ex2-functor* is just an exact functor between ex2-categories - it preserves kernels and cokernels. An *ex2-transformation* is a natural transformation between such functors. Analogously for the cases ex3, ex4, ex5.

This defines the sub-2-categories EX_2, EX_3, EX_4 and EX_5 of the 2-category EX_1 (see 1.7.0); they are 2-full in EX_1, i.e. full with respect to arrows (functors) as well as to cells (natural transformations).

Analogously we have the 2-categories EX and AB of p-exact and abelian $\mathcal{U}$-categories, that are (canonically embedded as) 2-full sub-2-categories of all the previous ones. (When convenient, we write EX as EX_6.)

Recall that a conservative ex1-functor is always N-reflective (by 1.7.2). As a partial converse, an N-reflective *ex4-functor* is conservative (because a morphism of a g-exact category is an isomorphism if and only if its kernel and cokernel are null).

A faithful ex5-functor is easily seen to be N-reflective, hence conservative. This need not be true for an ex4-functor (see 2.2.9(c)).

Each of the axioms (ex2), (ex3) and (ex4) - at least in some equivalent formulation - requires certain morphisms to be exact (or other morphisms to be iso). Therefore such a condition is reflected by any *conservative* ex1-functor $F \colon \mathsf{E} \to \mathsf{E}'$, in the sense that if E' satisfies it, then E too does. On the other hand, (ex5) is not of this nature, as shown by the conservative ex4-functor $\mathbf{G} \to \mathbf{1}$ from a 'null groupoid' to the p-exact category $\mathbf{1}$. However, (ex5) is reflected by every *faithful* conservative ex1-functor.

The following lemma is about ex2-,..., ex5-subcategories (defined in 1.3.7 and 1.7.3).

2.2.9 Lemma (Subcategories)

(a) A semiexact subcategory E' of an ex2-category E is necessarily ex2.

(b) A conservative semiexact subcategory of a homological, or g-exact category is of the same type.

(c) A semiexact subcategory of a componentwise p-exact category E is of the same type, and conservative in E. This need not be the case if E is only g-exact.

(d) An nsb-full semiexact subcategory of a homological category is homological.

(e) A g-exact subcategory of a g-exact category is necessarily conservative.

Inj_2 and Inc_2 *(1.7.4(b)) are thus homological subcategories of* Set_2; Cph_2 *(1.7.4(c)) is a homological subcategory of* Top_2; $\mathcal{T}_{LC}$ *(1.7.4(d)) is a homological subcategory of* $\mathcal{T}$; Ban_1 *(1.7.4(e)) is a homological (nsb-full) subcategory of* Ban.

Proof (a) A composition mn of normal monos of E' factorises $mn = h.i$, where h is the normal image of mn in E' (and in E), while i is a normal mono in E'; but mn is a normal mono in E, hence i is iso in E and N-epi in E'; therefore it is also iso in E'.

(b) By the last remark in 2.2.8.

(c) The ex5-case follows from the fact that a faithful ex5-functor is always conservative (Section 2.2.8). We have already seen in 1.4.4 that the g-exact category $\mathsf{E} = \mathsf{GrAb}$ has a homological subcategory E_+ which is not conservative in E and not g-exact.

(d) Obvious.

(e) By the characterisation of isomorphisms via kernels and cokernels, in 2.2.6.   □

2.3 Modular semiexact categories and diagram lemmas

We say that a semiexact category is modular (or distributive, or boolean) if its lattices $NsbA$ of normal subobjects are of this type *and* its transfer connections (f_*, f^*) are modular. We prove in Proposition 2.3.3 that every g-exact category is modular; of course, in this case $NsbA = SubA$.

The subcategories Inj_2 and Inc_2 of Set_2, consisting of pairs of sets and injections or inclusions are boolean homological categories (see 2.3.2, 2.3.4).

The modular property suffices to prove elementary diagram lemmas (cf. 2.3.5 - 2.3.7), by 'chasing' normal subobjects (instead of elements); but the interest of modularity will show more clearly in Chapter 3, in connection with the homology of chain complexes. The interest of distributivity has already been explored in the p-exact case (Part I, or [G5]-[G9]).

E is always a semiexact category.

2.3.1 Modular semiexact categories

An object A of the semiexact category E will be said to be *modular* if its lattice $NsbA$ of normal subobjects is modular (cf. 1.1.2).

A morphism $f: A \to B$ is *left modular*, or *right modular*, or *modular* if its associated connection $Nsb(f): NsbA \to NsbB$ is *left exact*, *right exact* or *exact* (see 1.2.5), i.e. satisfies the first condition below, or the second, or both

$$f^* f_* x = x \vee f^* 0, \qquad \text{for } x \in NsbA \qquad \text{(left modularity)},$$
$$f_* f^* y = y \wedge f_* 1, \qquad \text{for } y \in NsbB \qquad \text{(right modularity)}. \tag{2.12}$$

More particularly, we say that f is *left modular on x* (resp. *right modular on y*) when the above property holds for this particular normal subobject x of A (resp. y of B).

A semiexact category E is said to be *modular*, or an *exm-category*, if all its objects and morphisms are modular, or equivalently if its transfer functor $Nsb: E \to Ltc$ factorises through the (p-exact) subcategory Mlc of modular lattices and modular connections (Section 1.2.8). Then E is necessarily ex2 (use the condition (ex2b) in 2.1.3) and the transfer functor $Nsb: E \to Mlc$ is exact (by (ex2c)).

2.3.2 Examples

We show below that *every g-exact category is modular* (Proposition 2.3.3). In particular, Mlc itself is modular; this also follows, more simply, from the fact that the transfer functor of Mlc is isomorphic to its identity (cf. 1.5.9).

Plainly, *every nsb-faithful ex1-functor reflects modularity*; a fortiori, every conservative ex1-functor also does (by 1.7.2(c)).

In particular, the p-homological, not p-exact, category K Tvs of topological vector spaces over a topological field K is modular, because of its forgetful functor with values in the abelian category of vector spaces (over the underlying field); the latter is not conservative but is clearly nsb-faithful, since the normal subobjects of K Tvs are the linear subspaces equipped with the induced topology (see 1.6.3).

On the other hand, if K is the real or complex field, the homological categories K Hvs, Ban and Hlb of Hausdorff vector spaces, Banach spaces and Hilbert spaces, are not modular. The normal subobjects are now the *closed* linear subspaces, that generally do not obey the modular law; namely, the lattice of closed linear subspaces of the classic Hilbert space L^2 (hence of every infinite dimensional Hilbert space) is just orthomodular ([Bi], II.14).

The semiexact categories Gp and Rng (examined in 1.6.1) have modular lattices of normal subobjects, but are *not* modular (otherwise they would be ex2).

Analogously, the categories Set_2 and Top_2 have lattices of normal subobjects that are distributive, and even complete boolean algebras

$$\mathrm{Nsb}(X, A) \cong \{M \in \mathcal{P}X \mid M \supset A\}, \qquad (2.13)$$

yet these categories are not modular. Indeed, given a map $f \colon (X, A) \to (Y, B)$ and $A \subset M \subset X$, $B \subset N \subset Y$

$$f^* f_*(M) = f^{-1}(f(M) \cup B) = f^{-1}(f(M)) \cup f^{-1}(B), \qquad (2.14)$$

$$\begin{aligned} f_* f^*(N) &= f(f^{-1}(N) \cup B = (N \cap f(X)) \cup B \\ &= N \cap (f(X) \cup B) = N \cap f_*(X), \end{aligned} \qquad (2.15)$$

so that the connection $\mathrm{Nsb}(f)$ is always *right* exact, but it is exact (or modular) if and only if f is injective outside of its kernel $f^{-1}(B)$. But the subcategories Inj_2 and Inc_2 of Set_2, consisting of pairs of sets and injections or inclusions (Section 1.7.4), *are modular homological* (and even boolean, cf. 2.3.4).

Other examples of modular or non-modular homological categories are considered in 5.1.5.

The examples above show that a notion of *right modular* semiexact category, having modular lattices of normal subobjects and right exact transfer connections, may be of interest (but will not be developed here). It includes Gp, Rng, Set$_2$, Top$_2$, Set$_\bullet$, Top$_\bullet$,... but not Gp$_2$ (a lattice of subgroups need not be modular). Such a category is necessarily *right ex2*, i.e. its normal quotients are closed under composition. Its transfer functor is right exact (cf. 2.1.2) and takes values in the category of modular lattices and *right exact* connections, a p-homological subcategory of Ltc, that of course is right modular.

2.3.3 Proposition (Exactness and modularity)

Every g-exact category is modular.

For a semiexact category E, *the following conditions are equivalent:*

(a) E *is modular;*

(b) every morphism $f\colon A \to B$ *is modular, i.e. every connection* $\mathrm{Nsb}(f)$ $= (f_*, f^*)\colon \mathrm{Nsb}A \to \mathrm{Nsb}B$ *is exact*

$$f^* f_* x = x \vee f^* 0, \quad f_* f^* y = y \wedge f_* 1 \quad (x \in \mathrm{Nsb}A,\ y \in \mathrm{Nsb}B); \quad (2.16)$$

(c) for every $f\colon A \to B$, $x \in \mathrm{Nsb}A$ *and* $y \in \mathrm{Nsb}B$

$$f^*(f_* x \vee y) = x \vee f^* y, \qquad f_*(f^* y \wedge x) = y \wedge f_* x. \quad (2.17)$$

Proof (a) $\Rightarrow$ (c) By 1.2.8. (c) $\Rightarrow$ (b) Trivial. (b) $\Rightarrow$ (a) Let $x, a, b \in \mathrm{Nsb}A$, with $x \leqslant a$, and set

$$p = \mathrm{cok}\, b, \qquad f = pa, \qquad x' = a^* x.$$

Thus, $x = a_* x'$ and

$$(x \vee b) \wedge a = (a_* x' \vee p^* 0) \wedge a_* 1 = a_* a^* (p^* p_* (a_* x'))$$
$$= a_* f^* f_* (x') = a_* (x' \vee f^* 0) = a_* (x') \vee a_* a^* p^* (0)$$
$$= x \vee a_* a^* (b) = x \vee (b \wedge a).$$

The first assertion follows now easily: given a g-exact category, its (exact) transfer functor necessarily takes every (exact) morphism to an exact connection. □

2.3.4 *Distributive and boolean semiexact categories*

More particularly, a semiexact category E will be said to be *distributive*, or an *exd-category*, if its transfer functor factorises through the subcategory Dlc of distributive lattices and modular connections (Section 1.2.8).

In other words, this means that all the lattices of normal subobjects Nsb$A$ are distributive and all the transfer connections Nsb$(f) = (f_*, f^*)$ are modular. An nsb-full ex1-functor, again, reflects distributivity.

A *modular* semiexact category E is distributive if and only if it satisfies the following equivalent conditions:

(a) every lattice Nsb$A$ is distributive,

(a*) every lattice Nqt$A$ is distributive,

(b) every mapping f_* preserves binary meets (equivalently, f_* is a quasi homomorphism of lattices),

(b*) every mapping f^* preserves binary joins,

(c) every mapping p_* preserves binary meets, for each normal epi p,

(c*) every mapping m^* preserves binary joins, for each normal mono m.

Indeed we already know (by 1.2.8) that (a) implies (b*), which implies (c*), so that it suffices to show that (c*) implies (a) and achieve the argument by duality. Now, if $a, b, c \in$ Nsb$A$

$$(a \vee b) \wedge c = c_* c^* (a \vee b) = c_* c^* (a) \vee c_* c^* (b)$$
$$= (a \wedge c) \vee (b \wedge c). \tag{2.18}$$

Because of the functorial isomorphism Nsb$X \cong X$ (in 1.5.9), the category Dlc (p-exact, whence modular) is distributive, while Mlc is not. Other examples are in 2.4.5.

Similarly, we say that a semiexact category E is *boolean* if it is modular and all its lattices of normal subobjects are boolean algebras. We already know that this is the case of the subcategories Inj$_2$ and Inc$_2$ of Set$_2$ (see 2.3.2).

2.3.5 *Elementary diagram chasing*

Diagram lemmas in abelian categories are a well-known tool for homological algebra.

Loosely speaking, and in order to extend them to the present setting, let us distinguish between *elementary* lemmas - whose thesis just involves the morphisms assigned in the hypothesis - and *non-elementary* lemmas - that state the existence of some new arrow. In this sense, the Five Lemma and the 3×3-Lemma are elementary while the Snake Lemma - stating the existence (and properties) of the connecting morphism - is not.

As we show below, an elementary lemma can generally be proved by a sort of abstract diagram chasing based on *direct and inverse images of normal subobjects*, instead of the elements of the objects of concrete categories - a procedure already used by Mac Lane ([M3], XII.3), in the abelian case. The modular properties of direct and inverse images (in (2.12)) will often be used; for instance, the use of the property $f^* f_*(x) = x \vee f^* 0$ (where x is a normal subobject of A, the domain of f) systematically replaces the following standard argument of diagram chasing in concrete categories: knowing that $f(a') = f(a'')$, with $a', a'' \in A$, one introduces $a = a' - a'' \in \operatorname{Ker} f$.

Therefore these lemmas can be extended to *modular semiexact* categories, or even to *semiexact* categories by adding suitable hypotheses of modularity on specific morphisms.

On the other hand, the construction of new arrows - such as the connecting morphism - generally requires induction on subquotients in a *homological* category and is deferred to Chapter 3.

2.3.6 Five Lemma

In a modular semiexact category E*, a commutative diagram with exact rows is given*

$$
\begin{array}{ccccccccc}
A' & \xrightarrow{f} & B' & \xrightarrow{g} & C' & \xrightarrow{h} & D' & \xrightarrow{k} & E' \\
\downarrow{\scriptstyle a} & & \downarrow{\scriptstyle b} & & \downarrow{\scriptstyle c} & & \downarrow{\scriptstyle d} & & \downarrow{\scriptstyle e} \\
A'' & \xrightarrow[u]{} & B'' & \xrightarrow[v]{} & C'' & \xrightarrow[w]{} & D'' & \xrightarrow[x]{} & E''
\end{array}
\qquad (2.19)
$$

(a) If a is N-epi, while b and d are N-mono, then c is N-mono.

(a^*) If e is N-mono, while b and d are N-epi, then c is N-epi.

(b) If E is g-exact, a is epi, e is mono, b and d are isomorphisms, then c is an isomorphism.

Proof It suffices to prove (a). Since $c^* 0 \leqslant c^* w^* 0 = h^* d^* 0 = h^* 0 = g_* 1$, we have

$$
c^* 0 = g_* g^* c^* 0 = g_* b^* v^* 0 = g_* b^* u_* 1 = g_* b^* u_* a_* 1
$$
$$
= g_* b^* b_* f_* 1 = g_* f_* 1 = 0.
$$

$\square$

2.3.7 3×3 *Lemma*

In a modular semiexact category E, *a commutative diagram with short exact rows is given*

$$
\begin{array}{ccccc}
A' & \xrightarrow{\ n\ } & A & \xrightarrow{\ q\ } & A'' \\
{\scriptstyle a}\downarrow & & {\scriptstyle b}\downarrow & & {\scriptstyle c}\downarrow \\
B' & \xrightarrow{\ m\ } & B & \xrightarrow{\ p\ } & B'' \\
{\scriptstyle f}\downarrow & & {\scriptstyle g}\downarrow & & {\scriptstyle h}\downarrow \\
C' & \xrightarrow[\ r\ ]{} & C & \xrightarrow[\ s\ ]{} & C''
\end{array}
\qquad\qquad (2.20)
$$

(a) If c is an exact morphism and the first two columns are short exact, so is the third.

(a*) If f is an exact morphism and the last two columns are short exact, so is the first.

Proof Again, it suffices to prove (a). Recall that an exact morphism is a normal mono if and only if its kernel is null (see 1.5.5(c)). By using the modularity of our category

$$
c^*0 = q_*q^*c^*0 = q_*b^*p^*0 = q_*b^*m = q_*b^*(m \wedge b_*1)
$$
$$
= q_*b^*(m \wedge g^*0) = q_*b^*(m_*m^*(g^*0)) = q_*b^*(m_*f^*0)
$$
$$
= q_*b^*(m_*a_*1) = q_*b^*b_*n_*1 = q_*n = 0,
$$

$$
h^*0 = p_*p^*h^*0 = p_*g^*s^*0 = p_*g^*r = p_*g^*(r_*f_*1) = p_*g^*g_*m_*(1)
$$
$$
= p_*(m \vee g^*0) = p_*b = c_*q_*1 = c_*1.
$$

Finally, p and sg are normal epis (by 2.3.1, since E is ex2). Then h too is a normal epi, by 1.5.7(b*).   □

2.4 Additivity and monoidal structures

Semiadditive or additive homological categories have already been encountered in Sections 1.2 and 1.6. The fact that distributivity is incompatible with additivity (Proposition 2.4.4) shows why 'distributive homological algebra' cannot be developed in the context of abelian categories.

Monoidal structures consistent with null morphisms are briefly considered in 2.4.6.

E is always a semiexact category, and C a category.

2.4.1 Additivity

First, let us briefly recall some well-known topics on the sum of maps, in order to fix terminology. (These points are treated in more detail in I.2.1, and can also be found in most texts on categories.)

An N-*linear category* is a category C equipped with a sum of parallel maps, so that each hom-set is an abelian monoid and composition distributes over sum. In other words (and provided that all hom-sets are small), C is a category enriched over the category Abm of abelian monoids (Section 1.6.2), with respect to its obvious monoidal closed structure. Such a category is *preadditive*, or Z-*linear* (i.e. enriched over Ab) if all maps have opposites.

A category C is said to be *semiadditive* if it satisfies the following equivalent conditions (the equivalence is proved in I.2.1.3):

(a) C has a zero-object and binary *biproducts* $A \oplus B$, i.e. objects equipped with maps

$$A \underset{p}{\overset{i}{\rightleftarrows}} A \oplus B \underset{q}{\overset{j}{\leftrightarrows}} B \qquad pi = 1, \quad qj = 1, \quad qi = 0, \quad pj = 0, \qquad (2.21)$$

that make $(A \oplus B, p, q)$ the product of A and B, and $(A \oplus B, i, j)$ their sum;

(b) C has an N-linear structure and finite products;

(b*) C has an N-linear structure and finite sums.

If these conditions hold, the sum of maps is uniquely determined by the categorical structure (cf. I.2.1.3)

$$A \overset{d}{\rightarrow} A \oplus A \overset{f \oplus g}{\longrightarrow} B \oplus B \overset{\partial}{\rightarrow} B, \qquad f + g = \partial.(f \oplus g).d, \qquad (2.22)$$

where $d \colon A \to A \oplus A$ is the *diagonal* (of components id$A$), while $\partial \colon B \oplus B \to B$ is the *codiagonal* (defined by $\partial i' = \partial j' = $ id$B$, where i' and j' are the injections of $B \oplus B$).

On the other hand, the biproduct (2.21) is additively characterised by the following relations (cf. I.2.1.2)

$$pi = 1, \quad qj = 1, \quad pj = 0, \quad qi = 0, \quad ip + jq = 1. \qquad (2.23)$$

The category C is *additive* if it is semiadditive and each map has an opposite. (Then, the relations $pj = 0$, $qi = 0$ are redundant in (2.23)).

An *additive* functor $F \colon \mathsf{C} \to \mathsf{D}$ between N-linear categories preserves the sum of maps. If C and D are also preadditive, or semiadditive, or additive, the additional items of the structure (i.e. the opposite maps or the finite biproducts) are automatically preserved.

2.4.2 Additivity and exactness

Now, a *semiadditive semiexact* (resp. *homological*) category E will be a semiadditive category that is *pointed* semiexact (resp. pointed homological). Then the null maps coincide with the zero maps of the sum; moreover, the biproduct functor $- \oplus - \colon \mathsf{E}^2 \to \mathsf{E}$ is exact (being left and right adjoint to the diagonal functor $\mathsf{E} \to \mathsf{E}^2$). Similarly one defines an *additive* semiexact or homological category.

Notice that all these notions are determined by the mere categorical structure of E.

The p-exact case reduces here to the abelian one, since a p-exact category with finite products is always abelian (as already recalled in 1.3.6).

The following categories are semiadditive homological: Ltc (with biproducts $X \times Y$, see 1.2.7), pAb (Section 1.6.2), Abm, R Smd, K Tvs, K Hvs, Ban, Hlb. The last four are also additive, while Ltc, pAb and Abm are not.

It is easy to see that R Smd is additive if and only if the unital semiring R is actually a ring: in this case R Smd coincides with the abelian category R Mod of R-modules.

Ban_1 is not an additive subcategory of Ban. Moreover the product and sum in Ban_1 of two Banach spaces X, Y (described in 1.7.4(e)) are respectively equipped with the l_∞-norm and the l_1-norm

$$||(x,y)||_\infty = \max(||x||, ||y||), \qquad ||(x,y)||_1 = ||x|| + ||y||.$$

They are thus isomorphic in Ban but not in Ban_1, and the latter cannot have an $\mathbb{N}$-linear structure.

The p-homological category Set. (or the equivalent category $\mathcal{S}$ of sets and partial mappings) is not semiadditive, since its (obvious) sums and products are different. The p-exact category Mlc, again, has no semiadditive structure (cf. 1.2.8).

Cyclic groups form a preadditive p-exact category that is not additive. The same is true of abelian groups of order bounded by a fixed integer, or vector spaces (on a given field) of dimension similarly bounded.

2.4.3 The transfer functor in the semiadditive case

Let E be a semiadditive semiexact category. Then the transfer functor $\mathsf{Nsb}\colon \mathsf{E} \to \mathsf{Ltc}$ is *lax-additive*, in the sense that it comes equipped with a natural transformation

$$\varphi_{AB} \colon \mathrm{Nsb}A \times \mathrm{Nsb}B \to \mathrm{Nsb}(A \oplus B), \quad (x,y) \mapsto x \oplus y, \tag{2.24}$$

where $x \oplus y$ is a normal subobject of $A \oplus B$, because of the exactness of the biproduct. (More precisely, we should write $\mathrm{nim}\,(x \oplus y)$, to single out *the subobject* associated to the normal mono $x \oplus y$.)

We now prove that, for $f, g \colon A \to B$ in E, we have the following relation, in the *semiadditive ordered* category Ltc (Sections 1.2.1, 1.2.7)

$$\mathrm{Nsb}(f + g) \leqslant \mathrm{Nsb}(f) + \mathrm{Nsb}(g)$$

$$((f + g)_* \leqslant f_* \vee g_*, \quad (f + g)^* \geqslant f^* \wedge g^*). \tag{2.25}$$

Indeed, let us consider first the diagonal $d \colon A \to A \oplus A$, the codiagonal $\partial \colon A \oplus A \to A$ and prove that $d_*(x) \leqslant x \oplus x$ and $\partial_*(x \oplus y) = x \vee y$. The left diagram below proves the first inequality and $\partial_*(x \oplus x) \leqslant x$

The right diagram shows that $\partial_*(x \oplus y) \geqslant x$ (and $\geqslant y$).

On the other hand, if $x, y \leqslant z$ then $x \oplus y \leqslant z \oplus z$ and $\partial_*(x \oplus y) \leqslant \partial_*(z \oplus z) \leqslant z$, by our previous argument. It follows that:

$$(f + g)_*(x) = \partial_*(f \oplus g)_* d_*(x) \leqslant \partial_*(f \oplus g)_*(x \oplus x)$$
$$= \partial_*(f_*(x) \oplus g_*(x)) = f_*(x) \vee g_*(x).$$

Finally, if E is additive, the isomorphism -1_A induces the identity of $\mathrm{Nsb}A$

$$(-1)_*(x) = x_*(\mathrm{nim}\,(-1_X)) = x_*(1) = x.$$

It is now clear that, in order to extend the notion of semiadditive category and define an $\mathbb{N}$-*linear* or *preadditive semiexact* category, one should at least assume that the null maps coincide with the zero-maps of the sum *and* the property (2.25).

Abelian monoids of order bounded by a fixed integer form an $\mathbb{N}$-*linear homological category*, in this sense.

2.4.4 Proposition (Additivity versus distributivity)

An additive semiexact category is distributive if and only if it is trivial, i.e. all its objects are zero. In particular, a non-trivial abelian category cannot be distributive.

Proof Given any object A in the additive semiexact category E, we prove that the lattice $\mathrm{Nsb}(A \oplus A)$ is distributive (if and) only if A is the zero object.

With respect to the biproduct $A \oplus A$, write again i and j for the injections, p and q for the projections, $d = i + j \colon A \to A \oplus A$ for the diagonal. Now, i, j and d are normal subobjects of $A \oplus A$ (they are the kernels of q, p and $p - q$).

But $i \vee j = 1$ and $i \wedge d = 0$, because of the following arguments
- if $m \geqslant i, j$ then $i = mf$, $j = mg$ and $m(fp + gq) = ip + jq = 1$, whence m is epi: $m = 1$,
- if $m \leqslant d, i$ then $pm = qm = 0$, whence $m = 0$.

Finally, the distributivity condition implies that the normal mono $d \colon A \to A \oplus A$ is null, and then A too is null.

$$d = (i \vee j) \wedge d = (i \wedge d) \vee (j \wedge d) = 0.$$

$$\square$$

2.4.5 Examples of p-exact, non-abelian categories

Some examples of this kind have already been considered here: the categories Mlc and Dlc in 1.4.1 (the second is distributive), and the category $\mathcal{I}$ of sets and partial bijections in 1.4.3 (distributive and even boolean). We recall below other examples already listed in I.1.5.6, among others.

Obvious examples are the categories of cyclic groups (preadditive and distributive) and of vector spaces of dimension lower than a fixed integer (preadditive, generally non-distributive).

Other examples come from categorical constructions that can be performed in EX (the 2-category of p-exact categories, exact functors and natural transformations) and not in AB (abelian categories).

For instance, the *distributive expansion* Dst E of a p-exact category (see I.2.8.4, or here 5.1.7) is distributive p-exact, whence never abelian, unless E is trivial (in the sense of 2.4.4); in particular, the p-exact category Dst Ab, of interest for the study of spectral sequences of abelian groups, is not abelian.

Similarly, the projective category Pr E of a p-exact category is p-exact but non-abelian, in general, as already recalled in 2.1.4.

2.4.6 Monoidal homological categories

Let us say that a *monoidal N-category* is an N-category E equipped with a monoidal structure (see A4.1) *consistent with the N-structure*, in the sense that if f or g is a null morphism, $f \otimes g$ is also. Equivalently, if A or B is a null object, $A \otimes B$ is also.

Similarly, a semiexact, or ex2, or homological, or g-exact category will be said to be *monoidal* when equipped with a consistent monoidal structure.

The pointed categories Ab, R Mod, Abm, Ban have well-known tensor products $A \otimes B$, that make them monoidal p-homological categories. (Notice that their 'cartesian' structure $A \oplus B$ is *not* consistent with zero-maps, in the previous sense.)

We have already seen that Set_2 is a monoidal homological category, with the tensor product defined in 1.4.5

$$(X, A) \otimes (Y, B) = (X \times Y, (X \times B) \cup (A \times Y)). \tag{2.26}$$

Top_2 and Gp_2 have a similar structure; of course, in the second case, the union of subsets is replaced with the join of subgroups.

$\mathsf{Set}_\bullet$ is a monoidal p-homological category, with the usual smash product

$$(X, x_0) \wedge (Y, y_0) = (X \times Y)/(X \times \{y_0\} \cup \{x_0\} \times Y), [(x_0, y_0)]). \tag{2.27}$$

The exact functor $P \colon \mathsf{Set}_2 \to \mathsf{Set}_\bullet$ defined in 1.7.5(b) is strongly monoidal, i.e. there is a natural isomorphism

$$P((X, A) \otimes (Y, B)) \cong P(X, A) \wedge P(Y, B). \tag{2.28}$$

2.5 Abstract categories of pairs

The construction of a category of pairs C_2 (e.g. Top_2), is based on a category C (e.g. Top), equipped with some distinguished subobjects (e.g. the inclusions of subspaces) and follows a pattern that we formalise and extend here. A similar construction can be found in [LaV], Section 3, for a 'pre-regular' category.

Every variety of algebras gives a homological category of pairs (see 2.5.1). Categories of augmented rings, supplemented algebras and Lie algebras can also be dealt with in this way, viewing an augmented ring as a pair (Λ, I) formed of a ring and a left ideal of the latter (see 2.5.7).

2.5.1 *Categories with distinguished subobjects*

Let a category C be equipped with a set $\mathbf{d}$ of *distinguished* subobjects, also called *d-subobjects*, for short. For every object X, the ordered subset of $\mathrm{Sub}X$ formed by its d-subobjects will be written as d_X.

We say that C (more precisely the pair $(\mathsf{C}, \mathbf{d})$) is a *ds-category* if:

(ds1) for each object X, d_X is a small lattice containing 1_X and a least element, written as $n_X \colon N_X \to X$,

(ds2) the d-subobjects are stable under binary meets and inverse images (pullbacks along arbitrary maps),

(ds3) every map factorises through a smallest d-subobject of its codomain,

(ds4) if $a \colon A \to X$ is a d-subobject, then $a n_A \colon N_A \to X$ is *equivalent* to a d-subobject of X.

We say that C is a *normal* ds-category if it also satisfies the condition

(ds5) the d-subobjects are stable under composition (again, up to equivalence of subobjects),

that implies (ds4) and makes the meet-assumption in (ds2) redundant.

The category of topological spaces equipped with subspaces (or closed subspaces) is normal ds.

The category of groups equipped with subgroups (resp. invariant subgroups) is a normal (resp. *not* normal) ds-category.

More generally, *every variety of algebras* $\mathsf{C} = \mathrm{Alg}(\Omega, \Sigma)$, formed of all algebras of signature Ω that satisfy the equational laws of Σ, is easily seen to be a normal ds-category, when all subobjects (i.e. subalgebras) are distinguished. This fact will be extended to a large setting of concrete categories, in A2.7-A2.8. (But note that, with respect to the usual conventions of Universal Algebra, as for instance in Cohen's text [Co], *one should not exclude a priori the empty algebra*, since this exclusion can 'destroy' the lattices of subobjects and their counterimages: see A1.9.)

Let C be a ds-category. The meets of d_X subsist in $\mathrm{Sub}X$, by (ds2), while the joins need not. If $b \leqslant a$ in $\mathrm{Sub}X$ and $b \in \mathbf{d}$, the monomorphism c such that $b = ac$ belongs to $\mathbf{d}$, by an obvious pullback (again, up to identifying c with the equivalent d-subobject $a^*(b)$).

A morphism $f \colon X \to Y$ of C induces a Galois connection $(f_*, f^*) \colon d_X \to d_Y$, where $f_*(a)$ is the smallest d-subobject of Y through which fa factorises; the (contravariant) transformation $f \mapsto f^*$ is functorial (by pullback pasting), whence the (covariant) transformation $f \mapsto f_*$ is also (because of the adjunction).

The least d-subobject $n_X \colon N_X \to X$ is preserved by direct images, because of the adjunction. Therefore, if $a \colon A \to X$ is a d-subobject, it follows from (ds4) that

$$n_X = a_*(n_A) \sim a.n_A. \tag{2.29}$$

A *ds-functor* $F \colon \mathsf{C} \to \mathsf{D}$ will be a functor between ds-categories that preserves d-subobjects and the least d-subobjects.

2.5.2 General categories of pairs

Given a ds-category (C, d), we construct the semiexact category *of pairs* $C_2 = \mathrm{Pair}(C, d)$ as a full subcategory of the category of morphisms of C.

Its objects are the morphisms in d, its morphisms $f = (f, f') \colon a \to b$ are the (solid) commutative diagrams (2.30), with obvious composition

$$
\begin{array}{ccc}
X & \xrightarrow{\ f\ } & Y \\
{\scriptstyle a}\big\uparrow & \searrow{\scriptstyle g} & \big\uparrow{\scriptstyle b} \\
A & \xrightarrow[\ f'\ ]{} & B
\end{array}
\qquad (2.30)
$$

This morphism f is assumed to be *null* if there is a map g (necessarily unique) that makes (2.30) commutative. The null objects are the identities of C and $f \colon a \to b$ is null if and only if it factorises through $1 \colon B \to B$, or equivalently through $1 \colon X \to X$

$$
\begin{array}{ccccc}
X & \xrightarrow{\ g\ } & B & \xrightarrow{\ b\ } & Y \\
{\scriptstyle a}\big\uparrow & & \big\| & & \big\uparrow{\scriptstyle b} \\
A & \xrightarrow[\ f'\ ]{} & B & =\!=\!= & B
\end{array}
\qquad
\begin{array}{ccccc}
X & =\!=\!= & X & \xrightarrow{\ f\ } & Y \\
{\scriptstyle a}\big\uparrow & & \big\| & & \big\uparrow{\scriptstyle b} \\
A & \xrightarrow[\ a\ ]{} & X & \xrightarrow[\ g\ ]{} & B
\end{array}
\qquad (2.31)
$$

Kernels and cokernels exist and are computed as follows

$$
\begin{array}{ccccccc}
f^{*}B & \xrightarrow{\ m\ } & X & \xrightarrow{\ f\ } & Y & =\!=\!= & Y \\
{\scriptstyle k}\big\uparrow & & {\scriptstyle a}\big\uparrow & & \big\uparrow{\scriptstyle b} & & \big\uparrow{\scriptstyle h} \\
A & =\!=\!= & A & \xrightarrow[\ f'\ ]{} & B & \xrightarrow[\ n\ ]{} & B \vee f_{*}X
\end{array}
\qquad (2.32)
$$

where $m = f^{*}(b)$, $k \in d$ (because the left square is trivially a pullback) and $h = b \vee f_{*}(1_X)$ is a join in d_X, by (ds1).

In order to characterise the normal subobjects and quotients, note that in (2.32) $m = f^{*}(b) \in d$ (and $n = h^{*}(b) \in d$).

Therefore, given a pair $b \colon B \to X$, the general short exact sequence of central term b is bijectively determined by a d-subobject $a \colon A \to X$ such that $a \geqslant b$

$$
\begin{array}{ccccc}
A & \xrightarrow{\ a\ } & X & =\!=\!= & X \\
{\scriptstyle c}\big\uparrow & & \big\uparrow{\scriptstyle b} & & \big\uparrow{\scriptstyle a} \\
B & =\!=\!= & B & \xrightarrow[\ c\ ]{} & A
\end{array}
\qquad (2.33)
$$

In other words, for every pair $b \colon B \to X$, we have established an order-isomorphism between the quasi sublattice $\uparrow b = \{a \in d_X \mid a \geqslant b\}$ of d_X

and the lattice of normal subobjects of b in C_2

$$\uparrow b \to \text{Nsb}(b),$$
$$a \mapsto ((a, 1): c \rightarrowtail b) \qquad (c = a^*(b) \in \mathbf{d}). \tag{2.34}$$

The least normal subobject of the object $b: B \to X$ corresponds to b, the greatest to 1_X. Since all lattices d_X are assumed to be small, the local smallness condition holds (see 1.5.1) and C_2 is indeed semiexact (in the restricted sense).

Note that, even if all the lattices d_X are distributive, the category C_2 need not be modular: the modularity of the transfer connections (f_*, f^*) can fail, as it happens in Set_2 and Top_2 (cf. 2.3.2).

2.5.3 Subquotients

If C is a normal ds-category, its category of pairs C_2 is homological.

Indeed, the previous description of normal subobjects and quotients shows that these are stable under composition, i.e. C_2 satisfies the axiom (ex2) if and only if C is a normal ds-category.

On the other hand, the construction of subquotients works in the general case. Let $m: a \rightarrowtail x$ and $q: x \twoheadrightarrow b$ be given in C_2, as in the left diagram below

$$
\begin{array}{ccccccc}
A & \xrightarrow{m} & X & = & X & \qquad A & = & A & \xrightarrow{m} & X \\
{\scriptstyle a}\uparrow & & {\scriptstyle x}\uparrow & & {\scriptstyle b}\uparrow & {\scriptstyle a}\uparrow & & {\scriptstyle h}\uparrow & & {\scriptstyle b}\uparrow \\
X' & = & X' & \xrightarrow{q'} & B & \qquad X' & \xrightarrow{q'} & B & = & B
\end{array}
\tag{2.35}
$$

and assume that the monomorphism $m: a \rightarrowtail x$ is greater than

$$\ker (q: x \twoheadrightarrow b) = ((b, 1): q' \rightarrowtail x).$$

This means that $b = mh$ and $hq' = a$, for some mono $h: B \to A$. Then the morphism $qm: a \to b$ factorises as in the right diagram above, where h is a distinguished subobject (because the right square of monomorphisms is, trivially, a pullback).

2.5.4 The canonical embedding

Extending the concrete situations that we already know, a ds-category C embeds in C_2 sending each object X to its least d-subobject $n_X: N_X \to X$ (cf. 2.5.1)

$$n: C \to C_2, \qquad X \mapsto (n_X: N_X \to X),$$
$$(f: X \to Y) \mapsto n_f = (f, N_f): n_X \to n_Y, \tag{2.36}$$

where $N_f\colon N_X \to N_Y$ is determined by the condition $f.n_X = n_Y.N_f$ (and exists because $f_*(n_X) = n_Y$).

This functor n

(a) takes values in a semiexact category,

(b) takes each d-subobject $a\colon A \to X$ to a normal subobject in C_2 (namely $n(a)\colon n_A \rightarrowtail n_X$),

(c) takes the least d-subobject $n_X\colon N_X \to X$ to the null normal subobject.

Indeed, as to (b), there is a short exact sequence $\mathrm{Sh}(a)$, showing that, also here, every pair $a\colon A \to X$ can be presented as a cokernel n_X/n_A in C_2, of objects belonging to the image $n(\mathsf{C})$

$$
\begin{array}{ccccc}
A & \xrightarrow{\ a\ } & X & \xrightarrow{\ 1\ } & X \\
{\scriptstyle n_A}\big\uparrow & & {\scriptstyle n_X}\big\uparrow & & \big\uparrow{\scriptstyle a} \\
N_A & \xrightarrow[N_a]{} & N_X & \xrightarrow[v]{} & A
\end{array}
\tag{2.37}
$$

since Na is an isomorphism ($n_X = a_*(n_A) \sim an_A$, by (ds4)) and thus the morphism n_X factorises (uniquely) through a. Last, as to (c), if $a = n_X$ (and $A = N_X$), n_A is also an isomorphism and $n(a)$ is the least normal subobject of n_X (by (2.34)).

As usual for concrete categories of pairs, we shall often identify the object X of C with its image $n_X\colon N_X \to X$ in C_2; consistently, an arbitrary 'pair' $a\colon A \to X$ will also be written as X/A, when there is no ambiguity about the distinguished subobject a.

The sequence $\mathrm{Sh}(a)$ will be called the *canonical resolution* of $a = X/A$. It is functorial: each map $f\colon a \to b$ of C_2 produces a morphism $\mathrm{Sh}(f)\colon \mathrm{Sh}(a) \to \mathrm{Sh}(b)$, in a way that respects identities and composition.

This sequence will also be useful for calculating satellites, in Section 4.4; we shall see that $\mathrm{Sh}(a)$ is a normally projective resolution of a, and actually the only one up to isomorphism.

2.5.5 The universal property

Let C *be a ds-category.*

(a) The functor $n\colon \mathsf{C} \to \mathsf{C}_2$ *satisfies the properties 2.5.4(a) - (c), and every functor* $F\colon \mathsf{C} \to \mathsf{E}$ *that also satisfies them has a unique short exact extension* $G\colon \mathsf{C}_2 \to \mathsf{E}$ *to the category of pairs (satisfying* $Gn = F$*).* G *is defined as*

$$
G(a) = \mathrm{Cok}_{\mathsf{E}}(Fa), \qquad G(f) = \mathrm{Cok}_{\mathsf{E}}(Ff\colon Fa \to Fb). \tag{2.38}
$$

(b) If C *is a normal ds-category, the same universal property holds for functors with values in homological categories.*

(c) A ds-functor $F\colon$ C $\to$ D *between ds-categories has a natural extension to their semiexact categories of pairs*

$$F_2\colon \mathsf{C}_2 \to \mathsf{D}_2, \qquad F_2(a) = Fa, \quad F_2(f, f') = (Ff, Ff'). \tag{2.39}$$

The latter is short exact. If F *also preserves direct and inverse images of d-subobjects (hence their meets) and their joins, then* $F_2\colon \mathsf{C}_2 \to \mathsf{D}_2$ *preserves all kernels and cokernels, i.e. it is exact.*

Proof (a) If such an extension exists, given an object $a\colon A \to X$ consider its canonical resolution $\mathrm{Sh}(a) = (n_a, p_a)$ and let $\iota_a\colon \mathrm{Cok}\,(Fa) \to G(a)$ be the unique isomorphism of E that makes the following diagram commutative

$$
\begin{array}{ccccc}
FA & \xrightarrow{\ Fa\ } & FX & \xrightarrow{\ Gp_a\ } & G(a) \\
\big\| & & \big\| & & \big\uparrow{\scriptstyle \iota_a} \\
FA & \xrightarrow[\ Fa\ ]{} & FX & \xrightarrow[\ p'\ ]{} & \mathrm{Cok}\,(Fa)
\end{array}
\qquad .
$$

This determines the functor G, up to a unique functorial isomorphism.

Conversely, starting from the functor F, define

$$
\begin{aligned}
G\colon \mathsf{C}_2 \to \mathsf{E}, \qquad & G(a) = \mathrm{Cok}\,Fa, \\
G(f\colon a \to b) = {}& (\mathrm{Cok}\,(Ff)\colon G(a) \to G(b)).
\end{aligned}
\tag{2.40}
$$

G extends F, because of our hypotheses on the latter

$$
\begin{aligned}
G(n_A) = \mathrm{Cok}\,F(n_A) = \mathrm{Cok}\,(0_{FA}\colon (FA)_0 \to FA) = FA, \\
G(n_a) = Fa.
\end{aligned}
\tag{2.41}
$$

Now G preserves the short exact sequences consisting of canonical resolutions, hence - *because of the lemma below* - all of them.

Finally, (b) and (c) are obvious, taking into account the description of kernels and cokernels in (2.32). $\square$

2.5.6 Lemma

Let C *be a ds-category and* $F\colon \mathsf{C}_2 \to \mathsf{E}$ *a functor with values in a semiexact category. Then* F *is short exact if and only if it transforms the canonical resolutions* $\mathrm{Sh}(a)$ *of the objects of* C_2 *into short exact sequences of* E.

Proof The necessity is obvious. The sufficiency is a straightforward application of the special 3×3-Lemma in the semiexact category E (Lemma 2.2.1).

In fact, given an arbitrary short exact sequence in C_2 (see (2.33))

$$A/B \rightarrowtail X/B \longrightarrow\!\!\!\!\rightarrow X/A \qquad (2.42)$$

the canonical resolutions of A/B and X/B give a commutative diagram of C_2

$$(2.43)$$

The latter is transformed by F into a commutative diagram of E of the same kind, with short exact central row and column; then its lower row is also short exact. □

2.5.7 Augmented rings

Cartan - Eilenberg's text [CE] dedicates various chapters to the study of the homology and cohomology theories of augmented rings - including supplemented algebras and Lie algebras viewed as particular instances of augmented rings. We briefly show here how the category of augmented rings can be embedded in a semiexact category of pairs.

A *left augmented ring* $\varepsilon \colon \Lambda \to Q$ is defined in [CE], Ch. VIII, as a unital ring Λ equipped with an epimorphism ε of left Λ-modules; the latter is determined, up to isomorphism, by the *augmentation left ideal* $I = \operatorname{Ker} \varepsilon$, so that the structure essentially amounts to the pair (Λ, I) formed of a unital ring and a left ideal of the latter. A morphism of such structures amounts to a homomorphism of unital rings that takes the augmentation ideal of the first ring into the corresponding ideal of the second.

Now, except in trivial cases, an ideal of a unital ring has no unit and is not a subobject of the ring, in the category of *unital* rings and their homomorphisms; therefore the category of the pairs (Λ, I) described above is not a 'category of pairs' in the present sense.

However, one can consider a slightly different framework consistent with the present analysis. We start from the category Rng of rings (without assuming units), equipped with the set **L** of left ideals as distinguished subobjects, and we get a semiexact (non homological) category of pairs

$\mathsf{Rng}_L = \mathrm{Pair}(\mathsf{Rng}, \mathbf{L})$, which contains the previous category of the pairs (Λ, I).

Supplemented algebras and Lie algebras can be treated in a similar way, following [CE], Ch. X and Ch. XIII, respectively.

2.5.8 *Pointed objects*

The categories of pointed objects, like **Set.** and **Top.** (Section 1.6.4), are *not* categories of pairs in the present sense. The *points* $\top \to X$ of **Set** or **Top** do not satisfy the conditions for distinguished subobjects, in Section 2.5.1; furthermore, the cokernels of **Set.** and **Top.** are constructed in a different way, by means of pushouts rather than by joins of distinguished subobjects.

A notion of *category with distinguished points* $\top \to X$ (not necessarily all the maps from $\top$) can be easily given, in order to get an associated p-semiexact category. But here we are more interested in the connection between pairs and pointed objects on the same ds-category, that will be dealt with in Section 5.3.4.

2.5.9 *Distinguished quotients*

Dually, we define a *category with distinguished quotients* as a pair $(\mathsf{C}, \mathbf{d})$ such that $(\mathsf{C}^{\mathrm{op}}, \mathbf{d}^{\mathrm{op}})$ is a category with distinguished subobjects.

There is then a semiexact category

$$\mathrm{Pair}^*(\mathsf{C}, \mathbf{d}) = (\mathrm{Pair}(\mathsf{C}^{\mathrm{op}}, \mathbf{d}^{\mathrm{op}}))^{\mathrm{op}}.$$

Its objects are the **d**-quotients $a \colon X \to A$; a morphism $f = (f, f') \colon a \to b$ is a commutative square, as in the left (solid) diagram below

$$
\begin{array}{ccccccc}
X & \xrightarrow{f} & Y & \qquad X & = X & \xrightarrow{f} Y & \xrightarrow{c} P \\
\end{array}
\tag{2.44}
$$

The morphism f is null if and only if one can insert in the square a (unique) map g, as above (preserving commutativity, of course).

The kernels and cokernel of f are shown in the right diagram above, where k is the meet in **d** of a with the inverse f-image of the quotient 1_Y, while c is the pushout of a along f.

2.6 The epi-mono completion of a category

In the last three sections of this chapter we construct *the epi-mono completion* of a category C, or *Freyd completion* Fr C, and study its exactness properties.

This construction was introduced by the author in [G15, G16], as a generalisation of the Freyd embedding of a *stable* homotopy category into an *abelian* category [F1]-[F3]. Under some hypotheses on the existence of weak kernels and cokernels, but *without assuming stability*, a pointed category produces an unrestricted *homological* category, as we shall prove in Theorem 2.7.6. Well-poweredness of Fr C is a complex problem, briefly analysed in Section 2.8.

The epi-mono completion Fr C can be described as the free category with epi-mono factorisation system, over C (cf. 2.6.3). It is related to the *regular* or *Barr-exact completions* of a category with limits or weak limits, studied in Carboni [Ca2], or Carboni and Vitale [CaV].

A reader interested in category theory will recall that the endofunctor C ↦ C² of CAT has an obvious 2-monad structure (with 'diagonal' multiplication) whose pseudo algebras C² → C correspond to the factorisation systems over C (see Coppey [Cp], Korostenski and Tholen [KT]). Similarly, it has been proved that the pseudo algebras for the induced monad C ↦ Fr C correspond to the *epi-mono* factorisation systems over C [G17].

2.6.1 Factorisation systems

A *category with factorisation system* C, or *fs-category* for short, is endowed with a pair (E, M) satisfying the usual axioms (cf. [FK, JT, KT]):

(fs.1) E, M are subcategories of C containing all the isomorphisms,

(fs.2) every morphism u has a factorisation $u = u''.u'$ with $u' \in E$, $u'' \in M$,

(fs.3) (*orthogonality*) given a commutative square $mf = ge$, with $e \in E$, $m \in M$, there is a unique morphism u making the following diagram commutative

$$
\begin{array}{ccc}
A & \xrightarrow{e} & B \\
f \downarrow & \swarrow u & \downarrow g \\
C & \xrightarrow{m} & D
\end{array}
\qquad . \tag{2.45}
$$

The factorisation $u = u''.u'$ (with $u' \in E$, $u'' \in M$), is thus determined up to a unique central isomorphism; it will be called the *structural*, or *fs-factorisation*, of u; its middle object will be written as $\mathrm{Im}\,(u)$. An *fs-functor* will be a functor between fs-categories that preserves their structure.

We also have to consider a *strict factorisation system* (E_0, M_0) over C, satisfying:

(i) E_0, M_0 are subcategories of C containing all the identities,

(ii) every morphism u has a *strictly unique* factorisation $u = u''.u'$ with $u' \in E_0$, $u'' \in M_0$.

Then, $E_0 \cap M_0$ is the subcategory of identities; and there is a unique factorisation system (E, M) containing the former (or *spanned* by it), where $u = u''.u'$ is in E (resp. M) if u'' (resp. u') is iso. To prove this, the only not obvious point is the closure of E, or M, under composition. First, one proves that, if $e \in E_0$, i is iso and $ei = n.f$ is the unique (E_0, M_0)-factorisation, then n is an iso. Now, if u and v are consecutive in E, with u'', v'' iso, we have $vu = v''v'.u''u' = v''.(v'u'').u' = v''.(a''.a').u' = v''a''.a'u'$, with $a' \in E_0$, $a'' \in M_0$ and a'' iso; thus $vu \in E$.

A factorisation system is said to be *epi-mono* if all E-maps are epi and all M-maps are monos. Then, the morphisms of E and M will be called *fs-epis* (or distinguished epis) and *fs-monos* (or distinguished monos), respectively. There is a decomposition property for E-maps: if $e \in E$ and $e = vu$ then $v \in E$; dually for M.

2.6.2 The factorisation completion

Let C be any category and C^2 its category of maps (already considered in 1.7.6, for a semiexact category).

Let us recall that an object of C^2 is a C-morphism $x \colon X' \to X''$, which we *may* write as $\hat{x}$ when viewed as an object of C^2; a morphism $f = (f', f'') \colon \hat{x} \to \hat{y}$ is a commutative square of C

$$
\begin{array}{ccc}
X' & \xrightarrow{f'} & Y' \\
{\scriptstyle x}\big\downarrow & & \big\downarrow{\scriptstyle y} \\
X'' & \xrightarrow[f'']{} & Y''
\end{array}
\qquad (2.46)
$$

The composition is obvious. C^2 has a canonical factorisation system (not epi-mono, generally), where the map $f = (f', f'')$ is in E (resp. in M) if and only if f' (resp. f'') is an isomorphism

$$
\begin{array}{ccccc}
X' & {=\!=} & X' & \xrightarrow{f'} & Y' \\
{\scriptstyle x}\big\downarrow & & {\scriptstyle \bar{f}}\big\downarrow & & \big\downarrow{\scriptstyle y} \\
X'' & \xrightarrow[f'']{} & Y'' & {=\!=} & Y''
\end{array}
\qquad (2.47)
$$

Inside this system, the morphism f has a unique *strict* factorisation (2.47), whose middle object is the diagonal $\overline{f} = f''x = yf'$ of the given square (2.46). In other words, our system is spanned by a canonical strict system, where (f', f'') is in E_0 (resp. in M_0) if and only if f' (resp. f'') is an identity.

C is fully embedded in C^2, identifying the object X with $\hat{1}_X$, and $f\colon X \to Y$ with $(f, f)\colon \hat{1}_X \to \hat{1}_Y$. Each object $\hat{x}$ can be viewed as the structural image of the corresponding morphism $x\colon X' \to X''$ of $C \subset C^2$, whose (strict) factorisation is

$$x = (x, 1).(1, x)\colon X' \to \hat{x} \to X'', \qquad \mathrm{Im}\,(x) = \hat{x}. \qquad (2.48)$$

One deduces easily that the category with factorisation system C^2 is the *factorisation completion*, or the (bi)*free fs-category* on C. Namely, every functor $F\colon C \to D$ with values in an fs-category has an fs-extension $G\colon C^2 \to D$, determined up to a unique natural isomorphism

$$G(\hat{x}) = \mathrm{Im}_D(Fx), \qquad (2.49)$$

$$
\begin{array}{ccccc}
FX' & \xrightarrow{(Fx)'} & G(\hat{x}) & \xrightarrow{(Fx)''} & FX'' \\
{\scriptstyle Ff'}\Big\downarrow & & {\scriptstyle Gf}\Big\downarrow & & \Big\downarrow{\scriptstyle Ff''} \\
FY' & \xrightarrow[(Fy)']{} & G(\hat{y}) & \xrightarrow[(Fy)'']{} & FY''
\end{array}
\qquad . \qquad (2.50)
$$

(The essential uniqueness of Gf follows from the orthogonality axiom.)

2.6.3 The epi-mono completion

Now, the *epi-mono completion*, or *Freyd completion*, $\mathrm{Fr}\,C$ is a quotient C^2/R of the category of morphisms of C: two parallel morphisms $f = (f', f'')\colon x \to y$ and $g = (g', g'')\colon x \to y$ of C^2 are R-equivalent whenever their diagonals $\overline{f}, \overline{g}$ coincide

$$
\begin{array}{ccc}
X' & \xrightarrow{f'} & Y' \\
{\scriptstyle x}\Big\downarrow & \searrow{\scriptstyle \overline{f}} & \Big\downarrow{\scriptstyle y} \\
X'' & \xrightarrow[f'']{} & Y''
\end{array}
\qquad . \qquad (2.51)
$$

The morphism of $\mathrm{Fr}\,C$ represented by f will be written as $[f]$ or $[f', f'']$. Plainly, if f' is epi (resp. f'' is mono) in C, so is $[f]$ in $\mathrm{Fr}\,C$.

Consider the previous *strict* factorisation (2.47) in C^2: its middle object is precisely $\overline{f}$, and coincides with the diagonals of both morphisms of the factorisation. Therefore, our strict factorisation system in C^2 induces a similar system in Fr C, which is now epi-mono: a *canonical epi* (resp. mono) in Fr C is a morphism which can be represented by a square whose upper (resp. lower) arrow is an identity, and every morphism $[f]$ has a strictly unique *canonical factorisation* as a canonical epi followed by a canonical mono.

By 2.6.1, this strict system spans an epi-mono factorisation system for Fr C, in the usual sense: the distinguished epis (or fs-epis, denoted by $\twoheadrightarrow$) are those maps $[f]$ whose factorisation presents an iso at the right-hand; dually for fs-monos (denoted by $\rightarrowtail$).

It is easy to show that $[f]\colon x \to y$ is an fs-epi if and only if there is some $u\colon Y' \to X'$ such that $y f' u = y$ (y *sees* f' *as a split epi*): then our fs-mono $[f', 1]\colon \overline{f} \to y$ has inverse $[u, 1]\colon y \to \overline{f}$. Dually, $[f]\colon x \to y$ is an fs-mono if and only if there is some $v\colon X'' \to Y''$ such that $v f'' x = x$.

The quotient induces a full embedding $\mathsf{C} \to \mathrm{Fr}\,\mathsf{C}$, which identifies X with $1_X\colon X \to X$, and $f\colon X \to Y$ with $[f, f]\colon 1_X \to 1_Y$.

Fr C *is the (bi)free category with epi-mono factorisation system over the category* C, in the sense that every functor $F\colon \mathsf{C} \to \mathsf{D}$ with values in an epi-mono fs-category has an essentially unique fs-extension $G\colon \mathrm{Fr}\,\mathsf{C} \to \mathsf{D}$.

The construction of G is the same as above, in (2.49), (2.50); now, $G[f]$ is well defined, independently of the representative f, because E- and M-maps of D are respectively epi and mono, and the diagonal of the rectangle (2.50) is determined by $[f]$.

2.6.4 *Variations and covariations*

In order to study the *fs-subobjects* of X in Fr C (pertaining to the fs-structure, i.e. determined by *fs-monos*, or equivalently by the canonical ones) we begin by briefly recalling the notion of 'weak subobject', studied in [G15].

A *variation* $[x]_A$, or *weak subobject*, of the object A in C denotes a class of morphisms with values in A, equivalent with respect to mutual factorisation

$$x \sim_A y \quad \Leftrightarrow \quad \text{there exist } u, v \text{ such that } x = yu,\ y = xv. \qquad (2.52)$$

$$
\begin{array}{ccc}
X & \xrightarrow{\ x\ } & A \\
{\scriptstyle u}\big\downarrow\big\uparrow{\scriptstyle v} & & \big\| \\
Y & \xrightarrow[\ y\ ]{} & A
\end{array}
\ .
$$

In other words, x and y generate the same principal right ideal of maps with values in A (or, also, are connected by morphisms $x \to y \to x$ in the slice-category C/A of *objects over* A).

By a standard abuse of notation, as for subobjects, a variation $[x]_A$ will generally be denoted by any of its representatives x. The *domain* of a variation $x\colon X \to A$ is only determined up to a pair of arrows $u\colon X \to Y$, $v\colon Y \to X$ such that $x.vu = x$; in other words, x *sees* its domain as a retract of Y; and symmetrically, $x' = xv\colon Y \to A$ sees its domain as a retract of X.

The variations of A form a - *generally large* - ordered set $\mathrm{Var}(A)$, with $x \leqslant y$ if and only if x factors through y. The *identity variation* 1_A is the maximum; if C has an initial object, $\mathrm{Var}(A)$ has also a minimum, i.e. $0_A\colon \perp \to A$. (The smallness of such sets is a difficult question, discussed in Section 2.8.)

Weak pullbacks, defined by the existence part of the usual universal property (see A2.2), give meets of variations. Sums give joins (but weak sums are sufficient)

$$\vee(x_i\colon X_i \to A) = x\colon \textstyle\sum_i X_i \to A. \tag{2.53}$$

A variation $x\colon X \to A$ is said to be *epi* if it has a representative which is epi in C, or equivalently if all of them are. It is equivalent to the identity if and only if it is a retraction; a split epi onto A should thus be viewed as giving the same *information with values in A* as 1_A, with redundant duplication.

A variation is said to be *a subobject* if it has some mono representative $m\colon M \to A$ (all the other representatives are then given by the split extensions of M, and *include* all monos equivalent to m). The ordered set $\mathrm{Sub}(A)$ of subobjects of A is thus embedded in $\mathrm{Var}(A)$.

Transformations of weak subobjects, induced by morphisms (direct and inverse images), or by adjoint functors, or by product decompositions of objects, are studied in [G15]; in particular, inverse images of variations are obtained by means of weak pullbacks (see 2.6.8).

Dually, the *covariations*, or *weak quotients*, of A form an ordered set $\mathrm{Cov}(A)$, containing its quotients.

The notion of a weak subobject has appeared in the literature under other forms and names. What would be called here a *normal* or *regular* variation has been considered by Eckmann and Hilton [EcH] or Freyd [F4, F5], under the equivalent form of a 'principal right ideal' of maps, in order to deal with weak kernels or weak equalisers.

More recently, in connection with proof theory, Lawvere [Law] has used a 'proof-theoretic power set $\mathcal{P}_{\mathsf{C}}(A)$', defined as the 'poset-reflection of the slice category C/A', which amounts to $\mathrm{Var}(A)$.

The author acknowledges helpful remarks on these points, by F.W. Lawvere and P. Freyd.

2.6.5 *Variations as subobjects*

The ordered set $\text{Sub}_{\text{Fr}\,\mathsf{C}}(X)$ of *subobjects* of X in $\text{Fr}\,\mathsf{C}$ (*pertaining* to the fs-structure, i.e. determined by *fs-monos*, or equivalently by the canonical monos) can be identified with the ordered set $\text{Var}_{\mathsf{C}}(X)$ of variations of X in C

$$
\begin{array}{ccc}
M & \xrightarrow{\;m\;} & X \\
{\scriptstyle m}\big\downarrow & & \big\downarrow{\scriptstyle 1} \\
X & =\!=\!= & X
\end{array}
\qquad (2.54)
$$

More generally, the ordered set $\text{Sub}_{\text{Fr}\,\mathsf{C}}(\hat{x})$ of an object $x\colon X' \to X''$ can be identified with the set of C-variations of X'' lesser than x. Indeed, consider two morphisms $m\colon M \to X'$, $n\colon N \to X'$ and the associated fs-monos; $[m,1] \leqslant [n,1]$ means that there exist a map $[f',f'']\colon xm \to xn$ forming a commutative triangle $[n,1].[f',f''] = [m,1]$ in $\text{Fr}\,\mathsf{C}$

$$
\begin{array}{ccccc}
M & \xrightarrow{\;f'\;} & N & \xrightarrow{\;n\;} & X' \\
{\scriptstyle xm}\big\downarrow & & {\scriptstyle xn}\big\downarrow & & \big\downarrow{\scriptstyle x} \\
X'' & \xrightarrow[f'']{} & X'' & =\!=\!= & X''
\end{array}
\qquad
\begin{array}{ccc}
M & \xrightarrow{\;m\;} & X' \\
{\scriptstyle xm}\big\downarrow & {\scriptscriptstyle \searrow xm} & \big\downarrow{\scriptstyle x} \\
X'' & =\!=\!= & X''
\end{array}
\qquad (2.55)
$$

This is equivalent to the existence of f', f'' in C making the following diagram commute

$$
\begin{array}{ccc}
M & \xrightarrow{\;f'\;} & N \\
{\scriptstyle xm}\big\downarrow & {\scriptscriptstyle \searrow xm} & \big\downarrow{\scriptstyle xn} \\
X'' & \xrightarrow[f'']{} & X''
\end{array}
\qquad (2.56)
$$

But f'' can always be replaced with $1_{X''}$, and all this reduces to $xm \leqslant_{X''} xn$ in $\text{Var}_{\mathsf{C}}(X'')$.

Every object UX is *fs-projective* in $\text{Fr}\,\mathsf{C}$ (i.e. satisfies the usual lifting property with respect to fs-epis) and, dually, fs-injective. Moreover, each object $\hat{x}$ can be viewed as the image of the morphism $x\colon X' \to X''$ of $\mathsf{C} \subset \text{Fr}\,\mathsf{C}$, whose canonical factorisation is

$$
x = [x,1].[1,x]\colon X' \twoheadrightarrow \hat{x} \rightarrowtail X'', \qquad \text{Im}\,(x) = \hat{x}. \qquad (2.57)
$$

The arbitrary object $\hat{x}$ is thus a quotient of an fs-projective (namely, X') and a subobject of an fs-injective (X''): $\text{Fr}\,\mathsf{C}$ *has sufficient projectives and injectives, as an fs-category, belonging to the same class* $U(\text{Ob}\mathsf{C})$ (a sort of 'Frobenius condition', according to [F1, F2]).

As an easy consequence, the fs-projectives of $\text{Fr}\,\mathsf{C}$ coincide with the retracts of such objects UX; the fs-injectives as well.

2.6.6 The construction of limits

We prove now various results, showing in particular that if C has all products (resp. finite products) and weak equalisers (see A2.2), then $\mathrm{Fr}\,C$ has all limits (resp. finite limits).

Notice that these hypotheses apply to homotopy categories like HoTop, which do have ordinary products and weak equalisers (see 2.6.9), but lack ordinary equalisers (a fact related to the notion of *flexible* limits in bicategories, in the sense of [BKPS]). Of course, limits and weak limits are always understood to be *small*.

(a) Products in C give products in $\mathrm{Fr}\,C$, in the obvious way (inherited from C^2)

$$
\begin{array}{ccc}
\Pi\,X_i' & \xrightarrow{\ p_i\ } & X_i' \\
x \downarrow & & \downarrow x_i \\
\Pi\,X_i'' & \xrightarrow[\ q_i\ ]{} & X_i''
\end{array}
\qquad x = \Pi x_i, \tag{2.58}
$$

and $U\colon C \to \mathrm{Fr}\,C$ preserves the existing products.

Note that the cancellation property of the family of projections $[p_i, q_i]$ comes from the similar property of the family (q_i) in C; generally, we cannot replace $\Pi X_i''$ with a *weak* product.

(b) Weak equalisers in C produce ordinary equalisers in $\mathrm{Fr}\,C$

$$
\begin{array}{ccc}
E \xrightarrow{\ e\ } X' & \overset{f'}{\underset{g'}{\rightrightarrows}} & Y' \\
xe \downarrow \qquad x \downarrow & & \downarrow y \\
X'' =\!= X'' & \overset{f''}{\underset{g''}{\rightrightarrows}} & Y''
\end{array}
\qquad . \tag{2.59}
$$

Indeed, given a pair $[f], [g]\colon x \to y$, if e is a *weak equaliser of the pair* (yf', yg'), the fs-mono $[e, 1_{X''}]\colon xe \to x$ is the equaliser of the pair.

Moreover, the embedding U *takes a weak equaliser* $e\colon E \to X$ (of a pair $f, g\colon X \to Y$) *to a cone* $Ue\colon UE \to UX$ *which factorises through the equaliser* $[e, 1]$ *by a distinguished epi* $[1, e]$

$$
\begin{array}{ccccc}
E =\!= E & \xrightarrow{\ e\ } & X & \overset{f}{\underset{g}{\rightrightarrows}} & Y \\
\| \qquad e \downarrow & & \| & & \| \\
E \xrightarrow[\ e\ ]{} X & =\!= & X & \overset{f}{\underset{g}{\rightrightarrows}} & Y
\end{array}
\qquad . \tag{2.60}
$$

(c) Our statement on the completeness of $\mathrm{Fr}\,C$ is now proved. But it is interesting and easy to give an explicit construction of a general limit (assuming that C has products and weak equalisers).

Let a diagram be given in Fr C, i.e. a functor

$$D = ((\hat{x}_i), ([f_u])) \colon \mathsf{S} \to \mathsf{Fr}\, \mathsf{C},$$

defined over a small category S, with $i \in \mathrm{Ob}S$ and $u \colon i \to j$ in $\mathrm{Mor}S$.

Choose a weak limit L in C, of the diagram formed of all the arrows $x_i \colon X_i' \to X_i''$ and all the arrows $\overline{f}_u \colon X_i' \to X_j''$. The object L comes equipped with arrows $a_i \colon L \to X_i'$ such that $\overline{f}_u.a_i = x_j a_j \colon L \to X_j''$ (in a weakly universal way); take a product $\prod X_i''$ with projections q_i

$$
\begin{array}{ccccc}
L & \xrightarrow{\;a_i\;} & X_i' & \xrightarrow{\;f_u'\;} & X_j' \\
{\scriptstyle a}\downarrow & & \downarrow{\scriptstyle x_i} & & \downarrow{\scriptstyle x_j} \\
\prod X_i'' & \xrightarrow[\;q_i\;]{} & X_i'' & \xrightarrow[\;f_u''\;]{} & X_j''
\end{array}
\qquad (2.61)
$$

Letting $a = (x_i a_i) \colon L \to \prod X_i''$, the cone $[a_i, q_i] \colon \hat{a} \to \hat{x}_i$ is the limit of D in Fr C.

(d) In particular, if C has products, U satisfies a property with respect to all the existing weak limits of C, that extends the one already considered for weak equalisers:

$$
\begin{array}{ccccc}
L & =\!=\!= & L & \xrightarrow{\;a_i\;} & X_i \\
\| & & \downarrow{\scriptstyle a} & & \| \\
L & \xrightarrow[\;a\;]{} & \prod X_i & \xrightarrow[\;q_i\;]{} & X_i
\end{array}
\qquad (2.62)
$$

(*) U takes any weak limit $(L, (a_i))$ of a diagram $X = ((X_i), (f_u)) \colon S \to \mathsf{C}$ to a cone which is connected to the limit-cone of UX in Fr C by a distinguished epi.

(e) As a marginal remark, one can note that, if in C weak equalisers exist and *every map is a weak equaliser of some pair*, the fs-monos of Fr C coincide with the regular monomorphisms and are categorically determined (in fact, in (2.59), $[e, 1_{X''}]$ is the equaliser of $[f]$, $[g]$ also when xe is a weak equaliser of (f'', g'')). Practically, this assumption on C is mostly of interest for triangulated categories and stable homotopy, when Fr C is even abelian (see 2.7.7).

2.6.7 Theorem (Completeness properties)

Let C *be a category,* $F \colon \mathsf{C} \to \mathsf{D}$ *a functor with values in an epi-mono fs-category, and* $G \colon \mathsf{Fr}\, \mathsf{C} \to \mathsf{D}$ *its fs-extension. In the following, one can also restrict everything to the* finite *case:* finite products, finite limits, finite *weak limits, and so on.*

(a) If C *has products,* Fr C *also does and* U *preserves them. Moreover,* F *preserves products if and only if* G *does (so that* Fr C *is the free epi-mono fs-category with products over* C*, as a category with products).*

(b) Let C *have products and weak equalisers. Then* Fr C *is complete. Its fs-monos and fs-epis are stable under pullbacks. The ordered sets of variations in* C *have small meets.* U *preserves products and satisfies the property (*) on weak limits (in 2.6.6(d)). Moreover, the following conditions are equivalent*

(1) F *preserves products and satisfies (*) on weak equalisers,*

(2) F *preserves products and satisfies (*) on weak limits,*

(3) G *preserves all limits.*

   The property (*) in (1) can be equivalently replaced with the following (and similarly for (2)):

(**) every diagram $X = ((X_i), (f_u)) \colon S \to$ C *has a weak limit* $(L, (a_i))$ *taken by* U *to a cone which is connected to the limit-cone of* UX *in* Fr C *by a distinguished epi.*

(c) If C *has finite products and weak equalisers, and every map is a weak coequaliser, then* Fr C *is a regular category and all distinguished epis are regular.*

Proof After the previous construction of limits in Fr C, we only need to check the stability property of the fs-factorisation under pullbacks, which is done below (in 2.6.8). The fact that (**) implies (*) is easy: we know that $U(a_i) = p_i.e$ with $e \in E$, where $(\overline{L}, (p_i))$ is the limit of UX in Fr C

$$(2.63)$$

If $(M, (b_i))$ is also a weak limit of X, take some $u \colon L \to M$ such that $a_i = b_i u$ (for all i) and let $U(b_i) = p_i.f$; cancelling the limit cone (p_i), we have that $e = f.Uu$, whence also $f \in E$.

   Note that the property (*) is not closed under composition, and does not lead - naturally - to a category of categories with weak limits; to express the universal property (b) as an adjunction would require artificial constructs, probably of little interest.   □

2.6.8 *Inverse images*

In particular, we are interested in the construction of pullbacks in $\mathrm{Fr\,C}$, by means of weak pullbacks and ordinary products $A \times B$ in C.

Given $[f]\colon x \to z$ and $[g]\colon y \to z$ in the diagram below

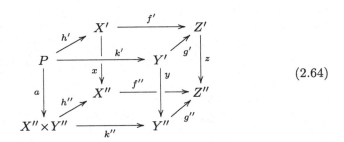

$$(2.64)$$

let h'', k'' be the projections of the product $X'' \times Y''$, let (P, h', k') be a weak pullback of (zf', zg') and $a = (xh', yk')\colon P \to X'' \times Y''$. Then the morphisms $[h]\colon a \to x$ and $[k]\colon a \to y$ are our solution.

Finally we prove that, in $\mathrm{Fr\,C}$, *fs-epis and fs-monos are stable under pullback*. For the first property, let $g' = 1$ in the previous diagram; then $[h]\colon a \to x$ is an fs-epi, by the characterisation given in 2.6.3 (h' is a split epi, because there exists some $u\colon X' \to P$ satisfying $h'u = 1$, $k'u = f'$).

The second fact is shown by another construction of the pullback, in the particular case $g'' = 1$, given in the following diagram

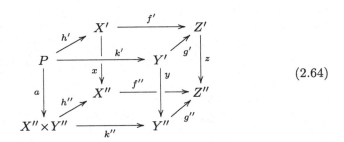

$$(2.65)$$

The stability of fs-monos also shows that the identification $\mathrm{Var_C}(X) = \mathrm{Sub_{Fr\,C}}(X)$ (in 2.6.5) is consistent with inverse images, given by weak pullbacks of variations in C ([G15], 3.1) and pullbacks of fs-monos in $\mathrm{Fr\,C}$. Take $f\colon X \to Z$ in C ($x = 1_X$, $z = 1_Z$) and $g' = y \in \mathrm{Var_C}(Z)$; the pullback of the fs-mono $[g', 1]\colon \hat{y} \to Z$ is realised as above, so that the C-inverse image $h' = f^*(y) \in \mathrm{Var_C}(X)$ corresponds to the $\mathrm{Fr\,C}$-inverse image $[h', 1] = [f]^*[y, 1] \in \mathrm{Sub_{Fr\,C}}(X)$.

2.6.9 Spaces and homotopy

Let us take $C = \mathsf{HoTop} = \mathsf{Top}/\simeq$, the homotopy category of spaces, where an object is a space and a morphism $[f]\colon X \to Y$ is an equivalence class of continuous mappings in $\mathsf{Top}(X, Y)$, up to homotopy. (It is well known, and easy to verify, that this quotient coincides with the category of fractions of Top which inverts homotopy equivalences.)

HoTop has small products and weak limits. In fact, Top has small products, satisfying the obvious 2-dimensional property with respect to homotopies (any family of homotopies $\alpha_i\colon f_i \to g_i\colon A \to X_i$ has a unique lifting $\alpha\colon f \to g\colon A \to \Pi X_i$). Moreover, HoTop has weak equalisers, because Top has *homotopy equalisers*, making a pair of parallel maps homotopic in a universal way

$$E \xrightarrow{\ e\ } X \underset{g}{\overset{f}{\rightrightarrows}} Y, \qquad \varepsilon\colon fe \to ge\colon E \to Y, \qquad (2.66)$$

$$E = \{(x, \eta) \in X \times PY \mid \eta(0) = f(x),\ \eta(1) = g(x)\},$$

$$e(x, \eta) = x, \qquad \varepsilon(x, \eta; t) = \eta(t).$$

(The path-space $PY = Y^{[0,1]}$ has the compact-open topology.)

Similarly, HoTop has small sums and weak colimits, because Top has small sums, consistent with homotopies, and *homotopy coequalisers*

$$X \underset{g}{\overset{f}{\rightrightarrows}} Y \xrightarrow{\ c\ } C, \qquad \kappa\colon cf \to cg\colon X \to C, \qquad (2.67)$$

where C is the quotient of $(X \times [0, 1]) + Y$ that identifies $[x, 0] = [fx]$ and $[x, 1] = [gx]$, for $x \in X$.

It follows that the Freyd completion $\mathsf{Fr\,HoTop}$ is complete and cocomplete. Every homotopy invariant functor $F\colon \mathsf{Top} \to \mathsf{D}$ with values in an epi-mono fs-category has a unique extension to an fs-functor $G\colon \mathsf{Fr\,HoTop} \to \mathsf{D}$. If F preserves (finite) products and satisfies the condition (*) with respect to weak equalisers (cf. 2.6.6) then G preserves (finite) limits; and dually.

In particular, for every space S, the S-homotopy functor

$$\pi_S = [S, -]\colon \mathsf{HoTop} \to \mathsf{Set},$$

has a unique extension to a limit-preserving fs-functor $\mathsf{Fr\,HoTop} \to \mathsf{Set}$. (Since Set has unique epi-mono factorisations, fs-functor means here to take fs-epis to epis and fs-monos to monos.)

Similarly, the S-cohomotopy functor

$$\pi^S = [-, S]\colon \mathsf{HoTop} \to \mathsf{Set}^{\mathrm{op}},$$

has a unique extension to a colimit-preserving fs-functor $\mathsf{Fr\,HoTop} \to \mathsf{Set}^{\mathrm{op}}$.

2.7 From homotopy categories to homological categories

If the category C has a zero-object, weak kernels and weak cokernels, as is the case of the homotopy category HoTop. of pointed spaces, then Fr C is an (unrestricted) *p-homological category*. Further hypotheses (in 2.7.6) make Fr C p-exact, or abelian as in the original Freyd's result [F2]; the latter case occurs, in particular, if C is triangulated (see 2.7.7).

In this section, semiexact (or homological, or p-exact) categories are always meant in the unrestricted sense, without assuming they are well-powered (see 1.5.1); a problem which will be briefly addressed in the next section.

2.7.1 Exactness

We consider now 'exactness properties' for pointed categories and *pointed functors* (that preserve the zero object).

At the 'weak level' (of C) we just need to consider *weak kernels* and *weak cokernels*, and the well-known notion of *triangulated category* (see Puppe [P3, P4], Verdier [Ve], Hartshorne [Har], Neeman [Ne]). At the 'strict level' (of Fr C) we use the notion of *p-exact* and *p-homological* category, always in the unrestricted sense.

The notation

$$k \in \mathrm{wker}(f), \qquad c \in \mathrm{wcok}(f), \tag{2.68}$$

will mean that k is *a* weak kernel of f and c a weak cokernel.

Let us also note that in a pointed epi-mono fs-category D, any normal mono is necessarily an fs-mono; in particular this holds for the null subobject $(0 \to A) = \mathrm{ker}\,(1_A)$. Similarly, all normal epis are fs-epis.

2.7.2 Kernels and cokernels

Assume now that the pointed category C has weak kernels and weak cokernels. We want to prove that Fr C is a p-homological category.

To begin with, the zero object of C is still so in Fr C. Moreover, every morphism $[f]$ in Fr C has kernel and cokernel, whose *canonical representatives* can be constructed choosing, in C, a weak kernel m and a weak cokernel p of the diagonal $\overline{f}$

$$
\begin{array}{ccccccc}
M & \xrightarrow{m} & X' & \xrightarrow{f'} & Y' & =\!=\!= & Y' \\
{\scriptstyle xm}\downarrow & & {\scriptstyle x}\downarrow & \searrow{\scriptstyle \overline{f}} & \downarrow{\scriptstyle y} & & \downarrow{\scriptstyle py} \\
X'' & =\!=\!= & X'' & \xrightarrow[f'']{} & Y'' & \xrightarrow[p]{} & P
\end{array}
\qquad
\begin{array}{l}
m \in \mathrm{wker}(\overline{f}), \\[2em]
p \in \mathrm{wcok}(\overline{f}).
\end{array}
\tag{2.69}
$$

(As in Section 2.6, the existence part of the universal properties comes from the analogue in C; the uniqueness part comes from the fact that $[m, 1]$ is mono in Fr C and $[1, p]$ is epi.)

Using the previous choice, we let

$$\ker[f] = [m, 1]\colon xm \rightarrowtail x, \qquad \cok[f] = [1, p]\colon y \twoheadrightarrow py. \tag{2.70}$$

We have thus proved that Fr C is p-semiexact. The normal factorisation $[f] = \operatorname{nim}[f].[g].\operatorname{ncm}[f]$ is given by the following three maps

$$
\begin{array}{ccccccc}
X' & =\!=\!= & X' & \xrightarrow{\ g'\ } & I & \xrightarrow{\ i\ } & Y' \\
{\scriptstyle x}\downarrow & & {\scriptstyle cx}\downarrow & & \downarrow{\scriptstyle yi} & & \downarrow{\scriptstyle y} \\
X'' & \xrightarrow{\ c\ } & C & \xrightarrow{\ g''\ } & Y'' & =\!=\!= & Y''
\end{array}
\tag{2.71}
$$

$$c \in \operatorname{wcok}(xm), \quad i \in \operatorname{wker}(py); \qquad ig' = f', \quad g''c = f''.$$

(The existence of g' and g'' is provided by the weak universal properties of i and c, since $py.f' = p\overline{f} = 0$ and $f''.xm = \overline{f}m = 0$).

Since $\operatorname{ncm}[f]$ is an fs-epi and $\operatorname{nim}[f]$ an fs-mono (cf. 2.7.1), the fs-factorisation of $[g]$ yields the one of $[f]$. The morphism $[f]$ is exact when this $[g]$ is iso; then, the normal factorisation of $[f]$ 'coincides' with its fs-factorisation.

2.7.3 Short exact sequences

We prove now that every commutative diagram in C with

$$
\begin{array}{ccccc}
M & \xrightarrow{\ m\ } & X' & =\!=\!= & X' \\
{\scriptstyle xm}\downarrow & & \downarrow{\scriptstyle x} & & \downarrow{\scriptstyle px} \\
X'' & =\!=\!= & X'' & \xrightarrow{\ p\ } & P
\end{array}
\qquad
\begin{array}{l}
m \in \operatorname{wker}(px), \\[12pt]
p \in \operatorname{wcok}(xm),
\end{array}
\tag{2.72}
$$

yields a short exact sequence in Fr C, and actually the general one (any such can be obtained in this way, up to isomorphism)

$$\bullet \xrightarrow{\ [m,1]\ } x \xrightarrow{\ [1,p]\ } \bullet \qquad [m, 1] = \ker[1, p], \ [1, p] = \cok[m, 1]. \tag{2.73}$$

In fact, take a normal subobject of $x\colon X' \to X''$, $[m, 1] = \ker[f', f'']$, with $m \in \operatorname{wker}(f''x)$. Take now its cokernel $[1, p]$, with $p \in \operatorname{wcok}(xm)$.

General properties of semiexact categories would ensure that $[m, 1] = \ker[1, p]$; but, directly, we can say more: the pair $(m, 1)$ is actually a *canonical representative for this kernel*, i.e. $m \in \operatorname{wker}(px)$; in fact, $f''xm = 0$, whence f'' factorises through p, so that any morphism v which annihilates px also annihilates $f''x$ and factorises through m

$$M \xrightarrow{m} X' \xrightarrow{x} X'' \xrightarrow{f''} \bullet \tag{2.74}$$

In particular, a normal subobject of $x\colon X' \to X''$ is always determined by an arrow $\mu = [m, 1]$, where $m\colon M \to X'$ is a weak kernel in C of px, for some $p\colon X'' \to \bullet$. As in 2.6.5, two weak kernels $m\colon M \to X'$, $n\colon N \to X'$ give equivalent normal monos $[m, 1] \sim [n, 1]$ (and determine the same normal subobject of x) if and only if xm and xn provide the same variation of X''.

Just because Fr C is (unrestricted) semiexact, the normal subobjects of any object $x\colon X' \to X''$ form a (possibly large) *lattice* Nsb(x). The normal quotients of x form a second lattice Nqt(x), *anti*-isomorphic to the former via kernel-cokernel duality. In particular, for $x = 1_X$, we get the lattice of *normal variations* of X in C, determined by weak kernels $m\colon M \to X$ up to mutual factorisation.

2.7.4 The axiom (ex2)

We prove now that normal monos in Fr C are closed under composition; by duality, the same holds for normal epis.

Consider the consecutive normal monos $[n, 1]$ and $[m, 1]$; by 2.7.3, we know that they are linked to their cokernels $[1, p]$ and $[1, q]$ by the following relations

$$\tag{2.75} \quad (a = qxm),$$

$$m \in \mathrm{wker}(px), \quad p \in \mathrm{wcok}(xm); \quad n \in \mathrm{wker}(qxm), \quad q \in \mathrm{wcok}(xmn).$$

Thus, p vanishes over xmn and factors as $p = uq$. It follows easily that

$[mn, 1] = \ker [q, 1_{X'}]$

$$
\begin{array}{ccccc}
N & \xrightarrow{mn} & X' & \rule[0.5ex]{2em}{0.4pt}\!\!= & X' \\
{\scriptstyle xmn}\downarrow & & {\scriptstyle x}\downarrow & & \downarrow{\scriptstyle qx} \\
X'' & \rule[0.5ex]{2em}{0.4pt}\!\!= & X'' & \xrightarrow{q} & Q
\end{array}
\qquad
\begin{array}{l}
m \in \mathrm{wker}(px), \\[3ex]
mn \in \mathrm{wker}(qx),
\end{array}
\qquad (2.76)
$$

(If $qx.v = 0$, also $px.v = 0$, whence $v = mv'$; now $qxm.v' = qx.v = 0$, and v' factors through n, which means that v factors through mn.)

2.7.5 The subquotient axiom

We finish proving that $\mathrm{Fr}\, \mathsf{C}$ is homological. We have a normal subobject and a normal quotient of $x\colon X' \to X''$, that satisfy the following relation

$$[m, 1]\colon m \rightarrowtail x, \qquad [1, q]\colon x \twoheadrightarrow q, \qquad [m, 1] \geqslant [n, 1] = \ker[1, q], \qquad (2.77)$$

and we have to show that the composite $[1, q].[m, 1]$ is exact.

Let $[1, p] = \mathrm{cok}\,[m, 1] \leqslant [1, q]$. By hypothesis, there exist morphisms f, g such that $xm.f = xn$, and $g.qx = px$. It suffices thus to consider the following commutative diagram (with $a = qxm$)

$$ (2.78) $$

Here $[1_M, q]$ is a normal epi (since $q \in \mathrm{wcok}(xn) = \mathrm{wcok}(xm.f)$) and $[m, 1_Q]$ a normal mono (since $m \in \mathrm{wker}(px) = \mathrm{wker}(g.qx)$).

2.7.6 Theorem (Exactness properties of the completion)

Let C be a pointed category, $F\colon \mathsf{C} \to \mathsf{E}$ a zero-preserving functor with values in a p-semiexact epi-mono fs-category and $G\colon \mathrm{Fr}\,\mathsf{C} \to \mathsf{E}$ its fs-extension.

(a) Let C have weak kernels and weak cokernels. Then $\mathrm{Fr}\,\mathsf{C}$ is a pointed homological epi-mono fs-category; the normal variations (Section 2.7.3) of X in C form a (possibly large) lattice, identified with the lattice of normal subobjects of X in $\mathrm{Fr}\,\mathsf{C}$. The functor G preserves kernels and cokernels if and only if

(**) every C-morphism $f\colon X \to Y$ has a weak kernel $k\colon K \to X$ and a

weak cokernel $c\colon Y \to C$ *such that, in the following commutative diagram of* E, *u is an fs-epi and v an fs-mono*

$$
\begin{array}{ccccccc}
FK & \xrightarrow{\;Fk\;} & FX & \xrightarrow{\;Ff\;} & FY & \xrightarrow{\;Fc\;} & FC \\
{\scriptstyle u}\Big\downarrow & & \Big\| & & \Big\| & & \Big\uparrow{\scriptstyle v} \\
\mathrm{Ker}\,(Ff) & \longrightarrow & FX & \xrightarrow[Ff]{} & FY & \longrightarrow & \mathrm{Cok}\,(Ff)
\end{array}
\qquad . \qquad (2.79)
$$

Equivalently, the same must happen for all *weak kernels and weak cokernels of* f. *In particular, condition (**) holds true for* $F = U$.

(a') *If, in the same hypotheses,* E *is p-exact, then* G *preserves kernels and cokernels if and only if every* C-*morphism* $f\colon X \to Y$ *has a weak kernel* $K \to X$ *and a weak cokernel* $Y \to C$ *such that the sequence*

$$
FK \to FX \to FY \to FC
$$

is exact in E.

(b) *Assume that* C *has weak kernels and weak cokernels, and moreover every morphism is a weak kernel and a weak cokernel.*

Then Fr C *is p-exact; all variations in* C *are normal and form a modular lattice* $\mathrm{Var}(A) = \mathrm{Sub}_{\mathrm{Fr}\,C}(A)$.

If also E *is p-exact, the functor* G *is exact if and only if the condition (**) is satisfied for its left-hand part (concerning* k *and* u), *if and only if it is satisfied for its right-hand part (concerning* c *and* v).

(c) *If* C *satisfies the hypotheses of (b) and has finite products (or sums), then* Fr C *is abelian.*

Proof (a) The first part has been proved above (in 2.7.2-2.7.5). The limit-colimit preserving properties are as in 2.6.7. (a') is a trivial consequence. Since (c) follows from (b), as already recalled in 2.7.1, we only need to verify the latter.

(b) First, we have to prove that an arbitrary morphism $f = [f', f'']$ factors as a normal epi followed by a normal mono. In fact, recall the distinguished factorisation $f = [f', 1].[1, f'']$ considered at the beginning, in (2.47).

Under the new hypotheses, f'' is a weak cokernel in C and $[1, f'']$ is a cokernel in Fr C, while f' is a weak kernel and $[f', 1]$ a kernel. Thus, Fr C is p-exact, by the general theory recalled above (cf. 2.7.1).

If also E is p-exact, the functor G is exact if and only if it preserves short exact sequences, if and only if it preserves kernels and epimorphisms; but the last condition is always satisfied by G, because all the epis of Fr C and E are distinguished. $\qquad\qquad\square$

2.7.7* Universal homology theories (Freyd)

It follows easily that, if C is a triangulated category, then Fr C is abelian and $U \colon \mathsf{C} \to \mathsf{Fr}\,\mathsf{C}$ is the universal homological functor over C. (See [F2], Lemma 4.1, where C is the stable homotopy category of spaces.)

In fact, the hypotheses of 2.7.6(c) are satisfied. First, if (u, v, w) is a (distinguished) triangle, it is easy to show that v is a weak cokernel of u (and dually a weak kernel of w)

$$
\begin{array}{ccccccc}
X & \xrightarrow{u} & Y & \xrightarrow{y} & Z & \xrightarrow{w} & \Sigma X \\
\downarrow & & \downarrow{\scriptstyle f} & & \downarrow & & \downarrow \\
0 & \longrightarrow & A & \xrightarrow[1]{} & A & \longrightarrow & 0
\end{array}
\qquad . \qquad (2.80)
$$

But any arrow can appear in a triangle, in any position, and the conclusion follows. Note that, in a triangle, any arrow is a weak kernel of the following, and a weak cokernel of the preceding one.

Moreover, the functor $U \colon \mathsf{C} \to \mathsf{Fr}\,\mathsf{C}$ is homological (i.e. the sequence $U_n \Sigma^{-n}$ is a homology theory), since it takes every triangle to an exact sequence; actually, it is the universal homological functor on C (by 2.7.6(a')): for every homological functor $H \colon \mathsf{C} \to \mathsf{E}$ (with values in an abelian category, or more generally in a p-exact one) there is an essentially unique exact functor $G \colon \mathsf{Fr}\,\mathsf{C} \to \mathsf{E}$ such that $GU = H$.

2.8 * A digression on weak subobjects and smallness

This last section of the present chapter briefly sketches without proofs some results of [G15] on variations, or weak subobjects, showing some cases where they form a small set and can be classified. It will not be used elsewhere.

2.8.1 Comments

As defined above, in 2.6.4, a *variation*, or *weak subobject*, of an object A in the category C is an equivalence class of morphisms with values in A, where $x \sim_A y$ if there exist maps u, v such that $x = yu$, $y = xv$. Among them, the *mono variations* (having some representative which is a monomorphism) can be identified with subobjects.

The variations of A form an ordered set $\mathrm{Var}(A)$, possibly large, which is a lattice under mild assumptions on C (see 2.6.4). Dually, a *covariation*, or *weak quotient*, of the object A is an equivalence-class of morphisms from A, extending the notion of a quotient.

As we have seen in the previous sections, variations are well connected with *weak* limits, much in the same way as subobjects are connected with limits; thus, they are of particular interest in homotopy categories and triangulated categories, which generally have ordinary products but only weak equalisers.

Nevertheless, the study of weak subobjects in *ordinary* categories, like abelian groups or groups, is interesting in itself and relevant to classify variations in homotopy categories of spaces, by means of homology and homotopy functors.

Various classifications are given in [G15]. The choice of the ground-category is crucial in order to obtain results of interest.

For instance, *finitely generated* abelian (co)variations always yield *countable* lattices, whereas any prime order group $\mathbb{Z}/p$ has *at least a continuum* of abelian variations and a *large set* of abelian covariations.

In the homotopy category of topological spaces, $\mathsf{HoTop} = \mathsf{Top}/ \simeq$, we get a distributive lattice $\mathrm{Var}_{\simeq}(A) = \mathrm{Fib}(A)$ of *types of fibrations* over the space A (see 2.8.3), which is hard to classify even in the simplest cases. But, restricting to CW-spaces, the *cw-variations* of the circle $\mathbb{S}^1$ are classified by the standard fibrations $y_n \colon \mathbb{S}^1 \to \mathbb{S}^1$ ($n > 0$), together with the universal covering $\mathbb{R} \to \mathbb{S}^1$, *consistently with what one might expect as 'homotopy subobjects' of the circle* (see 2.8.4).

2.8.2 Examples

In Set, every epi splits, by the axiom of choice, and the unique epi variation (cf. 2.6.4) of a set A is its identity: *variations and subobjects coincide.* The covariations of a non-empty set coincide with its quotients; but $\emptyset$ has two covariations, the identity and $0^{\emptyset} \colon \emptyset \to \{*\}$.

Similarly, in any category with epi-mono factorisations where all epis split, weak subobjects and subobjects coincide. This property, and its dual as well, hold in the category $\mathsf{Set}_{\bullet}$ of pointed sets, or in any category of vector spaces (over a fixed field), or also in the category of relations over any (well-powered) abelian category. In all these cases, the sets $\mathrm{Var}(A)$ and $\mathrm{Cov}(A)$ are small.

Consider now the category Ab of abelian groups and its full subcategory $\mathsf{Ab}_{\mathrm{fg}}$ of finitely generated objects (fg-abelian groups, for short). We have the lattice $\mathrm{Var}(A)$ of all *abelian variations* of A, and - if A is finitely generated - the sublattice $\mathrm{Var}_{\mathrm{fg}}(A)$ of *fg-variations* of A (having representatives in $\mathsf{Ab}_{\mathrm{fg}}$). By the structure theorem of fg-abelian groups, $\mathrm{Var}_{\mathrm{fg}}(A)$ and $\mathrm{Cov}_{\mathrm{fg}}(A)$ are always *countable*.

Since every subgroup of a free abelian group is free, it is easy to show that the abelian variations of a free abelian group F coincide with its subobjects (and are finitely generated whenever F is so). In particular, the weak subobjects of the group of integers $\mathbb{Z}$ form a noetherian distributive lattice, and can be represented by the 'positive' endomorphisms of $\mathbb{Z}$

$$x_n : \mathbb{Z} \to \mathbb{Z}, \quad x_n(a) = n.a \qquad (n \geqslant 0),$$
$$x_m \leqslant x_n \quad \Leftrightarrow \quad m\mathbb{Z} \subset n\mathbb{Z} \quad \Leftrightarrow \quad n \text{ divides } m. \tag{2.81}$$

The prime-order group $\mathbb{Z}/p$ has two subobjects and a totally ordered set of fg-variations, anti-isomorphic to the ordinal $\omega + 2$; it can be represented by the following natural homomorphisms x_n (including the natural projection $x_\infty : \mathbb{Z} \to \mathbb{Z}/p$)

$$x_n : \mathbb{Z}/p^n \to \mathbb{Z}/p, \quad x_n(\hat{1}) = \hat{1} \qquad (0 \leqslant n \leqslant \infty),$$
$$0 = x_0 < x_\infty < ... < x_3 < x_2 < x_1 = 1. \tag{2.82}$$

But $\mathbb{Z}/p$ has *at least a continuum* of *non-finitely generated* variations ([G15], 1.5). Similarly, $\mathbb{Z}/p$ has a totally ordered set of fg-abelian covariations, anti-isomorphic to $\omega + 1$, and a *large set* of abelian covariations ([G15], 1.6). The fg-variations of any cyclic group and of $\mathbb{Z}/p \oplus \mathbb{Z}/p$ are classified in [G15] (Section 4).

Also in the category Gp of groups, *the weak subobjects of a free group coincide with its subobjects*, by the Nielsen-Schreier theorem (any subgroup of a free group is free). But here a subgroup of a free group of finite rank may have countable rank ([Ku], Section 36). Thus, the set $\mathrm{Var}_{\mathrm{fg}}(G) \subset \mathrm{Var}(G)$ of fg-variations of an fg-group G is at most a continuum.

Consider now the full embedding Ab $\subset$ Gp. For an abelian group A, the set of abelian variations $\mathrm{Var}(A)$ is embedded in the set $\mathrm{Var}_{\mathrm{Gp}}(A)$ of its *group-variations*. Every group-variation $y : G \to A$ has an obvious *abelian closure* $y^{\mathrm{ab}} : \mathrm{ab}(G) \to A$, which is the least abelian variation of A following y; the latter is abelian if and only if it is equivalent to y^{ab}; $\mathrm{Var}(A) \subset \mathrm{Var}_{\mathrm{Gp}}(A)$ is a retract.

The group-variations of $\mathbb{Z}$, which is also free as a group, coincide with its subobjects and are all abelian. On the other hand, $\mathbb{Z}/2$ has also non-abelian fg-variations ([G15], 1.7).

2.8.3 Homotopy variations

Consider a quotient category $\mathsf{C}/\simeq$, modulo a congruence $f \simeq g$ (i.e. an equivalence relation between parallel morphisms, consistent with composition), which may be viewed as a sort of homotopy relation, since our main examples will be of this type.

A $\simeq$-*variation* of A in C is just a variation in the quotient $C/\simeq$. But it is simpler to take its representatives in C, as morphisms $x \colon \bullet \to A$ modulo the equivalence relation where $x \simeq_A y$ if and only if there are u, v such that $x \simeq yu$, $y \simeq xv$. The ordered set $\mathrm{Var}_{\simeq}(A)$ is thus a quotient of $\mathrm{Var}(A)$, often more manageable and more interesting. Similarly for covariations.

For a space X, we consider thus the ordered set $\mathrm{Var}_{\simeq}(X)$ of its *homotopy variations*, in the *homotopy category* $\mathsf{HoTop} = \mathsf{Top}/\simeq$ of topological spaces. This ordered set, invariant up to homotopy type, is a (possibly large) lattice. In fact, Top has (small) sums, consistent with homotopies, and homotopy pullbacks [Mat], whence the quotient $\mathsf{Top}/\simeq$ has sums and weak pullbacks.

Moreover, each homotopy variation *can be represented by a fibration*, because every map in Top factors as a homotopy equivalence followed by a fibration; we can thus view $\mathrm{Var}_{\simeq}(X) = \mathrm{Fib}(X)$ as the lattice of *types of fibrations* over X. Each homology functor $H_n \colon \mathsf{Top} \to \mathsf{Ab}$ can be used to represent homotopy variations as abelian variations and, possibly, distinguish them.

A homotopy class $\varphi = [f] \colon X \to Y$ acts on such lattices by direct and inverse images, giving a covariant connection (an adjunction between ordered sets)

$$\varphi_* \colon \mathrm{Fib}(X) \rightleftarrows \mathrm{Fib}(X) : \varphi^*,$$

$$\varphi_*[x] = [fx], \qquad \varphi^*[y] = \text{ class of a weak } f\text{-pullback of } y, \qquad (2.83)$$

$$1 \leqslant \varphi^* \varphi_*, \qquad \varphi_* \varphi^*[y] = [y] \wedge \varphi_*[1].$$

Globally, we get a homotopy-invariant functor Fib defined over Top, with values in the category of (possibly large) lattices and right exact connections [G15].

Dual facts hold for the lattice $\mathrm{Cov}_{\simeq}(X) = \mathrm{Cof}(X)$ of homotopy covariations, or *types of cofibrations* from X.

Most of these results can be extended to various other 'categories with homotopies', as pointed spaces, chain complexes, diagrams of spaces, spaces under (or over) a space, topological monoids, etc. (see [G14, G18] and references therein).

Finally, it is interesting to show that the lattice $\mathrm{Fib}(A) = \mathrm{Var}_{\simeq}(A)$ is *distributive*, and actually *binary meets distribute over small joins.*

First, note that pullbacks in Top distribute over sums (see [CaLW] for the notion of 'extensive' category): given a (small) topological sum with injections $u_i \colon X_i \subset X$ and a map $f \colon Z \to X$, the pullback-spaces $Z_i = f^{-1}(X_i)$ have topological sum Z.

Second, the sum-injections u_i are fibrations (every homotopy in X, starting from a map with values in X_i, has image contained in the latter), whence the previous pullbacks are homotopy pullbacks in Top and weak pullbacks in HoTop (pullbacks, actually, because $[u_i]$ is mono).

Now, given a family of variations $x_i \colon X_i \to A$, their join $x \colon X \to A$ and a variation $y \colon Y \to A$, we form the following commutative diagram

$$
\begin{array}{ccccc}
X_i & \underset{u_i}{\overset{x_i}{\rightrightarrows}} & X & \overset{x}{\longrightarrow} & A \\
{\scriptstyle f_i}\big\uparrow & & {\scriptstyle f}\big\uparrow & & {\scriptstyle y}\big\uparrow \\
Z_i & \underset{z_i}{\overset{v_i}{\rightrightarrows}} & Z & \overset{z}{\longrightarrow} & Y
\end{array}
\qquad (2.84)
$$

The right-hand square is a weak pullback in HoTop, whence $yz = y \wedge x$ in $\mathrm{Fib}(A)$, and the left-hand square too. Also the rectangle is a weak pullback, for every i; since (v_i) is the family of injections of a topological sum, the join of $y \wedge x_i = yz_i$ is $yz = y \wedge x$.

2.8.4 *CW-variations*

But it is important to restrict the class of spaces we are considering, to obtain more homogeneous sets of variations, which one might hopefully classify.

A first standard restriction is the category CW of *CW-spaces* (pointed spaces having the homotopy type of a connected CW-complex), with pointed maps. The variations of X in $\mathsf{CW}/\simeq$ will be called *cw-variations*; they form a *sublattice* $\mathrm{Var}_{cw}(X)$ of the lattice $\mathrm{Var}_{\simeq}(X)$ of all the homotopy variations of X.

Yet, $\mathrm{Var}_{cw}(X)$ may still be large (as follows from Freyd's results on the non-concreteness of homotopy categories [F4, F5]), and further restrictions should be considered.

The group variations of $\mathbb{Z}$, coinciding with the abelian variations $x_n \colon \mathbb{Z} \to \mathbb{Z}$ of (2.81), have corresponding cw-variations of the pointed circle $\mathbb{S}^1 = \mathbb{R}/\mathbb{Z}$

$$
\begin{aligned}
y_n \colon \mathbb{S}^1 \to \mathbb{S}^1, \qquad y_n[\lambda] = [n\lambda] \qquad (n \geqslant 0), \\
y_m \leqslant y_n \quad \Leftrightarrow \quad n \text{ divides } m,
\end{aligned}
\qquad (2.85)
$$

which realise them via π_1.

In fact, this sequence classifies all the cw-variations of the circle (as proved in [G15], 3.4). Note that, for $n > 0$, y_n is the covering map of $\mathbb{S}^1$ of degree n; the universal covering map $p \colon \mathbb{R} \to \mathbb{S}^1$ corresponds to the weak subobject y_0, also represented by $\{*\} \to \mathbb{S}^1$.

More generally, the cw-variations of any cluster of circles $\sum_{i \in I} \mathbb{S}^1$ are classified by the group-variations of the free group on the set I, i.e. its subgroups [G15], 3.4.

Cw-variations of the sphere and the projective plane are also studied in [G15], 2.4-2.5.

3

Subquotients, homology and exact couples

Subquotients and their induced morphisms in homological categories are studied in Sections 3.1, 3.2. The homology of a chain complex is treated in Section 3.3, together with the connecting morphism and the derived homology sequence.

Chain complexes over a *semi*additive homological category E have 'directed homotopies' with a very defective structure; but, in the *additive* case, these homotopies become reversible and acquire a rich structure based on the cylinder and cocylinder functors (Section 3.4).

Exact couples in homological categories and their associated spectral sequence are dealt with in Section 3.5.

3.1 Subquotients in homological categories

E is always a homological category. We define a subquotient of an object by a bicartesian square of normal monos and epis, and introduce 'regular induction' between subquotients, extending a notion already considered in Part I for p-exact categories (see I.2.6.9).

We have seen in Part I that, in a p-exact category, subquotients can be viewed as subobjects in the category of relations (I.2.6.5). This will be extended to g-exact categories in Section 5.4.

3.1.1 Subquotients and formal subquotients

As already observed in 2.2.1-2.2.2, the axiom (ex3) can be expressed in the following form (in the presence of (ex0) and (ex1)): given an object A and two normal subobjects $m \colon M \rightarrowtail A$ (the *numerator*), $n \colon N \rightarrowtail A$ (the *denominator*), with $m \geqslant n$, there is a bicartesian square (3.1), determined up to isomorphism (that can be embedded in a commutative diagram (2.8) with short exact rows and columns)

$$
\begin{array}{ccc}
M & \overset{m}{\rightarrowtail} & A \\
{\scriptstyle q'}\downarrow & & \downarrow{\scriptstyle q} \\
S & \underset{m'}{\rightarrowtail} & Q
\end{array}
\qquad
q = \operatorname{cok} n,
$$

$$
m = q^*(m'), \qquad q = m_\circ(q'). \tag{3.1}
$$

The object $S \cong \operatorname{Ncm}(qm) \cong \operatorname{Nim}(qm)$, determined up to isomorphism by $m \geqslant n$ in $\operatorname{Nsb}A$, will be called a *subquotient* of A and written as M/N, with the usual abuse of notation.

Of course it is equally well determined by two normal quotients of A, $p\colon A \twoheadrightarrow P$ and $q\colon A \twoheadrightarrow Q$, with $p \leqslant q$ (take $p = \operatorname{cok} m$) and will also be written as $Q//P$ when we want to stress some duality aspect. In particular, $P = A/M$ and $M = A//P$.

The subquotient M/N is null if and only if $M = N$ (more precisely, $m = n$); indeed, in (3.1), the object S is null if and only if $m' = 0_Q$ (the null normal subobject of Q), if and only if $m = \ker q = n$.

Of course, the mere object S does not contain sufficient information and we need the whole diagram (3.1).

Therefore we define a (formal) *subquotient* $s\colon S \dashrightarrow A$ to be such a bicartesian square of normal monos and normal epis, up to isomorphism in M and Q. Equivalently, one can give a diagram $S \leftarrow \bullet \rightarrowtail A$ up to a central isomorphism, determining (3.1) by pushout (cf. 2.2.2); or a diagram $S \rightarrowtail \bullet \leftarrow A$, determining (3.1) by pullback.

The formal subquotient s of diagram (3.1), with its numerator and denominator, will also be expressed as:

$$
s = mq'^\sharp = q^\sharp m',
$$

$$
\operatorname{num} s = m_*(1) = q^*(m'), \quad \operatorname{Num}(s) = M = \operatorname{Dom}(\operatorname{num} s), \tag{3.2}
$$

$$
\operatorname{den} s = m_* q'^*(0) = q^*(0), \quad \operatorname{Den}(s) = N = \operatorname{Dom}(\operatorname{den} s).
$$

The first equation can be justified using the the double category $\mathbb{Ind}\mathsf{E}$ and its vertical composition (see 3.2.7).

Furthermore, if E is p-exact (or g-exact, more generally), s can be identified with the *monorelation*

$$
mq'^\sharp = q^\sharp m'\colon S \to A
$$

(see I.2.6.5, or here in 5.4.4-5.4.6) determined by our bicartesian square; now $(\ )^\sharp$ denotes a reversed relation.

3.1.2 Definition and Proposition (The transfer mappings of a subquotient)

In the homological category E, *a subquotient*

$$s = mq'^{\sharp} = q^{\sharp}m' : S \dashrightarrow A$$

defines a covariant increasing mapping of normal subobjects

$$s_* : \mathrm{Nsb}S \to \mathrm{Nsb}A, \qquad s_*(y) = m_*q'^*(y) = q^*m'_*(y), \qquad (3.3)$$

and two contravariant increasing mappings

$$s^{\wedge}, s^{\vee} : \mathrm{Nsb}A \to \mathrm{Nsb}S, \quad s^{\wedge}(x) = q'_*m^*(x) \leqslant s^{\vee}(x) = m'^*q_*(x). \qquad (3.4)$$

Their main properties are (for $y \in \mathrm{Nsb}S$ and $x \in \mathrm{Nsb}A$):

$$s^{\wedge}s_*(y) = s^{\vee}s_*(y) = y, \qquad (3.5)$$

$$s_*s^{\wedge}(x) = (x \wedge \mathrm{num}\, s) \vee (\mathrm{den}\, s), \quad s_*s^{\vee}(x) = (x \vee \mathrm{den}\, s) \wedge (\mathrm{num}\, s), \quad (3.6)$$

$$s_*s^{\wedge}(x) = s_*s^{\vee}(x) = x, \quad for \quad \mathrm{den}\, s \leqslant x \leqslant \mathrm{num}\, s \;\; (in\; \mathrm{Nsb}A), \qquad (3.7)$$

$$\mathrm{num}\, s = s_*1, \qquad \mathrm{den}\, s = s_*0. \qquad (3.8)$$

As a consequence, s_* gives an isomorphism from the lattice $\mathrm{Nsb}S$ to the interval $[\mathrm{den}\, s, \mathrm{num}\, s] = [N, M]$ of $\mathrm{Nsb}A$; its inverse is the common restriction of $s^{\wedge}, s^{\vee}$ to that interval.

Since s_* is injective and because of (3.6), $s^{\wedge}$ and $s^{\vee}$ coincide if $\mathrm{Nsb}A$ is a modular lattice.

(Therefore, they always coincide for modular homological categories, and in particular for the g-exact ones; cf. 2.3.1, 2.3.3 and 5.4.5.)

More generally we have, for every $x \in \mathrm{Nsb}A$

$$s^{\wedge}(x) = s^{\vee}(x) \quad \Leftrightarrow \quad (x \wedge \mathrm{num}\, s) \vee (\mathrm{den}\, s) = (x \vee \mathrm{den}\, s) \wedge (\mathrm{num}\, s), \quad (3.9)$$

$$s^{\wedge}(x) = s^{\vee}(x) \quad if \;\; \mathrm{den}\, s \leqslant x \;\; or \;\; x \leqslant \mathrm{num}\, s. \qquad (3.10)$$

Proof The equalities and inequalities in (3.3) - (3.6) follow from the following identities

$$m_*q'^*(x) = m_*q'^*m'^*m'_*(x) = m_*m^*q^*m'_*(x)$$
$$= q^*m'_*(x) \wedge m = q^*m'_*(x) \wedge q^*m'_*(1) = q^*m'_*(x),$$

$$q'_*m^* = m'^*m'_*q'_*m^* = m'^*q_*m_*m^* \leqslant m'^*q_*,$$

$$s\hat{\ }s_* = q'_* m^* m_* q'^* = q'_* q'^* = 1,$$

$$s\check{\ }s_* = \dots = 1,$$

$$s_* s\hat{\ }(x) = m_* q'^* q'_* m^*(x) = m_*(m^* x \vee q'^* 0) = (m_* m^* x) \vee (m_* q'^* 0)$$
$$= (x \wedge m) \vee (m_* q'^* 0) = (x \wedge \mathrm{num}\, s) \vee (\mathrm{den}\, s).$$

$\square$

3.1.3 Regular induction

We shall often follow the common abuses of notation for subobjects (and quotients). Thus, a normal subobject $m \colon M \rightarrowtail A$ is often denoted by means of its domain M; if $N \leqslant M$ in $\mathrm{Nsb}\,A$ (i.e. we have a normal subobject $n \colon N \rightarrowtail A$ with $n \leqslant m$), N can also denote the corresponding normal subobject $m^*(n)$ of M, equivalent to the normal mono $n' \colon N \rightarrowtail M$ such that $mn' = n$.

Let a morphism $f \colon A \to B$ be given. If M and H are normal subobjects of A and B, respectively, and $f_*(M) \leqslant H$, we have a commutative diagram with short exact rows

$$
\begin{array}{ccccc}
M & \overset{m}{\rightarrowtail} & A & \overset{p}{\twoheadrightarrow} & A/M \\
{\scriptstyle f'}\downarrow & & {\scriptstyle f}\downarrow & & \downarrow{\scriptstyle f''} \\
H & \underset{h}{\rightarrowtail} & B & \underset{u}{\twoheadrightarrow} & B/H
\end{array}
\qquad (3.11)
$$

where f' comes from the fact that fm factorises through $\mathrm{Nim}\,(fm) = f_*(M) \leqslant H$, and f'' comes from the fact that uf annihilates $M = \mathrm{Ker}\,p$.

More generally, given two subquotients

$$s \colon M/N \dashrightarrow A, \qquad\qquad t \colon H/K \dashrightarrow B,$$

we say that *f has a regular induction* from M/N to H/K whenever

$$f_*(M) \leqslant H \quad \text{and} \quad f_*(N) \leqslant K. \qquad (3.12)$$

Then, we have a *regularly induced* morphism $g \colon M/N \to H/K$

$$
\begin{array}{ccccccc}
M & \overset{m}{\rightarrowtail} & A & \overset{q}{\dashrightarrow} & A/N & & \\
{\scriptstyle f'}\downarrow & & {\scriptstyle f}\downarrow & & \downarrow{\scriptstyle f''} & & \\
H & \underset{h}{\rightarrowtail} & B & \underset{v}{\twoheadrightarrow} & B/K & &
\end{array}
\qquad
\begin{array}{ccccc}
M & \overset{q'}{\twoheadrightarrow} & M/N & \overset{m'}{\rightarrowtail} & A/N \\
{\scriptstyle f'}\downarrow & & \downarrow{\scriptstyle g} & & \downarrow{\scriptstyle f''} \\
H & \underset{v'}{\twoheadrightarrow} & H/K & \underset{h'}{\rightarrowtail} & B/K
\end{array}
\qquad (3.13)
$$

first form the left diagram above, by applying twice (3.11), then the right one, by normal factorisation of the rows of the former and its naturality (Section 1.5.5).

We combine these data in an *inductive square* (that will be a cell of the double category $\mathbb{I}nd\mathsf{E}$, see 3.2.7)

$$
\begin{array}{ccc}
A & \xrightarrow{\ f\ } & B \\
{\scriptstyle s}\big\uparrow & & \big\uparrow{\scriptstyle t} \\
M/N & \xrightarrow{\ g\ } & H/K
\end{array}
\qquad\qquad f_*(M) \leqslant H, \quad f_*(N) \leqslant K. \qquad\qquad (3.14)
$$

For instance, the central morphism in the normal factorisation of any morphism f is induced by f, from the normal quotient $\mathrm{Ncm}\, f$ to the normal subobject $\mathrm{Nim}\, f$.

(We shall see in 5.4.6 that, if E is g-exact, the induced morphism g can be expressed in $\mathrm{Rel}\,\mathsf{E}$:

$$
g = t^\sharp f s = (v' h^\sharp) f(m q'^\sharp) = (h'^\sharp s) f(q^\sharp m'),
$$

as a composition of relations.)

Notice that, if f is an isomorphism (or even an identity) and the induced morphism g is an isomorphism, its inverse g^{-1} need *not* be regularly induced by f^{-1}. This problem is at the basis of the coherence problems dealt with in Part I, in the p-exact setting (see 5.4.6).

Here, we shall only deal with *regular* induction.

3.1.4 Basic properties

Regular induction is consistent with composition, identities and isomorphisms, in the following sense.

(a) If $f : A \to B$ regularly induces f' from M/N to H/K and $g : B \to C$ regularly induces g' from H/K to P/Q, then gf regularly induces $g'f'$ from M/N to P/Q.

(b) The identity of A regularly induces the identity from any subquotient M/N of A to itself.

(c) An isomorphism $f : A \to B$ regularly induces an isomorphism from any subquotient M/N of A to f_*M/f_*N.

For (c), it is sufficient to note that f and f^{-1} regularly induce two morphisms, inverse to each other. A related result is in 3.2.5(b).

3.1.5 Proposition (Direct and inverse images for an induced morphism)

The calculus of direct and inverse images along the *regularly* induced morphism $g \colon M/N \to H/K$ *provides a basic tool for diagram chasing. It is expressed by the following formulas (to be compared with (3.7))*

$$g_*(x) = t\check{\,}f_*s_*(x) = t\hat{\,}f_*s_*(x), \tag{3.15}$$

$$g^*(y) = s\check{\,}f^*t_*(y) = s\hat{\,}f^*t_*(y). \tag{3.16}$$

Proof With the notation of (3.13), we have:

$$g_*(x) = h'^*h'_*g_*(x) = h'^*f''_*m'_*(x) = h'^*f''_*q_*q^*m'_*(x)$$
$$= h'^*v_*f_*q^*m'_*(x) = t\check{\,}f_*s_*(x) = t\hat{\,}f_*s_*(x),$$

$$g^*(y) = g^*h'^*h'_*(y) = m'^*f''^*h'_*(y) = m'^*q_*q^*f''^*h'_*(y)$$
$$= m'^*q_*f^*v^*h'_*(y) = s\check{\,}f^*t_*(y) = s\hat{\,}f^*t_*(y).$$

The last equalities in the two equations above, concerning the contravariant transfer mappings, follow from the property (3.10), that holds because of the following inequalities

$$f_*s_*(x) \leqslant f_*s_*(1) = f_*(M) \leqslant H = \operatorname{Num} t,$$
$$f^*t_*(x) \geqslant f^*t_*(0) = f^*(K) \geqslant f^*f_*(N) \geqslant N = \operatorname{Den} s.$$

$\square$

3.1.6 Canonical morphisms

In particular, if $A = B$ and f is the identity, we have a *canonical* morphism

$$g \colon M/N \to M'/N', \quad \text{for } M \leqslant M', \ N \leqslant N' \text{ in } \mathsf{Nsb}A, \tag{3.17}$$

that is regularly induced by 1.

This morphism need not be exact (e.g. in $\mathsf{Gp_2}$); but it is a *normal mono* if $N = N'$ (then f'' is the identity in (3.13)), whereas it is a *normal epi* if $M = M'$

$$M/N \rightarrowtail M'/N \quad (N \leqslant M \leqslant M'),$$
$$M/N \twoheadrightarrow M/N' \quad (N \leqslant N' \leqslant M). \tag{3:18}$$

In the first case it is easy to verify that, for the subquotient $t \colon M'/N \dashrightarrow A$, we have:

$$t_*(M/N \rightarrowtail M'/N) = M \quad (N \leqslant M \leqslant M'). \tag{3.19}$$

Indeed, the canonical monomorphism $x\colon M/N \rightarrowtail M'/N$ is induced by 1_A, from $s\colon M/N \dashrightarrow A$ to $t\colon M'/N \dashrightarrow A$, whence (by (3.15)) $x = x_*(1) = t\check{\ }s_*(1) = t\check{\ }(M)$ and $t_*(x) = t_*t\check{\ }(M) = M$ (see (3.7)).

3.1.7 Proposition (Additivity and induction)

Let E *be a semiadditive homological category (as defined in 2.4.1). If the maps* $f, g\colon A \to B$ *regularly induce* $f', g'\colon M/N \to H/K$, *respectively, then their sum* $f + g$ *regularly induces* $f' + g'\colon M/N \to H/K$.

Moreover, if E *is additive,* $-f$ *regularly induces* $-f'$.

Proof Using the relations between sum and transfer mappings, in (2.25), we have:

$$
\begin{aligned}
(f + g)_*(M) &\leqslant f_*(M) \vee g_*(M) \leqslant H, \\
(f + g)_*(N) &\leqslant f_*(N) \vee g_*(N) \leqslant K.
\end{aligned}
\tag{3.20}
$$

Moreover, the induced map is $f' + g'$, as it follows easily from the diagrams (3.13) that define regular induction.

The last statement is an obvious consequence. $\qquad\square$

3.2 Induction, exactness and modularity

Further properties of subquotients and induced morphisms are studied here, also in connection with exactness and modularity conditions for morphisms.

E is always a homological category, $f\colon A \to B$ is a morphism that regularly induces $g\colon M/N \to H/K$ from $s\colon M/N \dashrightarrow A$ to $t\colon H/K \dashrightarrow B$, so that $f_*M \leqslant H$ and $f_*N \leqslant K$, as in the inductive square (3.14).

3.2.1 Normal subobjects of a subquotient

With respect to our subquotient $s\colon M/N \dashrightarrow A$, we shall frequently identify *the normal subobjects of* M/N *with the subquotients* X/N *of* A, *for* $N \leqslant X \leqslant M$ in $\mathrm{Nsb}A$, by means of the following isomorphism of lattices

$$
\begin{aligned}
\mathrm{Nsb}(M/N) &\to [N, M], \quad x \mapsto s_*(x), \\
[N, M] &\to \mathrm{Nsb}(M/N), \quad X \mapsto (x'\colon X/N \rightarrowtail M/N),
\end{aligned}
\tag{3.21}
$$

where $x'\colon X/N \rightarrowtail M/N$ is the canonical monomorphism, induced by 1_A (see (3.18)).

Indeed, we already know that $s_*\colon \mathrm{Nsb}(M/N) \to \mathrm{Nsb}A$ induces an isomorphism onto its image, the interval $[\mathrm{Den}\,s, \mathrm{Num}\,s] = [N, M]$ of $\mathrm{Nsb}A$ (cf.

3.1.2); further, the two mappings above are inverse to each other, because of the relation $s_*(x') = X$ (cf. (3.19)).

3.2.2 Kernel and cokernel of an induced morphism

With these identifications the normal factorisation of the morphism g, regularly induced by f, can be written as follows

$$(M \wedge f^*K)/N \xrightarrow{\;k\;} M/N \xrightarrow{\;g\;} H/K \xrightarrow{\;c\;} H/(K \vee f_*M)$$

$$(M \wedge f^*K)/N \xrightarrow{p} \qquad \qquad \xrightarrow{m} (K \vee f_*M)/K$$

$$M/(M \wedge f^*K) \xrightarrow{\qquad\qquad g' \qquad\qquad} (K \vee f_*M)/K \tag{3.22}$$

where *the kernel and cokernel, the normal image and coimage of g are canonical morphisms* between subquotients of A and B, while g' is induced by g (hence also by f, as it will be proved in 3.2.3).

As we already know that g induces g' (by 3.1.3), it suffices to calculate $\ker g$. With the notation of (3.13) and the results of 3.1.5

$$s_*(\ker g) = s_* s\check{\ } f^* t_*(0) = s_* s\check{\ }(f^* v^* h'_*(0)) = s_* s\check{\ }(f^*K)$$
$$= (f^*K \vee N) \wedge M = M \wedge f^*K. \tag{3.23}$$

Since $s_* \colon \mathrm{Nsb}(M/N) \to [N, M]$ is an isomorphism of lattices and $s_*(k)$ also coincides with $M \wedge f^*K$ (cf. 3.2.1), the conclusion follows.

3.2.3 Subquotients of subquotients and third Noether isomorphism

If $N \leqslant N' \leqslant M' \leqslant M$ in $\mathrm{Nsb}A$, the previous arguments give a square diagram, composed of canonical morphisms

$$\begin{array}{ccc} M'/N & \xrightarrow{\;m\;} & M/N \\ {\scriptstyle h}\downarrow & & \downarrow{\scriptstyle q} \\ M'/N' & \xrightarrow[k]{} & M/N' \end{array} \qquad\qquad M'/N' \cong (M'/N)/(N'/N). \tag{3.24}$$

Since $\ker q = (N'/N \rightarrowtail M/N) \leqslant m$ (see (3.22)), our square is bicartesian (by 3.1.1) and M'/N' *is* canonically a subquotient of M/N: in other words, we have the (third-type) *Noether isomorphism* displayed above.

Conversely, each subquotient $t \colon T \dashrightarrow M/N$ can be obtained in this form, letting $M' = s_*(\mathrm{num}\,t)$ and $N' = s_*(\mathrm{den}\,t)$.

Now, given a second chain $K \leqslant K' \leqslant H' \leqslant H$ in $\mathrm{Nsb}B$, if the morphism $f \colon A \to B$ is coherent with these filtrations, i.e. $f_*(N) \leqslant K$ and so on, the morphisms induced by f produce a natural transformation of the square (3.24) into the analogous square of subquotients of B.

This proves the *transitivity of regular induction*: if f induces $g: M/N \to H/K$ and the latter induces $h: M'/N' \to H'/K'$, then f induces h.

3.2.4 Exact objects and second Noether isomorphism

In any *ex2-category* we say that an object A is *exact* if the axiom (ex4) holds *locally* in A, in the sense that:

(a) for every $m \in \mathrm{Nsb}A$ and every $q \in \mathrm{Nqt}A$, the morphism qm is exact.

In our *homological* category E, we can rewrite this in terms of subquotients of A. Given two normal subobjects of any object A, $m: M \rightarrowtail A$ and $n = \ker q: N \rightarrowtail A$, the composed morphism $f = qm: M \rightarrowtail A \twoheadrightarrow A/N$ has the following normal factorisation, where g is induced by f, hence by 1_A (cf. (3.22) and Section 3.2.3)

$$
\begin{array}{ccccc}
M \wedge N & \rightarrowtail & M & \xrightarrow{\ \ f\ \ } & A/N & \twoheadrightarrow & A/(M \vee N) \\
& & \downarrow & & \uparrow & & \\
& & M/(M \wedge N) & \underset{g}{\to} & (M \vee N)/N & &
\end{array}
\qquad (3.25)
$$

Therefore the object A is exact if and only if

(a') for every $m: M \rightarrowtail A$ and $n: N \rightarrowtail A$, the canonical morphism g of the normal factorisation of $f = (M \rightarrowtail A \twoheadrightarrow A/N)$ is invertible, a *Noether isomorphism* of the second type.

The exact objects and exact morphisms of an ex2-category form its *exact centre* (a g-exact category, see 5.1.3, 5.1.4). Provided we already know that A is an exact object, the diagram above can also be drawn in the general ex2-case (where we do not have subquotients), *since it actually lives in the exact centre of* E.

3.2.5 Proposition (Induction and exact morphisms)

Let $f: A \to B$ be an exact *morphism, with $g: M/N \to H/K$ regularly induced by f.*

(a) If $f_*M \geqslant K$ and $f^*K \leqslant M$, the induced morphism g is exact.

(b) If $f_*M = H$ and $f^*K = N$, g is an isomorphism.

Note. Taking into account the regular induction, the hypothesis of (a) can be rewritten as:

$$
K \leqslant f_*M \leqslant H, \qquad N \leqslant f^*K \leqslant M.
$$

Proof (a) Let $f_*M \geqslant K$ and $f^*K \leqslant M$. First, note that the connection $\mathrm{Nsb}f = (f_*, f^*)$ is exact; since $K \leqslant f_*M$, it follows that

$$f_*f^*K = K \wedge f_*A = K. \tag{3.26}$$

Form the left diagram below, where the lower inductive square is given by the normal factorisation of g (see 3.2.2), under our hypotheses

$$
\begin{array}{ccc}
A & \xrightarrow{\;f\;} & B \\
\big\uparrow & & \big\uparrow \\
M/N & \xrightarrow{\;g\;} & H/K \\
\big\uparrow & & \big\uparrow \\
M/f^*K & \xrightarrow[g']{} & f^*M/K
\end{array}
\qquad
\begin{array}{ccc}
A & \xrightarrow{\;f\;} & B \\
{}^{r}\big\uparrow & & \big\uparrow {}^{i} \\
A/R & \xrightarrow{\;f'\;} & I \\
\big\uparrow & & \big\uparrow \\
M/f^*K & \xrightarrow[f'']{} & f^*M/K
\end{array}
\quad . \tag{3.27}
$$

Form now the right diagram above: the upper inductive square is the normal factorisation of f, with central *isomorphism* $f': A/R \to I$ (for $R = \mathrm{Ker}\, f$, $I = \mathrm{Nim}\, f$), while the lower inductive square exists and produces an isomorphism f'', because the conditions of 3.1.4(c) are satisfied (use (3.26))

$$f'_*(M/R) = i\hat{\ } f_* r_*(M/R) = i\hat{\ } f_*M = f_*M \wedge I = f_*M,$$

$$f'_*(f^*K/R) = i\hat{\ } f_* r_*(f^*K/R) = i\hat{\ } f_* f^*K = f_* f^*K \wedge I = f_* f^*K = K.$$

Therefore g' and f'' are both induced by f, because of the transitive property, and coincide so that g' is an isomorphism and g is exact.

(b) In these stronger hypotheses, g coincides with the isomorphism g' of the previous argument. $\square$

3.2.6 Proposition (Induction and modular morphisms)

Let $g: M/N \to H/K$ be regularly induced by $f: A \to B$ and let

$$X \in \mathrm{Nsb}A, \quad N \leqslant X \leqslant M; \qquad Y \in \mathrm{Nsb}B, \quad K \leqslant Y \leqslant H, \tag{3.28}$$

so that, according to the identification (3.21), $X/N \in \mathrm{Nsb}(M/N)$ and $Y/K \in \mathrm{Nsb}(H/K)$. Then:

(a) g is left modular on X/N (cf. 2.3.1) provided that: f is left modular on X, $f_*X \leqslant H$ and $f^*0 \leqslant M$,

(b) g is right modular on Y/K provided that: f is right modular on Y, $f^*Y \leqslant M$ and $f_*A \geqslant K$.

Proof We verify (a). Let us write:

$$s = mq^\sharp\colon M/N \dashrightarrow A, \qquad\qquad t = hv^\sharp\colon H/K \dashrightarrow B,$$

$$x\colon X \rightarrowtail A, \qquad\qquad x'\colon X/N \rightarrowtail M/N,$$

so that $s_*x' = m_*q^*(x') = x$.

By the calculus of direct and inverse images along the induced map g (see (3.15), (3.16))

$$g^*g_*(x') = q_*m^*f^*h_*v^*v_*h^*f_*m_*q^*(x') = q_*m^*f^*h_*v^*v_*h^*f_*(x)$$

$$\leqslant q_*m^*f^*h_*h^*f_*(x) = q_*m^*f^*(f_*x \wedge h) = q_*m^*f^*f_*x$$

$$= q_*m^*(x \vee f^*0) = q_*m^*(m_*q^*x' \vee m_*m^*f^*0)$$

$$= q_*m^*m_*(q^*x' \vee m^*f^*0) = q_*(q^*x' \vee m^*f^*0) = (q_*q^*x') \vee (q_*m^*f^*0)$$

$$= x' \vee (q_*m^*f^*0) \leqslant x' \vee (q_*m^*f^*h_*v^*0) = x' \vee g^*0.$$

The other inequality, $g^*g_*(x') \geqslant x' \vee g^*0$, is obvious. $\qquad\square$

3.2.7* The double category of inductive squares

Subquotients and regular induction can be further formalised by introducing the double category $\mathbb{Ind}E$ *of inductive squares of the homological category* E, that we sketch here but will not be explicitly used in the following.

The objects and the horizontal morphisms come from E. The vertical morphisms are the formal subquotients (see 3.1.1); the cells are the *inductive squares* (3.14), with f' induced by f, from s to t

$$\begin{array}{ccc} A & \xrightarrow{\ f\ } & B \\[2pt] {\scriptstyle s}\Big\uparrow\Big\downarrow & & {\scriptstyle t}\Big\uparrow\Big\downarrow \\[2pt] M/N & \xrightarrow{\ f'\ } & H/K \end{array} \qquad\qquad f_*(M) \leqslant H, \quad f_*(N) \leqslant K. \qquad (3.29)$$

The horizontal composition (of morphisms and cells) is defined by composition in E, that is consistent with regular induction.

The vertical composition of subquotients is computed by completing the following diagram with the dotted pullback and pushout, that are bicartesian

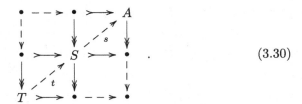

$$(3.30)$$

The vertical composition of cells follows from the fact that this construction is consistent with regular induction (by 3.2.3).

The formulas $s = mq'^\sharp = q^\sharp m'$, that we used in 3.1.2, are instances of this vertical composition, provided - of course - that we view $m\colon M \dashrightarrow A$ as the subquotient $M \rightarrowtail A = A$ of A defined by the normal subobject m and $q^\sharp\colon Q \dashrightarrow A$ as the subquotient $Q = Q \leftarrow A$ defined by the normal quotient q.

All the vertical morphisms are mono (with regard to vertical composition); the subquotient $s'\colon M'/N' \dashrightarrow A$ precedes $s\colon M/N \dashrightarrow A$ (in the canonical preorder of monos with values in A) if and only if $N \leqslant N' \leqslant M' \leqslant M$, i.e.

$$\text{num } s' \leqslant \text{num } s \quad \text{and} \quad \text{den } s' \geqslant \text{den } s, \quad \text{in Nsb}.A$$

3.3 Chain complexes and homology

We now prove the connecting morphism lemma and the homology sequence theorem, for homological categories, by means of regular induction on subquotients. In both cases we get *order-two* sequences, that are *exact* under some modularity conditions.

E is always a homological category, with (exact) transfer functor Nsb: $\mathsf{E} \to \mathsf{Ltc}$.

We generally treat the category $\mathsf{Ch}_\bullet\mathsf{E}$ of *unbounded* chain complexes $A_\bullet = A = ((A_n), (\partial_n))$, indexed on $\mathbb{Z}$, and occasionally the category $\mathsf{Ch}_+\mathsf{E}$ of *positive* chain complexes, with components indexed on $\mathbb{N}$. Notice that the subobject $B_n(A)$ of boundaries is defined as a *normal* image.

3.3.1 Complexes

An (unbounded) chain complex in E is a sequence $A_\bullet = ((A_n), (\partial_n))$, indexed on $\mathbb{Z}$

$$\dots A_{n+1} \xrightarrow{\partial_{n+1}} A_n \xrightarrow{\partial_n} A_{n-1} \dots \qquad (3.31)$$

where $\partial_{n+1}\partial_n$ is always null. In other words, we have

$$\operatorname{nim}\partial_{n+1} \leqslant \ker\partial_n \qquad (\operatorname{cok}\partial_{n+1} \geqslant \operatorname{ncm}\partial_n). \qquad (3.32)$$

As usual, A_n is called the *component* of $A_{\bullet}$ in degree n, and ∂_n the *differential* or *boundary.*

A *morphism* of complexes $f\colon A_{\bullet} \to C_{\bullet}$ is a sequence of morphisms $f_n\colon A_n \to C_n$ such that, for all $n \in \mathbb{Z}$, $\partial_n f_n = f_{n-1}\partial_n$; the differentials of $A_{\bullet}$ and $C_{\bullet}$ are distinguished by the context.

This forms the category $\mathrm{Ch}_{\bullet}\mathsf{E}$ of (unbounded) *chain complexes* of E. It has a natural structure of homological category, created by the conservative, faithful functor into the countable product $\mathsf{E}^{\mathbb{Z}}$

$$U\colon \mathrm{Ch}_{\bullet}\mathsf{E} \to \mathsf{E}^{\mathbb{Z}}, \qquad A_{\bullet} \mapsto (A_n). \qquad (3.33)$$

For this structure, a morphism f is null (or a normal mono, or an exact morphism) if and only if all its components are, and $\mathrm{Ker}\,(f) = ((\mathrm{Ker}\,f_n),(\partial_n))$ with differentials induced by those of $A_{\bullet}$.

In this chapter we shall always use unbounded chain complexes. But the homological category $\mathrm{Ch}_{+}\mathsf{E}$ formed by *positive* chain complexes (with components A_n indexed on $\mathbb{N}$) can be given a similar treatment. There is an embedding V and a truncation functor P

$$\mathrm{Ch}_{+}\mathsf{E} \xrightarrow{\ V\ } \mathrm{Ch}_{\bullet}\mathsf{E} \xrightarrow{\ P\ } \mathrm{Ch}_{+}\mathsf{E} \qquad\qquad PV = 1, \qquad (3.34)$$

where V completes the positive complex $A_{\bullet} = ((A_n),(\partial_n))$ by adding to its right *the null normal quotient* N of the component A_0

$$\ldots A_1 \xrightarrow{\ \partial_1\ } A_0 \xrightarrow{\ \partial_0\ } N \xrightarrow{\ \mathrm{id}\ } N \ldots \qquad (3.35)$$

$$A_n = N \ (n < 0), \qquad \partial_0 = 0^{A_0}\colon A_0 \twoheadrightarrow N, \qquad d_n = 1_N \ (n \leqslant -1).$$

It is useful to identify $A_{\bullet}$ with this unbounded complex, in order to obtain simpler formulas.

If E is pointed homological, N is the zero object and V is right adjoint to P (with the obvious unit $A_{\bullet} \to VPA_{\bullet}$). But this is not true in the general case: e.g. for Set_2, it is easy to see there may be no map $A_{\bullet} \to VPA_{\bullet}$.

Similarly one treats the homological category $\mathrm{Ch}^{\bullet}\mathsf{E}$ of (unbounded) *cochain complexes* $A^{\bullet} = ((A^n),(d^n))$ of E (indexed on $\mathbb{Z}$), and the homological category $\mathrm{Ch}^{+}\mathsf{E}$ of *positive cochain complexes*, with components indexed on $\mathbb{N}$.

3.3.2 Homology

The complex $A_\bullet = ((A_n), (\partial_n))$ determines, in each component A_n, the normal subobjects $B_n \leqslant Z_n$ of boundaries and cycles, and the homology subquotient

$$Z_n = Z_n A_\bullet = \operatorname{Ker} \partial_n, \qquad B_n = B_n A_\bullet = \operatorname{Nim} \partial_{n+1},$$
$$H_n = H_n A_\bullet = \operatorname{Cok}(B_n \rightarrowtail Z_n) = Z_n / B_n. \tag{3.36}$$

Equivalently, one can use the corresponding normal quotients $\hat{Z}_n \leqslant \hat{B}_n$ (in $\operatorname{Nqt} A_n$)

$$\hat{Z}_n = \hat{Z}_n A_\bullet = \operatorname{Ncm} \partial_n = A_n / Z_n,$$
$$\hat{B}_n = \hat{B}_n A_\bullet = \operatorname{Cok} \partial_{n+1} = A_n / B_n, \tag{3.37}$$
$$H_n A_\bullet \cong \operatorname{Ker}(\hat{B}_n \twoheadrightarrow \hat{Z}_n) = \hat{B}_n // \hat{Z}_n.$$

There is thus a commutative diagram, with a bicartesian subquotient square (cf. 3.1.1)

$$\tag{3.38}$$

Here $\delta_n \colon \hat{B}_n \to Z_{n-1}$ is induced by the differential ∂_n, so that $\operatorname{nim} \delta_{n+1} = (B_n \rightarrowtail Z_n)$ and

$$H_n A_\bullet = \operatorname{Cok}(B_n \rightarrowtail Z_n) = \operatorname{Cok} \delta_{n+1},$$
$$H_n A_\bullet \cong \operatorname{Ker}(\hat{B}_n \twoheadrightarrow \hat{Z}_n) = \operatorname{Ker} \delta_n. \tag{3.39}$$

It will be useful to remark that, in the commutative diagram

$$\tag{3.40}$$

$\underline{\partial}_n$ is the central morphism of the normal factorisations of ∂_n *and* δ_n, so that:

(a) ∂_n is exact $\Leftrightarrow$ δ_n is exact $\Leftrightarrow$ $\underline{\partial}_n$ is an isomorphism,

(b) ∂_n is modular $\Leftrightarrow$ δ_n is modular $\Leftrightarrow$ $\mathrm{Nsb}(\underline{\partial}_n)$ is an isomorphism. Clearly, we have defined functors

$$Z_n, \ B_n, \ \hat{Z}_n, \ \hat{B}_n, \ H_n \colon \mathrm{Ch}_\bullet\mathsf{E} \to \mathsf{E}.$$

We are going to show, by the usual argument based on the connecting morphism, that the functors H_n give raise to a connected sequence. Note that all these functors act on maps by regular induction; therefore, if E is semiadditive homological, they are additive (see 3.1.7).

Each functor $F \colon \mathrm{Ch}_\bullet\mathsf{E} \to \mathsf{E}$ of the above has a corresponding functor $\mathrm{Ch}_+\mathsf{E} \to \mathsf{E}$ that can be defined as PFV, by embedding $\mathrm{Ch}_+\mathsf{E}$ in $\mathrm{Ch}_\bullet\mathsf{E}$ and truncating the result of F (cf. (3.34)).

For cochain complexes, *we always speak of cycles, boundaries and homology* (or cochain-homology), rather than *cocycles and so on.*

3.3.3 Lemma (Snake Lemma, or the connecting morphism)

The two middle squares of the diagram below are given, commutative and with exact rows. We complete the diagram by kernels and cokernels (together with the induced morphisms of the other rows)

$$(3.41)$$

$$g = \mathrm{cok}\, f,$$
$$h = \ker f.$$

Then there is a connecting morphism $\partial \colon \mathrm{Ker}\, c \to \mathrm{Cok}\, a$, *regularly induced by* $b \colon B' \to B''$ *on the subquotients* $\mathrm{Ker}\, c$ *(of* B'*) and* $\mathrm{Cok}\, a$ *(of* B''*)*

$$s = (\mathrm{Ker}\, c \rightarrowtail C' \leftarrowtail B'), \qquad t = (\mathrm{Cok}\, a \leftarrowtail A'' \rightarrowtail B''). \qquad (3.42)$$

It forms a sequence of order two, *that is natural for natural transformations of the two middle squares (its exactness is analysed in the next proposition)*

$$(3.43)$$

Note. The initial data can be equivalently reduced to the morphisms f, b and k, under the condition that their composition kbf be null; then we take $g = \operatorname{cok} f$, $h = \ker f$ and define a and c by the universal property of kernels and cokernels.

Proof The morphism b does have a regular induction from s to t:

$$b_*(\operatorname{num} s) = b_* g^*(c') = b_* g^* c^*(0) = b_* b^* k^*(0) \leqslant k^*(0) = \operatorname{num} t, \quad (3.44)$$

$$b_*(\operatorname{den} s) = b_*(\ker g) = b_* f_*(1) = h_* a_*(1) = h_*(\ker a'') = \operatorname{den} t. \quad (3.45)$$

The sequence (3.33) is thus established, together with its naturality.

Moreover, $g'f'$ is null, because it is annihilated by c', an N-mono, and we just need to show that $\partial g'$ is null, as the rest will follow by duality. We show that $\partial_* g'_*(1) = 0$, where ∂_* is computed along the diagram, by means of the calculus of direct images along induced morphisms (see (3.15))

$$\partial_* g'_*(1) = a''_* h^* b_* g^* c'_* g'_*(1) = a''_* h^* b_* g^* g_* b'_*(1) = a''_* h^* b_*(b' \vee \ker g)$$
$$= a''_* h^* b_*(\ker b \vee f_*(1)) = a''_* h^* b_* f_*(1) = a''_* h^* h_* a_*(1) = a''_* a_*(1) = 0.$$

$\square$

3.3.4 Proposition (Exactness and modularity conditions)

With the same hypotheses and notation

(a) *if f is right modular on the subobject* $\ker b$ *(i.e. $f_* f^*(b^*0) = b^*0 \wedge f_*1$), then* $\operatorname{nim} f' = \ker g'$,

(b) *if b is left modular on the subobject* $\operatorname{nim} f$ *(i.e. $b^* b_*(f_*1) = f_*1 \vee b^*0$), then* $\operatorname{nim} g' = \ker \partial$,

(b*) *if b is right modular on the subobject* $\ker k$ *(i.e. $b_* b^*(k^*0) = k^*0 \wedge b_*1$), then* $\operatorname{nim} \partial = \ker h'$,

(a*) *if k is left modular on the subobject* $\operatorname{nim} b$ *(i.e. $k^* k_*(b_*1) = b_*1 \vee k^*0$), then* $\operatorname{nim} h' = \ker k'$,

(c) *if f is a normal mono, then* $f' \sim \ker g'$,

(c*) *if k is a normal epi, then* $k' \sim \operatorname{cok} h'$,

(d) *if b is exact, so is ∂ and the sequence (3.43) is exact in the 'central objects',* $\operatorname{Ker} c$ *and* $\operatorname{Cok} a$,

(e) *if $x \in \operatorname{Nsb}(\operatorname{Ker} b)$ and $(b'_* x \vee g^*0) \wedge b' = b'_* x \vee (g^*0 \wedge b')$ in* $\operatorname{Nsb} B'$, *then g' is left modular on x, i.e. $g'^* g'_*(x) = x \vee g'^*0$,*

(e*) *if $x \in \operatorname{Nsb}(\operatorname{Cok} b)$ and $(b''^* x \wedge h) \vee b''0 = b''^* x \wedge (h \vee b''^*0)$ in* $\operatorname{Nsb} B''$, *then h' is right modular on x, i.e. $h'_* h'^*(x) = x \wedge h'_*1$,*

(f) if f is right modular, b is modular and k is left modular, then the sequence (3.43) is exact,

(g) if f is a normal mono, b is modular and k is a normal epi, then the sequence (3.43) is exact, begins with a normal mono and ends with a normal epi.

Proof (a) By hypothesis

$$b'_*(f_*1) = f_*(a') = f_*a^*(0) = f_*a^*h^*(0) = f_*f^*b^*(0) = b^*0 \wedge f_*1,$$
$$b'_*(g'^*(0)) = b'_*g'^*c'^*(0) = b'_*b'^*g^*(0) = b'_*b'^*f_*(1)$$
$$= f_*(1) \wedge b' = b^*(0) \wedge f_*(1).$$

Since b' is a normal mono, the conclusion follows.

(b) By the calculus of inverse images along an induced morphism

$$\partial^*0 = c'^*g_*b^*h_*a''^*(0) = c'^*g_*b^*h_*a_*(1) = c'^*g_*b^*b_*f_*(1)$$
$$= c'^*g_*(f_*1 \vee b^*0) = c'^*g_*(g^*0 \vee b') = c'^*g_*(b') = c'^*c'_*g'_*(1) = g'_*(1).$$

(c) If f is a normal mono, so is f' (by the detection property 1.5.7(b)); hence $f' \sim \operatorname{nim} f' = \ker g'$.

(d) Because of (b) and (b*), we only have to prove that the morphism ∂ is exact. This follows from Lemma 3.2.5(a), on the exactness of the induced morphism, since the morphism b, that induces ∂ from s to t, is exact and moreover (using again the exactness of b)

$$b_*(\operatorname{num} s) \geqslant b_*(\operatorname{den} s) = \operatorname{den} t, \quad \text{(by (3.45))},$$
$$\operatorname{num} s = g^*(c') = g^*c^*(0) = b^*k^*(0) \geqslant \ker b,$$
$$b^*(\operatorname{den} t) = b^*(b_*(\operatorname{den} s)) = \operatorname{den} s \vee \ker b \leqslant \operatorname{num} s.$$

(e) We end by this verification, as the rest follows by duality or from the previous points:

$$g'^*g'_*(x) = g'^*c'^*c'_*g'_*(x) = b'^*g^*g_*b'_*(x) = b'^*g^*g_*b'_*(x)$$
$$= \bar{b}'^*(b'_*x \vee g^*0) = b'^*b'_*b'^*(b'_*x \vee g^*0) = b'^*((b'_*x \vee g^*0) \wedge b'))$$
$$= b'^*(b'_*x \vee (g^*0 \wedge b')) = b'^*(b'_*x \vee b'_*b'^*(g^*0))$$
$$= b'^*b'_*(x \vee b'^*g^*0) = x \vee b'^*g^*0 = x \vee g'^*0.$$

□

3.3.5 Theorem (The homology sequence)

A short exact sequence of chain complexes is given, in Ch.E

$$U \xrightarrow{\ m\ } V \xrightarrow{\ p\ } W \qquad m = \ker p, \quad p = \operatorname{cok} m. \qquad (3.46)$$

(a) There is a homology sequence of order two, natural for morphisms of short exact sequences

$$\ldots H_n(V) \xrightarrow{\hat{p}_n} H_n(W) \xrightarrow{\partial_n} H_{n-1}(U) \xrightarrow{\hat{m}_{n-1}} H_{n-1}(V) \ldots \qquad (3.47)$$

where $\hat{m}_n = H_n(m)$, $\hat{p}_n = H_n(p)$ and ∂_n is induced by the differential ∂_n^V of the complex V.

(b) Central exactness. If the differential ∂_n^V of the central complex V is an exact morphism, so is the differential ∂_n of the homology sequence; moreover, the sequence itself is exact in the domain of ∂_n (i.e. $H_n(W)$) and in its codomain (i.e. $H_{n-1}(U)$).

Note. This partial exactness result is the key to prove the universality of chain homology in non-exact cases (cf. Section 4.5).

(c) The modular theory. If the following conditions hold for every $n \geqslant 0$, the homology sequence is exact:

$$(B_n V \vee U_n) \wedge Z_n V = B_n V \vee (U_n \wedge Z_n V), \qquad (3.48)$$

$$\partial^* \partial_* (U_n) = U_n \vee Z_n V, \qquad \partial_* \partial^* (U_{n-1}) = U_{n-1} \wedge B_{n-1} V, \qquad (3.49)$$

where (3.49) means that the differential $\partial = \partial_n^V$ is left modular on U_n and right modular on U_{n-1} (see 2.3.1).

(d) Conditions (3.48) and (3.49) are automatically satisfied when the homological category E *is modular.*

Proof (a) The short exact sequence (3.46) produces the following commutative diagrams (where we omit the degree of the morphisms)

$$
\begin{array}{ccccc}
Z_n U & \xrightarrow{\ m'\ } & Z_n V & \xrightarrow{\ p'\ } & Z_n W \\
\downarrow & & \downarrow & & \downarrow \\
U_n & \xrightarrow{\ m\ } & V_n & \xrightarrow{\ p\ } & W_n \\
\partial \downarrow & & \downarrow \partial & & \downarrow \partial \\
U_{n-1} & \xrightarrow{\ m\ } & V_{n-1} & \xrightarrow{\ p\ } & W_{n-1} \\
\downarrow & & \downarrow & & \downarrow \\
\hat{B}_{n-1} U & \xrightarrow{\ m''\ } & \hat{B}_{n-1} V & \xrightarrow{\ p''\ } & \hat{B}_{n-1} W
\end{array}
\qquad . \qquad (3.50)
$$

The middle rows are exact by hypothesis, the first row by the left exactness of the functor Ker, and the last by the right exactness of Cok (cf. 1.7.6). This yields the two central rows of the next diagram, that are exact, and we can connect them with three morphisms δ_n induced by the corresponding ∂_n (by (3.38))

$$
\begin{array}{ccccc}
H_nU & \xrightarrow{\hat{m}} & H_nV & \xrightarrow{\hat{p}} & H_nW \\
\downarrow & & \downarrow & & \downarrow \\
\hat{B}_nU & \xrightarrow{m''} & \hat{B}_nV & \xrightarrow{p''} & \hat{B}_nW \\
\delta\downarrow & & \delta\downarrow & & \delta\downarrow \\
Z_{n-1}U & \xrightarrow{m'} & Z_{n-1}V & \xrightarrow{p'} & Z_{n-1}W \\
\downarrow & & \downarrow & & \downarrow \\
H_{n-1}U & \xrightarrow{\hat{m}} & H_{n-1}V & \xrightarrow{\hat{p}} & H_{n-1}W
\end{array}
\qquad (3.51)
$$

∂_H

Completing the diagram by the Snake Lemma 3.3.3, we get the order-two sequence (3.47), with differential ∂_H induced by δ_n^V, hence also by ∂_n^V (by the transitivity of regular induction, 3.2.3).

(b) If $\partial_n^V \colon V_n \to V_{n-1}$ is exact, so is $\delta_n \colon \hat{B}_nV \to Z_{n-1}V$ (by 3.3.2(a)) and the thesis follows immediately from 3.3.4(d).

(c) We want to apply Proposition 3.3.4. We omit the degrees of morphisms, write $\partial = \partial^V$, $\delta = \delta^V$ and use the canonical morphisms q, r, z of the following commutative diagram

$$
\begin{array}{ccccc}
U_n & \xrightarrow{m} & V_n & \xrightarrow{\partial} & V_{n-1} \\
q\downarrow & & r\downarrow & & \uparrow z \\
\hat{B}_nU & \xrightarrow{m''} & \hat{B}_nV & \xrightarrow{\delta} & Z_{n-1}V
\end{array}
\qquad (3.52)
$$

With this notation, we check the hypotheses (a), (b) of 3.3.4 (whereas (a*) and (b*) follow by duality):

$$
m''_*m''^*(\delta^*0) = m''_*q_*q^*m''^*(\delta^*0) = r_*m_*m^*r^*\delta^*0
$$
$$
= r_*(r^*\delta^*0 \wedge m) = r_*(\partial^*0 \wedge m) = r_*((\partial^*0 \wedge m) \vee r^*0)
$$
$$
= r_*((Z_nV \wedge U_n) \vee B_nV) = r_*(Z_nV \wedge (U_n \vee B_nV)) =
$$
$$
= r_*(\partial^*0 \wedge (m \vee r^*0)) = r_*(r^*\delta^*0 \wedge r^*r_*(m))
$$
$$
= r_*r^*(\delta^*0 \wedge r_*(m)) = \delta^*0 \wedge r_*(m) = \delta^*0 \wedge m''_*1,
$$

$$\delta^*\delta_* m_*''(1) = r_*\partial^* z^* z_*\partial_* r^* m_*''1 = r_*\partial^*\partial_* r^* m_*''1 =$$
$$r_*\partial^*\partial_* r^* m_*'' q_* 1 = r_*\partial^*\partial_* (r^* r_* m) = r_*\partial^*\partial_* (m \vee r^* 0) = \dots$$
$$= r_*(m \vee \partial^* 0) = r_*(m) \vee r_*(\partial^* 0)$$
$$= m_*''1 \vee r_*\partial^* z_* 0 = m_*''1 \vee \delta^* 0.$$

$\square$

3.3.6 Theorem (Homology and modularity)

The following conditions on the homological category E *are equivalent:*

(a) E *is modular;*

(b) the connected sequence (H_n, ∂_n): $\mathrm{Ch}_\bullet\mathsf{E} \to \mathsf{E}$ *of chain homology functors of* E *is exact, or, in other words, for every short exact sequence of chain complexes of* E, *the homology sequence* (3.47) *is exact;*

(c) the connecting-morphism sequence (3.43) *is always exact, for every commutative diagram* (3.41) *whose central rows are exact;*

(d) as before, for every commutative diagram (3.41) *whose central rows are short exact.*

Proof We already know that (a) $\Rightarrow$ (b), by 3.3.5, and (a) $\Rightarrow$ (c), by 3.3.4(f); since (d) is a particular instance of (c) and also of (b), it suffices to show that (d) $\Rightarrow$ (a).

Assuming (d), we must prove that for every morphism $f: A \to B$ in E the associated connection $\mathrm{Nsb}(f): \mathrm{Nsb}A \to \mathrm{Nsb}B$ is exact, i.e. we always have $f^* f_*(x) = x \vee f^* 0$ and $f_* f^*(y) = y \wedge f_* 1$.

For the first condition, let $x: X \rightarrowtail A$ be a normal subobject, take $y = f_* x: Y \rightarrowtail B$ and form the following commutative diagram with short exact rows

$$
\begin{array}{ccccc}
X & \xrightarrow{\ x\ } & A & \xrightarrow{\ p\ } & A/X \\
{\scriptstyle g}\downarrow & & {\scriptstyle f}\downarrow & & \downarrow{\scriptstyle h} \\
Y & \xrightarrow[\ y\ ]{} & B & \xrightarrow[\ q\ ]{} & B/Y
\end{array}
\qquad (3.53)
$$

By hypothesis we have the exact sequence

$$\mathrm{Ker}\, f \xrightarrow{\ p'\ } \mathrm{Ker}\, h \xrightarrow{\ d\ } \mathrm{Cok}\, g \xrightarrow{\ y'\ } \mathrm{Cok}\, f . \qquad (3.54)$$

Here p' is N-epi (i.e. $\mathrm{nim}\, p' = 1$) because $\mathrm{Cok}\, g$ is null:

$$g_* 1 = y^* y_* g_* 1 = y^* f_* x = y^*(y) = 1.$$

Let $k = \ker f$, $k'' = \ker h$, so that $pk = k''p'$. Then

$$f^* f_*(x) = f^* y = f^* q^* 0 = p^* h^* 0 = p^*(k'') = p^*(k''_*(1))$$
$$= p^*(k''_* p'_*(1)) = p^*(p_* k_*(1)) = p^* p_*(k) = k \vee p^* 0 = f^* 0 \vee x.$$

Analogously, starting from a normal subobject $y\colon Y \rightarrowtail B$, one takes $x = f^* y\colon X \rightarrowtail A$ and constructs a diagram (3.53) with $\mathrm{Ker}\, h$ null, whence $\ker y' = 0$. $\qquad\square$

3.3.7 Homology of chain complexes of lattices

In $\mathrm{Ch}_\bullet(\mathsf{Ltc})$, we can identify the subquotient $H_n(X)$ of a chain complex $X = ((X_n), (\partial_n))$ with a *closed interval* of the lattice X_n (see 1.4.1)

$$H_n(X) = [\partial_{n+1\bullet}(1), \partial_n{}^\bullet(0)] \subset X_n, \tag{3.55}$$

identifying, for every n-cycle $x \leqslant \partial_n{}^\bullet(0)$, the homology class $[x]$ with the element $x \vee \partial_{n+1\bullet}(1)$ of X_n.

A morphism $f = (f_\bullet, f^\bullet)\colon X \to Y$ of chain complexes of lattices (and connections) induces a connection in homology, that can be described as follows, using the previous identification

$$H_n f\colon H_n(X) \to H_n(Y),$$
$$(H_n f)_\bullet(x) = f_{n\bullet}(x) \vee \partial_{n+1\bullet}(1), \tag{3.56}$$
$$(H_n f)^\bullet(y) = f_n{}^\bullet(y) \wedge \partial_n{}^\bullet(0).$$

Indeed, for the covariant part, if $\partial_{n+1\bullet}(1) \leqslant x \leqslant \partial_n{}^\bullet(0)$, then

$$f_{n\bullet}(x) \leqslant f_{n\bullet}\partial_n{}^\bullet(0) \leqslant f_{n\bullet}\partial_n{}^\bullet f_{n-1}{}^\bullet(0) = f_{n\bullet}f_n{}^\bullet\partial_n{}^\bullet(0) \leqslant \partial_n{}^\bullet(0).$$

3.4 Chain homotopies, semiadditivity and additivity

E is always a semiadditive homological category, whose biproducts are written as $A \oplus B$, unless - in a particular case - a specific notation is more appropriate.

We now define homotopies $\varphi\colon f \to g$ in the category $\mathrm{Ch}_\bullet\mathsf{E}$ of chain complexes; these have a rich structure in the additive case, but a very defective one when E lacks opposites.

3.4.1 Review of semiadditive homological categories

Recall that a semiadditive (resp. additive) category has a zero-object and binary biproducts, that produce an enriched structure over abelian monoids (resp. abelian groups; see 2.4.1). A semiadditive (or additive) homological category E has been defined in 2.4.2, as a semiadditive (or additive) category that is *pointed* homological.

A typical example is the category Ltc of lattices and connections, studied in Section 1.2. Its biproducts are ordinary cartesian products $X \times Y$ (cf. 1.2.7). The sum of maps is computed as $f + g = (f_\bullet \vee g_\bullet, f^\bullet \wedge g^\bullet)$; it is idempotent and lacks opposites.

The categories pAb (preordered abelian groups, see 1.6.2) and Abm (abelian monoids, see 1.6.2) are again semiadditive homological categories where opposite maps are missing.

The categories K Tvs, K Hvs, Ban, Hlb (introduced in 1.6.3) are additive homological. The category R Smd of semimodules over the unital semiring R is always semiadditive homological, and is additive if and only if R is actually a ring; but then R Smd = R Mod is even abelian.

Let us also recall that the transfer functor Nsb: E $\to$ Ltc of a semiadditive homological category E satisfies the property: $(f+g)_* \leqslant f_* + g_*$ (cf. (2.25)), and that regular induction on the subquotients of E preserves the sum of maps (cf. 3.1.7).

3.4.2 Homotopy for chain complexes

Let us consider now the p-homological category Ch$_\bullet$E of (unbounded) chain complexes $A_\bullet = ((A_n), (\partial_n))$ over E (see 3.3.1). Also Ch$_\bullet$E is semiadditive homological, with the obvious biproducts

$$((A_n), (\partial_n)) \oplus ((B_n), (\partial_n)) = ((A_n \oplus B_n), (\partial_n^A \oplus \partial_n^B)), \qquad (3.57)$$

and is additive if and only if E is.

The homology functors H_n: Ch$_\bullet$E $\to$ E are additive, as already remarked in 3.3.2.

Given two morphisms $f, g \colon A_\bullet \to B_\bullet$, a (directed) *homotopy* $\varphi = (\varphi_n)$: $f \to g \colon A_\bullet \to B_\bullet$ will consist of a sequence of maps of E such that

$$\varphi_n \colon A_n \to B_{n+1}, \quad g_n = f_n + \partial_{n+1}\varphi_n + \varphi_{n-1}\partial_n \quad (n \in \mathbb{Z}). \qquad (3.58)$$

Notice that the sequence

$$h_n = \partial_{n+1}\varphi_n + \varphi_{n-1}\partial_n \colon A_n \to B_n,$$

defines a morphism of chain complexes $h \colon A_\bullet \to B_\bullet$.

The existence of a homotopy $f \to g$ defines a *preorder* of categories on Ch$_\bullet$E (i.e. a preorder relation on arrows with the same domain and codomain, that is consistent with composition). To prove transitivity, we take a second homotopy $\psi\colon g \to h\colon A_\bullet \to B_\bullet$ and define

$$\varphi + \psi\colon f \to h\colon A_\bullet \to B_\bullet, \quad (\varphi + \psi)_n = \varphi_n + \psi_n.$$

Let $f \simeq g$ denote the congruence of categories generated by this preorder; it amounts to the existence of a finite sequence of directed homotopies $f \to f_1 \leftarrow f_2 \to \ldots \leftarrow f_{n-1} \to g$ (and coincides with our preorder when E is additive).

The quotient of $\mathrm{Ch}_\bullet\mathsf{E}$ modulo the homotopy congruence defines the *homotopy category*

$$\mathrm{HoCh}_\bullet(\mathsf{E}) = \mathrm{Ch}_\bullet\mathsf{E}/\simeq, \tag{3.59}$$

with the same objects and homotopy classes $[f]\colon A_\bullet \to B_\bullet$.

As ususal, two chain complexes $A_\bullet, B_\bullet$ are said to be *homotopy equivalent* if they are isomorphisc objects in the homotopy category, or equivalently if there exists maps $f\colon A_\bullet \rightleftarrows B_\bullet\colon g$ in $\mathrm{Ch}_\bullet\mathsf{E}$, with $gf \simeq \mathrm{id}A_\bullet$ and $fg \simeq \mathrm{id}B_\bullet$.

3.4.3 Proposition (Homotopy invariance)

Two homotopic maps $f \simeq g\colon A_\bullet \to B_\bullet$ in $\mathrm{Ch}_\bullet\mathsf{E}$ always induce the same morphisms $f_{*n} = g_{*n}\colon H_n(A_\bullet) \to H_n(B_\bullet)$ in homology.

As a consequence, a homotopy equivalence $A_\bullet \rightleftarrows B_\bullet$ induces a sequence of isomorphisms $H_n(A_\bullet) \cong H_n(B_\bullet)$.

Proof Plainly, we can assume that there exists a homotopy $\varphi\colon f \to g\colon A_\bullet \to B_\bullet$. Then there is a map h of chain complexes with $g = f + h$

$$h = \partial\varphi + \varphi\partial\colon A_\bullet \to B_\bullet, \quad h_n = \partial_{n+1}\varphi_n + \varphi_{n-1}\partial_n \quad (n \in \mathbb{Z}),$$

$$\partial_n h_n = \partial_n(\partial_{n+1}\varphi_n + \varphi_{n-1}\partial_n) = \partial_n\varphi_{n-1}\partial_n$$
$$= (\partial_n\varphi_{n-1} + \varphi_{n-2}\partial_{n-1})\partial_n = h_{n-1}\partial_n,$$

But $h_*(Z_n(A_\bullet)) \leqslant B_n(B_\bullet)$, so that h induces the zero morphism $h_{*n} = 0$ in homology.

Homology is an additive functor and the thesis follows: $g_{*n} = f_{*n} + h_{*n} = f_{*n}$. $\qquad\square$

3.4.4 Remarks

If D is an *additive* category (not necessarily homological), homotopies in $\mathrm{Ch}_\bullet\mathsf{D}$ are defined as above. They have very regular properties, studied from a structural point of view in a text on 'directed algebraic topology' [G18] (Section 4.4).

This study is based on two adjoint endofunctors $I \dashv P$ of $\mathrm{Ch}_{\bullet}\mathsf{D}$, the *cylinder* and the *path* functor, defined as follows on a complex $A_{\bullet} = ((A_n), (\partial_n))$ (letting u_i denote the canonical injections)

$$(IA_{\bullet})_n = A_n \oplus A_{n-1} \oplus A_n,$$
$$\partial = (\partial u_1 - u_2, -\partial u_2, \partial u_3 + u_2), \qquad (3.60)$$

$$(PA_{\bullet})_n = A_n \oplus A_{n+1} \oplus A_n,$$
$$\partial = (\partial u_1, -u_1 - \partial u_2 + u_3, \partial u_3). \qquad (3.61)$$

Then a homotopy $\varphi \colon f \to g \colon A_{\bullet} \to B_{\bullet}$ can be viewed as a chain morphism:

$$(f, \varphi, g) \colon A_{\bullet} \to PB_{\bullet}, \qquad (3.62)$$

or equivalently, because of the adjunction, as the morphism $IA_{\bullet} \to B_{\bullet}$ of co-components $f_n \colon A_n \to B_n$, $\varphi_{n-1} \colon A_{n-1} \to B_n$ and $g_n \colon A_n \to B_n$.

Together with various natural transformations between the powers of the endofunctors I or P (faces, degeneracy, connections, reversion, transposition, etc.), this provides $\mathrm{Ch}_{\bullet}\mathsf{D}$ with a very rich structure, explored in [G18] and called a *reversible, permutable dIP4-homotopical category.*

Now, more generally, homotopies can be defined as above in $\mathrm{Ch}_{\bullet}\mathsf{D}$, when D is just a *semi*additive category. As we have seen, we get a 'directed' notion: homotopies *cannot be reversed* (if D lacks opposites).

Notice that this new structure is *much more defective* than the various frameworks of directed homotopy theory investigated in [G18]. First, there is now no cylinder or cocylinder functor representing homotopies, because we need opposites to construct the differentials of $IA_{\bullet}$ and $PA_{\bullet}$, in the formulas above. Second, the category $\mathsf{C} = \mathrm{Ch}_{\bullet}\mathsf{D}$ lacks a basic element of those frameworks, namely a 'reversor' $R \colon \mathsf{C} \to \mathsf{C}$ turning an object into the 'opposite one', so that the existence of a directed homotopy $f \to g$ implies the existence of a directed homotopy $Rg \to Rf$.

3.5 The exact couple

The exact couple and its associated spectral sequence, introduced by Massey [Mas1], are here extended to homological categories. This extension will be applied to *homotopy* spectral sequences, whose terms do not live in any abelian context but can be seen as objects of a suitable homological category (Sections 6.3-6.5).

E is always a homological category and $\mathrm{Bgr}\mathsf{E}$ the homological category of bigraded objects over E, with bidegree in $\mathbb{Z} \times \mathbb{Z}$ and morphisms of every bidegree (r, s) (see 1.4.4)

$$f = (f_{hk}) \colon A \to B, \quad f_{hk} \colon A_{hk} \to B_{h+r,k+s}.$$

Then, f is assumed to be null when all its components are. Using BgrE will make the exposition simpler, allowing us to defer to the end the issue of bigraduation.

Notice that, even if E is pointed, BgrE has no zero-object (generally): the object whose components are zero is just *quasi* zero (as defined in 1.3.2); thus, if E is p-exact, or even abelian, the category BgrE is just g-exact, in the present sense.

In Section 6.5 we shall investigate a generalised notion of 'quasi exact' couple.

3.5.1 *Definition*

An *exact couple* $C = (D, E, u, v, \partial)$ in the homological category E is a system of objects and morphisms

$$(3.63)$$

so that

(a) this triangle is exact, i.e. $\operatorname{nim} u = \ker v$, $\operatorname{nim} v = \ker \partial$ and $\operatorname{nim} \partial = \ker u$,

(b) all the endomorphisms $u^r = u...u \colon D \to D$ are exact, for $r \geqslant 1$,

(c) v is left modular on $\operatorname{Ker} u^r$, for $r \geqslant 1$,

(d) ∂ is right modular on $\operatorname{Nim} u^r$, for $r \geqslant 1$.

Note that the endomorphism $d = v\partial \colon E \to E$ has a null square, $dd = v(\partial v)\partial$, since ∂v is null.

The conditions (b) - (d) are trivially satisfied in a g-exact category and can be viewed as a sort of C^∞-condition on our couple; indeed, if k-*exact* couple means that (a) - (d) are satisfied for $1 \leqslant r \leqslant k$, it is easy to see that the derived couple C' is $(k-1)$-*exact* (following the argument of 3.5.2 and 3.5.3).

3.5.2 The derived exact couple

The derived couple $C' = (D', E', u', v', \partial')$ of the exact couple C has the following objects

$$D' = \operatorname{Nim} u \in \operatorname{Nsb}(D), \qquad E' = H(E, d), \qquad (3.64)$$

$$E' = \partial^*(\operatorname{Ker} v)/v_*(\operatorname{Nim} \partial) = \partial^*(\operatorname{Nim} u)/v_*(\operatorname{Ker} u).$$

In order to define its morphisms we also use an isomorphic copy of D' *coming from the exactness of the morphism u*. It is a normal quotient of D (instead of a normal subobject), and the isomorphism i is regularly induced by $u \colon D \to D$

$$\hat{D}' = \operatorname{Ncm} u = D/\operatorname{Ker} u, \qquad i \colon \hat{D}' \cong D'. \qquad (3.65)$$

Now we can set

$$u' \colon D' \to D', \quad \text{regularly induced by } u$$
$$(\text{since } u_*(D') = \operatorname{Nim} u^2 \leqslant D'),$$

$$\partial' \colon E' \to D', \quad \text{regularly induced by } \partial$$
$$(\partial_*(\partial^*(D')) \leqslant D', \;\; \partial_*(v_*(\operatorname{Ker} u)) = 0), \qquad (3.66)$$

$$\hat{v}' \colon \hat{D}' \to E', \quad \text{regularly induced by } v$$
$$(v_*(1) = \partial^*0 \leqslant \operatorname{Num} E', \;\; v_*(\operatorname{Ker} u) = \operatorname{Den} E').$$

Finally, we define the last morphism as

$$v' = \hat{v}'.i^{-1} \colon D' \to \hat{D}' \to E', \qquad (3.67)$$

but notice that i^{-1} and v' are *not* regularly induced morphisms, generally (cf. 3.1.3).

3.5.3 Theorem

The derived couple of an exact couple is exact. If the morphisms v, ∂ are also exact, so are v' and ∂'.

Proof (a) The exactness in E' is trivial, by the normal factorisation of induced morphisms (cf. 3.2.2)

$$\operatorname{Ker} \partial' = (\operatorname{Num} E' \wedge \partial^*(0))/\operatorname{Den} E' = \operatorname{Ker} \partial/\operatorname{Den} E',$$
$$\operatorname{Nim} v' = (\operatorname{Den} E' \vee v_*1))/\operatorname{Den} E' = \operatorname{Ker} \partial/\operatorname{Den} E'. \qquad (3.68)$$

The exactness in the 'left occurrence' of D' follows from the right modularity of ∂ on Nim u

$$\text{Nim } \partial' = \partial_*(\partial^*(\text{Nim } u) = \text{Nim } u \wedge \text{Nim } \partial = D' \wedge \text{Ker } u = \text{Ker } u'. \quad (3.69)$$

Similarly, the exactness in the 'right occurrence' of D' follows from the left modularity of v over Ker u; indeed, in the following diagram of inductive squares (where $v' = \hat{v}'.i^{-1} \colon D' \to E'$)

$$
\begin{array}{ccccc}
D & \xleftarrow{\ u\ } & D & \xrightarrow{\ v\ } & E \\
{\scriptstyle m}\big\uparrow & & \big\downarrow{\scriptstyle q} & & \big\uparrow{\scriptstyle s} \\
D' & \xleftarrow{\ i\ } & \hat{D}' & \xrightarrow{\ \hat{v}'\ } & E'
\end{array}
$$

$$\text{Ker } \hat{v}' = (\text{Num } \hat{D}' \wedge v^*(\text{Den } E'))/\text{Den } \hat{D}'$$
$$= v^*(v_* \text{Ker } u)/\text{Den } \hat{D}' = (\text{Ker } u \vee \text{Ker } v)/\text{Ker } u,$$
$$\text{Ker } v' = i_*(\text{Ker } \hat{v}') = u_*(\text{Ker } u \vee \text{Ker } v)/u_*(u^*(0))$$
$$= u_*(\text{Ker } v) = u_*(u_*(1)) = u_*(D') = \text{Nim } u'.$$

(b) In order to prove that all the composites of the endomorphism $w = u'$ are exact, consider the commutative diagram

$$
\begin{array}{ccccc}
D & \xrightarrow{\ u\ } & D & \xrightarrow{\ u^r\ } & D \\
{\scriptstyle q}\big\downarrow & & \big\uparrow{\scriptstyle m} & & \big\uparrow{\scriptstyle m} \\
\hat{D}' & \xrightarrow{\ i\ } & D' & \xrightarrow{\ w^r\ } & D'
\end{array}
$$

where $(u^r m)iq = u^r u = u^{r+1}$ is exact, by hypothesis.

Applying two 'detection' properties for exact morphisms (cf. 1.5.7(c), (c*)), so is $u^r m = m w^r$ and finally so is w^r. It should be noted that the exactness of w^r follows from the exactness of u^{r+1}.

(c) We now prove that v' is left modular over Ker u^r; similarly, one verifies that ∂' is right modular on Nim u^r. Consider the following diagram of inductive squares

$$
\begin{array}{ccccc}
D & \xleftarrow{\ u^r\ } & D & \xrightarrow{\ v\ } & E \\
{\scriptstyle q}\big\downarrow & & \big\downarrow{\scriptstyle q} & & \big\uparrow{\scriptstyle s} \\
\hat{D}' & \xleftarrow{\ \hat{u}^r\ } & \hat{D}' & \xrightarrow{\ \hat{v}^r\ } & E'
\end{array} .
$$

Plainly, it is sufficient to prove that $\hat{v}'$ is left modular over Ker $\hat{u}^r$

$$\text{Ker } \hat{u}^r = (\text{Ker } u^r \wedge \text{Num } \hat{D}')/\text{Den } \hat{D}' = \text{Ker } u^r/\text{Den } \hat{D}'. \quad (3.70)$$

This follows from 3.2.6, as v is left modular on $\operatorname{Ker} u^r$, by hypothesis, and

$$v_*(\operatorname{Ker} u^r) \leqslant \partial^*(\operatorname{Nim} u) = \operatorname{Num} E', \qquad v^*0 \leqslant 1 = \operatorname{Num} \hat{D}'.$$

(d) Finally, we assume that v and ∂ are exact and prove that the induced morphisms $\hat{v}' \colon \hat{D}' \to E'$, $\partial' \colon E' \to D'$ are also, by means of Lemma 3.2.5

$$v_*(\operatorname{Num} \hat{D}') = v_*D \geqslant v_*(\operatorname{Nim} \partial) = \operatorname{Den} E',$$

$$\operatorname{Num} \hat{D}' = D \geqslant v^*(\operatorname{Den} E'),$$

$$\partial_*(\operatorname{Num} E') \geqslant 0 = \operatorname{Den} D',$$

$$\operatorname{Num} E' = \partial^*(\operatorname{Ker} v) \geqslant \partial^*(0) = \partial^*(\operatorname{Den} D').$$

It follows that $v' = \hat{v}'.i^{-1}$ is also exact $\square$

3.5.4 Iterated derivation

We have thus proved that an exact couple C has derived (exact) couples of any order.

They can be directly calculated in the usual way. We let $C = C^1$, $C' = C^2$ and define the r-th derived couple $C^r = (D^r, E^r)$, for $r \geqslant 1$

$$D^r = \operatorname{Nim} u^{r-1} \in \operatorname{Nsb}(D), \quad D_r = \operatorname{Ncm} u^{r-1} \in \operatorname{Nqt}(D),$$

$$i \colon D_r \cong D^r \quad \text{(isomorphism induced by } u^{r-1} \colon D \to D), \qquad (3.71)$$

$$E^r = \partial^*(\operatorname{Nim} u^{r-1})/v_*(\operatorname{Ker} u^{r-1}) = \partial^*(D^r)/v_*(\operatorname{Ker} u^{r-1}).$$

The object E^r is a subquotient of E, since $v_*(\operatorname{Ker} u^{r-1}) \leqslant v_*(1) \leqslant \partial^*(0) \leqslant \partial^*(D^r)$. The morphisms of the r-th derived couple are, by definition:

$\qquad u^{(r)} \colon D^r \to D^r, \quad$ induced by u

$$\qquad (u_*(D^r) = D^{r+1} \leqslant D^r),$$

$\qquad \partial^{(r)} \colon E^r \to D^r, \quad$ induced by ∂

$$\qquad (\partial_*(\partial^*(D^r)) \leqslant D^r, \quad \partial_*(v_*(\operatorname{Ker} u^{r-1})) = 0), \qquad (3.72)$$

$\qquad v_{(r)} \colon D_r \to E^r, \quad$ induced by v

$$\qquad (v_*(1) \leqslant \partial^*(0) \leqslant \operatorname{Num} E^r, \quad v_*(\operatorname{Ker} u^{r-1}) = \operatorname{Den} E^r),$$

$$\qquad v^{(r)} = v_{(r)}.i^{-1} \colon D^r \cong D_r \to E^r.$$

Now, assuming the exactness of C^r, it suffices to show that the couple C^{r+1} is isomorphic to the derived couple $(C^r)'$. First:

$$(D^r)' = \operatorname{Nim} u^{(r)} = u_*^{(r)}(D^r) = u_*^{(r)}(\operatorname{Nim} u^{r-1}) = \operatorname{Nim} u^r = D^{r+1}.$$

Second, $(E^r)'$ is the homology of E^r with respect to $d^r = \partial^{(r)}.v^{(r)}\colon E^r \to E^r$. Using also (3.70), we have

$$\operatorname{Ker} d^r = \partial^{(r)*}(\operatorname{Ker} v^{(r)}) = \partial^{(r)*}(\operatorname{Nim} u^{(r)}) = \partial^{(r)*}(D^{r+1})$$
$$= (\operatorname{Num} E^r \wedge \partial^*(D^{r+1}))/\operatorname{Den} E^r = \partial^*(D^{r+1})/\operatorname{Den} E^r,$$
$$\operatorname{Nim} d^r = v_*^{(r)}(\operatorname{Nim} \partial^{(r)}) = v_*^{(r)}(\operatorname{Ker} u^{(r)}) = v_{(r)*}(\operatorname{Ker} \hat{u}^{(r)})$$
$$= v_{(r)*}(\operatorname{Ker} u^r/\operatorname{Den} E^r) = (v_*(\operatorname{Ker} u^r) \vee \operatorname{Den} E^r)/\operatorname{Den} E^r$$
$$= v_*(\operatorname{Ker} u^r)/\operatorname{Den} E^r,$$

and the thesis follows, by a Noether isomorphism:

$$(E^r)' = H(E^r, d^r) = \operatorname{Ker} d^r/\operatorname{Nim} d^r = \partial^*(D^{r+1})/v_*(\operatorname{Ker} u^r) \cong E^{r+1}.$$

3.5.5 The bigraded case

A *bigraded exact couple* $C = (D, E, u, v, \partial)$ *of type 1* in the homological category E is an exact couple in the homological category BgrE recalled above, where u, v and ∂ have bidegree $(0, 1), (0, 0)$ and $(-1, -1)$, respectively.

In other words, it is a system of morphisms in E, indexed on $n, p \in \mathbb{Z}$

$$u = u_{np}\colon D_{n,p-1} \to D_{np}, \qquad v = v_{np}\colon D_{np} \to E_{np},$$
$$\partial = \partial_{np}\colon E_{np} \to D_{n-1,p-1}, \tag{3.73}$$

such that:

(a) the following sequences are exact

$$\ldots E_{n+1,p} \xrightarrow{\partial} D_{n,p-1} \xrightarrow{u} D_{np} \xrightarrow{v} E_{np} \xrightarrow{\partial} D_{n-1,p-1} \ldots \tag{3.74}$$

(b) all the morphisms $u_{n,p}^r = u_{np}\ldots u_{n,p-r+1}\colon D_{n,p-r} \to D_{np}$ are exact, for $r \geqslant 1$,

(c) v_{np} is left modular on $\operatorname{Ker}(u_{n,p+r}^r\colon D_{np} \to D_{n,p+r})$, for $r \geqslant 1$,

(d) ∂_{np} is right modular on $\operatorname{Nim}(u_{np}^r\colon D_{n,p-r} \to D_{np})$, for $r \geqslant 1$.

3.5.6 The spectral sequence

Therefore the r-th derived couple $C^r = (D^r, E^r, u^{(r)}, v^{(r)}, \partial^{(r)})$ of a bigraded exact couple C consists of the following bigraded objects (where

$r \geqslant 1$, $D^1_{np} = D_{np}$ and $E^1_{np} = E_{np}$):

$$D^r_{n,p} = \operatorname{Nim}(u^{r-1} \colon D_{n,p-r+1} \to D_{np}) \in \operatorname{Nsb}(D_{np}),$$

$$D^{np}_r = \operatorname{Ncm}(u^{r-1} \colon D_{np} \to D_{n,p+r-1}) \in \operatorname{Nqt}(D_{np}),$$

$$i \colon D^{np}_r \cong D^r_{n,p+r-1} \tag{3.75}$$

(isomorphism induced by $u^{r-1} \colon D_{np} \to D_{n,p+r-1}$),

$$E^r_{n,p} = \partial^*(D^r_{n-1,p-1})/v_*(\operatorname{Ker}(u^{r-1} \colon D_{np} \to D_{n,p+r-1})), \tag{3.76}$$

and the following morphisms:

$$u^{(r)}_{np} \colon D^r_{n,p-1} \to D^r_{n,p} \quad \text{(induced by } u\text{)},$$

$$\partial^{(r)}_{np} \colon E^r_{n,p} \to D^r_{n-1,p-1} \quad \text{(induced by } \partial\text{)},$$

$$v^{np}_{(r)} \colon D^{np}_r \to E^r_{n,p} \quad \text{(induced by } v\text{)}, \tag{3.77}$$

$$v^{(r)}_{np} = v^{np}_{(r)}.i^{-1} \colon D^r_{n,p+r-1} \to D^{np}_r \to E^r_{n,p}.$$

Note that the couple C^r has morphisms $u^{(r)}, v^{(r)}, \partial^{(r)}$ of bidegrees $(0,1)$, $(0,1-r)$, $(-1,-1)$; it can be called a bigraded exact couple *of type r*.

This gives a spectral sequence $(E^r_{np}, d_{n,p}r)$, with differential of bidegree $(-1,-r)$

$$d^r_{n,p} = v^{(r)}.\partial^{(r)} \colon E^r_{np} \to E^r_{n-1,p-r}. \tag{3.78}$$

3.5.7 Convergence

In the abelian case, the convergence of spectral sequences and the relative 'mapping theorems' are usually studied by additive constructions that transform properties of maps into properties of objects; typically, the mapping cylinder [EM].

For p-exact categories, similar results can be obtained without using additivity, by means of convergence properties of subquotients (see [G8]). Likely, this approach can be extended to *homological* categories.

4

Satellites

Connected sequences of functors and satellites can be extended to semiexact categories; the main definitions are given (or adapted) in Section 4.1. The construction of satellites by means of the existence of sufficient normally injective objects is dealt with in Section 4.2; their identification by means of effacements in Section 4.3.

Additivity is here important, in order to get some existence results (see 4.2.5), but not necessary; and indeed the applications we give here concern the calculus of left satellites for (non-additive) categories of pairs (Section 4.4) and the discussion of the universality of chain homology (Section 4.5), in non-additive cases too. Section 4.4 also shows that the categories of pairs have 'left satellite dimension' $\leqslant 1$, in the obvious sense; for instance, this dimension is 0 for Set_2 and 1 for Top_2.

4.1 Connected sequences and satellites

Satellites, i.e. universal connected sequences of functors, can be constructed by limits or colimits. (It is a particular case of a pointwise Kan extension, that - in order to leave the background from category theory as simple as possible - will be dealt with directly.)

We always consider functors $F\colon \mathsf{E} \to \mathsf{B}$ defined on a semiexact category E and taking values in an N-category B, i.e. a category with an assigned ideal of null maps (see 1.3.1); recall that F is an N-functor if it preserves the null maps, or equivalently the null objects (1.3.1, 1.3.2). A short exact sequence of E will be written in the form $a = (a', a'') = (A' \rightarrowtail A \twoheadrightarrow A'')$. Moreover, two 'extended integers' $h < k$, in the ordinal sum $\overline{\mathbb{Z}} = \{-\infty\} + \mathbb{Z} + \{+\infty\}$, are given.

4.1.1 Connected sequences of functors

A *connected* (resp. *exact*) ∂-sequence (F_n, ∂_n) of functors $\mathsf{E} \to \mathsf{B}$ (for $h < n < k$), from a semiexact category E to an N-category B, is given by the following data:

(a) a sequence of functors $F_n \colon \mathsf{E} \to \mathsf{B}$, that generally are not exact, for $h < n < k$,

(b) for every short exact sequence $a = (A' \rightarrowtail A \twoheadrightarrow A'')$ of E, a sequence of morphisms $\partial_n = \partial_n^a \colon F_n A'' \to F_{n-1} A'$ (for $h < n - 1 < n < k$), natural for morphisms of short exact sequences and such that the following sequence is of order two (resp. exact) in B

$$\ldots F_n(A') \longrightarrow F_n(A) \longrightarrow F_n(A'') \xrightarrow{\partial_n} F_{n-1}(A') \ldots \qquad (4.1)$$

The functors F_n of a connected sequence are necessarily N-functors, since the pair $(1_Z, 1_Z)$ is a short exact sequence for every null object Z. If the sequence is exact, each functor F_n is *half exact*, i.e. takes the short exact sequence a to the exact sequence $F_n(A') \to F_n(A) \to F_n(A'')$.

The differential ∂_n can be seen as a natural transformation

$$\partial_n \colon F_n P'' \to F_{n-1} P' \colon \mathsf{ShE} \to \mathsf{B}, \qquad (4.2)$$

where ShE is the category of short exact sequences of E, while $P', P'' \colon \mathsf{ShE} \to \mathsf{E}$ are the first and last projection. Notice that, even if E is abelian, ShE is just an N-category: the Snake Lemma 3.3.3 shows that it is not closed under kernels and cokernels (unless E is trivial).

These sequences of functors are clearly preserved by composition with a short exact ex1-functor $H \colon \mathsf{E}' \to \mathsf{E}$ (i.e. a functor that preserves short exact sequences, see 1.7.0) or with an N-functor $K \colon \mathsf{B} \to \mathsf{B}'$ (assumed to preserve exact sequences, if our connected sequence of functors is exact).

Note that the numeration of indices in (4.1) grows backwards; this could cause misunderstandings in terminology. We shall always use the terms *lower* and *upper*, *first* and *last* with reference to indices, the terms *left* and *right* with reference to (4.1). Therefore our sequence is *upper* (or *left*) *bounded* if $k < +\infty$, in which case its last (or left-hand) functor is F_{k-1}.

Similarly one treats *connected d-sequences*, with a differential

$$d^n = d_a^n \colon F^n A'' \to F^{n+1} A',$$

of degree 1 (for $h < n < n + 1 < k$); in this case the numeration grows onwards.

4.1.2 Null functors

Recall that a *null* functor $N: \mathsf{E} \to \mathsf{B}$ is a functor that takes all the maps of E to null maps of B, or equivalently all the objects of E to null objects of B (cf. 1.3.1).

Notice that, if B is not pointed, there can be various such functors, non-isomorphic and also non-isomorphic to constant functors. For instance, Set_2 (or Top_2) has the following remarkable null endofunctors (and infinitely many others; see 1.7.5 for notation)

$$\bot: (X, A) \mapsto \bot = (\emptyset, \emptyset), \quad \top: (X, A) \mapsto \top = (\{*\}, \{*\}),$$
$$N_0: (X, A) \mapsto (A, A), \qquad N^0: (X, A) \mapsto (X, X). \tag{4.3}$$

We assume now that B has an initial object $\bot$, that is null (if B is semiexact, the latter condition follows from the former, by 1.5.3(g)). We write $\bot: \mathsf{E} \to \mathsf{B}$ for the constant functor at $\bot$, that is plainly initial in the category of all functors $\mathsf{E} \to \mathsf{B}$.

Then every connected sequence (F_n, ∂_n) that is left bounded ($k < +\infty$) can be extended to the left, by setting $F_n = \bot$ for $n \geqslant k$. It is easy to see that the extended sequence is exact if and only if the original sequence is exact and *moreover* its last (or left-hand) functor F_{k-1} takes normal monos to N-monos (cf. 1.3.1)

$$\dots \bot \to \bot \to \bot \to F_{k-1}(A') \to F_{k-1}(A) \to F_{k-1}(A'') \dots \tag{4.4}$$

Dually, the existence of a null terminal object for B allows one to extend a *right bounded* connected ∂-sequence to the right.

4.1.3 Ker, Cok and homology

If E is semiexact we already know (by 1.7.6) that the functors

$$\mathrm{Ker}: \mathsf{E}^2 \to \mathsf{E}, \qquad \mathrm{Cok}: \mathsf{E}^2 \to \mathsf{E}, \tag{4.5}$$

defined on the category of morphisms of E, are respectively left exact and right exact, i.e. preserve - respectively - kernels and cokernels.

If E is homological, the connecting map (introduced in 3.3.3) yields a connected sequence $(\mathrm{Ker}, \mathrm{Cok}, \partial)$ that is also exact if E is modular (by 3.3.4). The sequence can be canonically extended to the left (right) whenever E has an initial (terminal) object, without prejudice to exactness.

Analogously, if E is a homological category, the homology functors of positive chain complexes and their differentials form a right-bounded connected sequence (H_n, ∂_n) from $\mathsf{Ch}_+\mathsf{E}$ to E, whereas the homology functors of positive cochain complexes give a left-bounded connected sequence

(H^n, d^n) from Ch^+E to E; these sequences of functors are exact if and only if E is modular (by 3.3.5).

Connected ∂-sequences of functors $E \to B$, where E is semiexact and B is homological, are often obtained (in accord with an obvious remark in 4.1.1), by a composition

$$E \xrightarrow{\ C\ } Ch_+B \xrightarrow{\ H_n\ } B \tag{4.6}$$

of a *short exact* functor C, associating to each object of E a chain complex of B, with the connected sequence (H_n, ∂_n) of 'algebraic' homology functors over the chain complexes of B. If B is modular (a fortiori if it is g-exact, or abelian), the sequence is exact. For instance, singular homology for pairs of spaces and the relative homology of groups are both defined in this way.

4.1.4 Left satellites

We assume from now on that the N-category B has a null initial object and that h is a fixed integer. We consider right-bounded connected ∂-sequence (F_n, ∂_n) of functors from E to B with $h \leqslant n < k$, so that the functor F_h is the first of the sequence.

Such a sequence will be said to be *left universal* if

(ls) for every connected sequence (G_n, ∂_n) of the same indexing type and every natural transformation of the first (or right-hand) functors $\varphi_h \colon G_h \to F_h \colon E \to B$, there exists precisely one *natural transformation* of connected sequences $(\varphi_n) \colon (G_n, \partial_n) \to (F_n, \partial_n)$ that extends φ_h ($h \leqslant n < k$).

Plainly, the naturality of (φ_n) means that each $\varphi_n \colon G_n \to F_n$ is natural and moreover for every short exact sequence $a = (a', a'') = (A' \rightarrowtail A \twoheadrightarrow A'')$ of E the following diagram commutes in B

$$
\begin{array}{ccc}
F_n(A'') & \xrightarrow{\ \partial_n\ } & F_{n-1}(A') \\
\varphi\big\uparrow & & \big\uparrow\varphi \\
G^n(A'') & \xrightarrow[\ \partial_n\ ]{} & G_{n-1}(A')
\end{array}
\qquad . \tag{4.7}
$$

Note the following properties.

(a) The solution, if extant, is determined by the first functor F_h up to a unique isomorphism of connected sequences.

(b) Any 'coinitial' subsequence of a left universal sequence is again left universal; therefore, given F_h, the functor F_n does not depend on the length $k - h$ of the sequence we are using (provided that $k > n$, of course); this fact depends on the possibility of extending the connected sequences to the left, with the initial object.

(c) (*Pasting*) If $h < k < k'$ and the sequences

$$(F_n, \partial_n)_{h \leqslant n \leqslant k}, \qquad (F_n, \partial_n)_{k \leqslant n < k'},$$

are both left universal, then the sequence $(F_n, \partial_n)_{h \leqslant n < k'}$ obtained by pasting them is again left universal.

(d) All the functors F_n are N-functors (see 4.1.1).

The functors F_n are called *left satellites* of F_h and we let $S_p F_h = F_{h+p}$.

4.1.5 The first left satellite

Consider now an N-functor $F \colon \mathsf{E} \to \mathsf{B}$; plainly, $S_0 F = F$.

The functor $S_1 F$, or more precisely the pair $(S_1 F, \partial_1)$, is called *the (first) left satellite* of F (if it exists). Because of the pasting property, we want to construct this satellite and iterate the procedure.

Therefore, we are looking for a functor $S = S_1 F \colon \mathsf{E} \to \mathsf{B}$, equipped with a natural family $\partial_a \colon SA'' \to FA'$ (indexed on the short exact sequences of E), so that all the sequences

$$
\begin{array}{ccccc}
S(A') & \xrightarrow{\ Sa'\ } & S(A) & \xrightarrow{\ Sa''\ } & S(A'') \\
 & & & \nearrow{\scriptstyle \partial} & \\
F(A') & \underset{Fa'}{\rightrightarrows} & F(A) & \underset{Fa''}{\rightarrow} & F(A'')
\end{array}
\tag{4.8}
$$

are of order two in B; and we want the universal solution. The problem can be reduced to a simpler form, as in the following lemma.

4.1.6 Lemma

Part A (*Characterisation of connected sequences*). *Assume we have two functors* $F, S \colon \mathsf{E} \to \mathsf{B}$ *and, for every short exact sequence* $a = (A' \rightarrowtail A \twoheadrightarrow A'')$ *of* E, *a natural* B-*map* $\partial_a \colon SA'' \to FA'$.

Then these data form a connected sequence, i.e. all sequences (4.8) *are of order two, if and only if* F *and* S *are N-functors, i.e. preserve the null objects.*

Part B (*Characterisation of universal connected sequences*). *If* (F, S, ∂) *is a connected sequence, then* (S, ∂) *is the left satellite of* F *if and only if for every pair* (T, ∂') *forming a connected sequence with* F *there is a unique natural transformation* $\varphi \colon T \to S$ *such that, for every short exact sequence* a *in* E

$$\varphi.\partial_a = \partial'_a. \tag{4.9}$$

Both statements can be easily rewritten for sequences of functors of arbitrary length. The second says that in the universal property for connected sequences (F_n, ∂_n) one can equivalently assume that the new sequence (G_n, ∂_n) and the natural transformation φ start with $G_h = F_h$ and $\varphi_h = 1$.

Proof (A) The fact that F and S be N-functors is necessary, by 4.1.1. Conversely, assume that this holds. Then, for every short exact sequence a of E, the functors F and S annihilate the composite $a''a'$ and it suffices to prove that $\partial.S(a'')$ and $F(a').\partial$ are also null.

Consider in E the left commutative diagram below

$$
\begin{array}{ccccc}
A_0 & \rightarrowtail & A & \overset{1}{\twoheadrightarrow} & A \\
& & \downarrow{\scriptstyle 1} & & \downarrow{\scriptstyle a''} \\
A' & \overset{a'}{\rightarrowtail} & A & \overset{a''}{\twoheadrightarrow} & A'' \\
\downarrow{\scriptstyle a'} & & \downarrow{\scriptstyle 1} & & \downarrow \\
A & \overset{1}{\rightarrowtail} & A & \twoheadrightarrow & A^0
\end{array}
\qquad
\begin{array}{ccc}
SA & \overset{\partial}{\longrightarrow} & FA_0 \\
\downarrow{\scriptstyle Sa''} & & \downarrow \\
SA'' & \overset{\partial}{\longrightarrow} & FA' \\
\downarrow & & \downarrow{\scriptstyle Fa'} \\
SA^0 & \overset{\partial}{\longrightarrow} & FA
\end{array}
\qquad (4.10)
$$

with short exact rows. The null normal subobject A_0 and the null normal quotient A^0 are annihilated by F and S, and the right diagram above shows that the previous composed maps are indeed null.

(B) The general property follows easily from the reduced one. Given a connected sequence (G_0, G_1, ∂') and a natural transformation $\varphi_0 \colon G_0 \to F$, form a new 'intermediate' connected sequence (F, G_1, ∂'') with $\partial a'' = \varphi_0 \partial a' \colon G_1 A'' \to F A'$ and apply the reduced property to the new sequence

$$
\begin{array}{ccc}
S(A'') & \overset{\partial}{\longrightarrow} & F(A') \\
\varphi \uparrow & & \uparrow{\scriptstyle 1} \\
G_1(A'') & \overset{\partial''}{\longrightarrow} & F(A') \\
{\scriptstyle 1}\uparrow & & \uparrow{\scriptstyle \varphi_0} \\
G_1(A'') & \underset{\partial'}{\longrightarrow} & G_0(A')
\end{array}
\qquad . \qquad (4.11)
$$

$\square$

4.1.7 First satellite as a Kan extension

Assume, from now on, that B too is a semiexact category.

We write $\mathrm{Ch}_2\mathsf{E}$ for the semiexact category of 2-truncated chain complexes of the semiexact category E

$$a = (a', a'') = (A' \to A \to A'') \qquad (a''.a' \text{ null}), \qquad (4.12)$$

and ShE for the full N-subcategory of Ch_2E consisting of the short exact sequences (cf. 4.1.1).

Let P', P'' denote the projections on the first or third object of the complex and H' the homology in the first position

$$P', P'' : Ch_2E \to E, \qquad P'(a) = A', \quad P''(a) = A'', \qquad (4.13)$$

$$H' : Ch_2E \to E, \qquad H'(a) = \text{Ker } a'. \qquad (4.14)$$

The same constructions apply to the semiexact category B. Let $F: E \to B$ be an N-functor and consider its obvious extension $\hat{F}: \text{ShE} \to Ch_2B$.

To assign an N-functor $S: E \to B$ and a natural transformation $\delta: SP'' \to H'\hat{F}$

$$
\begin{array}{ccc}
\text{ShE} & \xrightarrow{\hat{F}} & Ch_2B \\
{\scriptstyle P''}\Big\downarrow & {\nearrow}\atop{\scriptstyle \delta} & \Big\downarrow{\scriptstyle H'} \\
E & \dashrightarrow[S] & B
\end{array}
\qquad (4.15)
$$

amounts to giving a connected ∂-sequence (F, S, ∂) starting with F, where ∂_a is associated to δ_a in the following obvious way (and the sequence is connected, by Lemma 4.1.6(A))

$$SA' \to SA \to SA'' \to FA' \to FA \to FA'',$$
$$\partial_a = (SA'' \to \text{Ker } Fa' \rightarrowtail FA'). \qquad (4.16)$$

(It would seem simpler to use $P': Ch_2B \to B$ in (4.15), instead of H'. But the present construction will lead more directly to a limit procedure for the satellite.)

Lemma 4.1.6(B) now says that the left satellite (S, ∂) of F can be equivalently expressed in this way:

(a) $S: E \to B$ is an N-functor and $\delta: SP'' \to H'\hat{F}$ a natural transformation,

(b) for every N-functor $T: E \to B$ and natural transformation $\delta_T: TP'' \to H'\hat{F}$ there is a unique natural transformation $\varphi: T \to S$ such that $\delta_T = \delta.(\varphi P'')$

$$
\begin{array}{ccc}
\text{ShE} & \xrightarrow{\hat{F}} & Ch_2B \\
{\scriptstyle P''}\Big\downarrow & {\nearrow}\atop{\scriptstyle \delta} & \Big\downarrow{\scriptstyle H'} \\
E & \dashrightarrow[{\underset{T}{\overset{S}{\uparrow\varphi}}}] & B
\end{array}
\qquad (4.17)
$$

In other words, (S, ∂) is the left satellite of F if and only if (S, δ) is the *right Kan extension* of $H'\hat{F}$ along P'', *within N-functors* (see [M5], Chapter X).

4.1.8 Calculus by limits

This suggests us to give a construction of the left satellite by means of limits in the category B (see A2.2).

Given an object X in E, let us 'lift' it in ShE by means of the *comma category* $(X \downarrow P'')$ of the functor P'': ShE $\to$ E. An object of $(X \downarrow P'')$ is a pair (a, x) where a is an object of ShE and $x\colon X \to A''$ is a map of E with values in $A'' = P''(a)$, as in the upper row of the following diagram

$$
\begin{array}{ccccccc}
A' & \rightarrowtail & A & \xrightarrow{a''} & A'' & \xleftarrow{x} & X \\
{\scriptstyle f'}\downarrow & & \downarrow{\scriptstyle \bar{f}} & & \downarrow{\scriptstyle f''} & & \| \\
C' & \underset{c'}{\rightarrowtail} & C & \xrightarrow{c''} & C'' & \xleftarrow{y} & X
\end{array}
\qquad (4.18)
$$

A morphism $f\colon (a, x) \to (c, y)$ of $(X \downarrow P'')$ is defined by a map $f = (f', \bar{f}, f'')\colon a \to c$ in ShE such that $(P''f).x = f''.x = y$, as in the (commutative) diagram above. The composition is the 'same' of ShE.

Consider also the functor

$$
F_X = H'\hat{F}U_X\colon (X \downarrow P'') \to \text{B}, \qquad (a, x) \mapsto \operatorname{Ker} F(a'), \qquad (4.19)
$$

where $U_X\colon (X \downarrow P'') \to$ ShE is the forgetful functor $(a, x) \mapsto a$.

Assume now that

(i) (*limit condition*) for every X, this functor F_X has a limit $S(X)$ in B, with the following cone of 'projections' (for $f\colon (a, x) \to (c, y)$)

$$
S(X) = \lim F_X, \qquad p(a, x)\colon S(X) \to \operatorname{Ker} F(a'), \qquad (4.20)
$$

$$
\begin{array}{ccc}
SX & \xrightarrow{p(a,x)} & F_X(a, x) = \operatorname{Ker} F(a') \\
 & {\scriptstyle p(c,y)}\searrow & \downarrow{\scriptstyle F_X(f)=\operatorname{Ker}(Ff',F\bar{f})} \\
 & & F_X(c, y) = \operatorname{Ker} F(c')
\end{array}
$$

(ii) (*annihilation condition*) for every null object Z of E, $S(Z)$ is null in B.

Note that the completeness of B is neither sufficient nor necessary to ensure the existence of the limits above; it does not suffice, because the indexing categories $(X \downarrow P'')$ are not small, in general; it is not necessary, since these categories (or at least the diagrams F_X) can be very simple in particular situations - they can even have an initial object (see 4.2.2, 4.2.5 and 4.4.5-4.4.7).

Then the satellite (S, ∂) exists and will be called a *pointwise* satellite, because of its 'local computation'. (It is indeed a pointwise Kan extension, cf. [M5], Ch. X, Thm. 5.3). The functor $S \colon \mathsf{E} \to \mathsf{B}$, already defined for objects, takes the map $u \colon X \to Y$ of E to the unique map $Su \colon SX \to SY$ such that

$$p(a, y).Su = p(a, yf) \colon SX \to \operatorname{Ker} F(a'), \tag{4.21}$$

$$
\begin{array}{ccc}
SX & \xrightarrow{\;Su\;} & SY \\
& \searrow{\scriptstyle p(a,yu)} & \Big\downarrow{\scriptstyle p(a,y)} \\
& & \operatorname{Ker} F(a')
\end{array}
\qquad (A' \overset{a'}{\rightarrowtail} A \overset{a''}{\twoheadrightarrow} A'' \overset{y}{\leftarrow} Y \overset{u}{\leftarrow} X).
$$

The cone (4.20) yields the components of the natural transformations $\delta \colon SP'' \to H'\hat{F}$ and $\partial \colon SP'' \to FP'$

$$
\begin{aligned}
\delta_a &= p(a, 1_{A''}) \colon SA'' \to \operatorname{Ker} Fa', \\
\partial_a &= (SA'' \to \operatorname{Ker} Fa' \rightarrowtail FA').
\end{aligned}
\tag{4.22}
$$

Note that, as in (4.22), every short exact sequence a of E determines an object $(a, 1_{A''})$ of the category $(A'' \downarrow P'')$; we shall often identify these things in the sequel.

4.1.9 Proposition (The additive case)

Let $F \colon \mathsf{E} \to \mathsf{B}$ be an additive functor between additive semiexact categories. Then ShE is an additive category and $H'\hat{F} \colon \mathsf{ShE} \to \mathsf{B}$ an additive functor. If F has a pointwise satellite S, the latter is also additive.

Proof Using the definition of S on maps (see (4.21)), it is sufficient to show that the projections $p(a, x) \colon S(X) \to \operatorname{Ker} Fa'$ are *additive* in the second variable

$$p(a, x) + p(a, x') = p(a, x + x') \qquad (A' \twoheadrightarrow A \twoheadrightarrow A'' \leftarrow X). \tag{4.23}$$

Using the exactness of the biproduct functor (cf. 2.4.2), we consider the short exact sequence $a \oplus a$, together with the map $z \colon X \to A'' \oplus A''$ of components x, x', and form the commutative diagram:

$$\begin{array}{ccccccc}
A' \oplus A' & \xrightarrow{a' \oplus a'} & A \oplus A & \xrightarrow{a'' \oplus a''} & A'' \oplus A'' & \xleftarrow{z} & X \\
\scriptstyle f' \downarrow & & \scriptstyle \bar{f} \downarrow & & \scriptstyle f'' \downarrow & & \| \\
A' & \xrightarrow{a'} & A & \xrightarrow{a''} & A'' & \xleftarrow{w} & X
\end{array} \qquad (4.24)$$

where the map $f : a \oplus a \to a$ is *either* the first projection $p = (p', \bar{p}, p'')$, *or* the second projection q, *or* the codiagonal $\partial = p + q$; accordingly, $w = f''z$ is *either* $p''z = x$, *or* $q''z = x'$, *or* $\partial''z = (p + q).z = x + x'$.

Then, by the consistency of the cone (cf. (4.20))

$$p(a, x + x') = F_X(\partial).p(a \oplus a, z)$$
$$= F_X(p).p(a \oplus a, z) + F_X(q).p(a \oplus a, z) = p(a, x) + p(a, x').$$

$\square$

4.2 Calculus of satellites

Sufficient conditions for the existence of pointwise left satellites and for their calculus are given, using 'sufficient' normally projective objects.

E and B are always semiexact categories, and B has an initial object, necessarily null (cf. 1.5.3(h)).

4.2.1 Normally projective objects

The notion we now introduce corresponds to the classical definition of a projective object, replacing everywhere 'epimorphism' with 'normal epi'; if the category is g-exact, the two notions coincide (by 2.2.6).

Accordingly, a *normally projective* object P of the semiexact category E is such that, in the left diagram below, for every *normal* epi p, every map f has some lifting g $(pg = f)$

$$\begin{array}{ccc}
 & P & \qquad A \xrightarrow{m} B \\
\scriptstyle g \nearrow & \downarrow \scriptstyle f & \qquad \quad \searrow \downarrow \scriptstyle g \\
A \xrightarrow{p} B & & \scriptstyle f \qquad Q
\end{array} \qquad (4.25)$$

The dual definition of a *normally injective* object Q is described at the right. Normally projective (resp. injective) objects are closed under existing sums (resp. products).

Given a short exact sequence (i, q) whose third object B is normally projective, the normal epi q is a retraction, i.e. there exists some j such that $qj = 1$

$$A \underset{p}{\overset{i}{\rightleftarrows}} C \underset{j}{\overset{q}{\rightleftarrows}} B \quad . \tag{4.26}$$

Now, if E is additive semiexact, it is easy to see that the sequence *splits* in a biproduct (cf. (2.23)). (Define $f = 1_C - jq \colon C \to C$; then $qf = 0$ and $f = ip$ for a unique $p \colon C \to A$; finally $ipj = fj = 0$, and the rest follows trivially.)

The semiexact category E will be said to have *sufficient normally projective objects* if every object X is a *normal quotient* of some normally projective object, $P \twoheadrightarrow X$.

Equivalently, every object X can be embedded in a short exact sequence $p = (p', p'')$, with P normally projective

$$M \overset{p'}{\rightarrowtail} P \overset{p''}{\longrightarrow} X. \tag{4.27}$$

This sequence will be called a *left semiresolution* of X, and a *normally projective resolution* (of length 1) if M too is normally projective.

4.2.2 Theorem

Let $F \colon \mathsf{E} \to \mathsf{B}$ be an N-functor between semiexact categories. We assume that E has sufficient normally projective objects and that

(a) given two maps $f, g \colon a \to c$ of short exact sequences of E, with $P''f = P''g$

$$\begin{array}{ccccc}
A' & \overset{a'}{\rightarrowtail} & A & \overset{a''}{\longrightarrow} & A'' \\
\Vert\!\Vert & & \Vert\!\Vert & & \downarrow{\scriptstyle P''f = P''g} \\
C' & \underset{c'}{\rightarrowtail} & C & \underset{c''}{\longrightarrow} & C''
\end{array} \tag{4.28}$$

if A is normally projective then $H'\hat{F}(f) = H'\hat{F}(g)$.

Then F has a pointwise left satellite, that can be computed via any left semiresolution $p = (p', p'')$ of X (with P normally projective, see (4.27))

$$S(X) = \operatorname{Ker}(Fp' \colon FM \to FP) \qquad (M \rightarrowtail P \twoheadrightarrow X). \tag{4.29}$$

The projection relative to an object (a, x) of $(X \downarrow P'')$ is computed on any map $f \colon (p, 1_X) \to (a, x)$ in $(X \downarrow P'')$

$$p(a, x) = H'\hat{F}(f \colon p \to a) \colon SX \to \operatorname{Ker} Fa', \tag{4.30}$$

$$SX \rightarrowtail FM \xrightarrow{Fp'} FP \xrightarrow{Fp''} FX$$

$$p(a,x) \downarrow \qquad \downarrow Ff' \qquad \downarrow \qquad \downarrow \qquad .$$

$$\mathrm{Ker}\,Fa' \rightarrowtail FA' \xrightarrow{Fa'} FA \xrightarrow{Fa''} FA''$$

The natural transformations $\delta\colon SP'' \to H'\hat{F}$ and $\partial\colon SP'' \to FP'$ are defined as follows, for $f\colon (p,1_{A''}) \to (a,1_{A''})$

$$\delta_a = p(a, 1_{A''})\colon SA'' \to KerFa', \tag{4.31}$$

$$SA'' \rightarrowtail FM \longrightarrow FP \longrightarrow FA''$$

$$\delta_a \downarrow \quad \diagdown{}_{\partial_a} \quad \downarrow Ff' \qquad \downarrow \qquad \parallel \qquad .$$

$$\mathrm{Ker}\,Fa' \rightarrowtail FA' \xrightarrow{Fa'} FA \xrightarrow{Fa''} FA''$$

Proof Given X in E, every left semiresolution $p = (p',p'')$ of X yields an object $(p,1_X)$ of the comma category $(X \downarrow P'')$, that is *weakly initial* in the latter and *initial* for the functor F_X.

Indeed, for any object (a,x) of $(X \downarrow P'')$, there is always *some* map $f\colon (p,1_X) \to (a,x)$

$$M \overset{p'}{\rightarrowtail} P \xrightarrow{p''} X \overset{\displaystyle=\!=}{} X$$

$$P'f \downarrow \qquad \downarrow \qquad \downarrow P''f \qquad \parallel \qquad P''f = x\,.$$

$$A' \overset{}{\underset{a'}{\rightarrowtail}} A \xrightarrow{a''} A'' \overset{x}{\longleftarrow} X$$

Moreover, if g is another such map, then $P''g = x = P''f$ and P is normally projective; by hypothesis (a)

$$F_X(f) = H'\hat{F}U_X(f) = H'\hat{F}(f\colon p \to a) = H'\hat{F}(g\colon p \to a)$$
$$= H'\hat{F}U_X(g) = F_X(g).$$

Therefore the limit condition 4.1.8(i) is satisfied by taking for $S(X)$ the value of F_X on any initial object $(p,1_X)$ and for projection $p(a,x)$ the value of F_X on any map $f\colon (p,1_X) \to (a,x)$, as specified in the statement.

The annihilation condition 4.1.8(ii) also holds: the transformation S preserves the null objects, since if X is null, the normal monomorphism $p'\colon M \rightarrowtail P$ is an isomorphism. By 4.1.8, S is the left satellite of F, with map-action consistent with the projections (cf. (4.21)) and differential as in (4.31). $\square$

4.2.3 Corollary

In the same hypotheses, the left satellite $S = S_1F$ annihilates on every normally projective object. Moreover, if E *satisfies the following condition:*

(np1) every normal subobject of a normally projective object is normally projective,

then all the satellites of F after the first are null functors.

Proof If P is normally projective, its semiresolution $P_0 \rightarrowtail P = P$ gives $S(P) = \mathrm{Ker}\,(FP_0 \to FP)$, a null object of B.

If (np1) is satisfied, the satellite of any N-functor F that annihilates on normally projective objects is null, by formula (4.29) (where M is normally projective); in particular, this holds if F itself is a left satellite, by the previous argument. $\qquad\square$

4.2.4 Theorem (The satellite of a right exact functor)

(a) In the same hypotheses of Theorem 4.2.2, assume moreover that E *is homological,* B *is modular semiexact and F is right exact. Then, for every short exact sequence $a = (a', a'')$ in* E, *the associated six-term sequence of* B

$$
\begin{array}{ccc}
S(A') \xrightarrow{\;Sa'\;} & S(A) \xrightarrow{\;Sa''\;} & S(A'') \\
& \searrow{\scriptstyle \partial} & \\
F(A') \xrightarrow[Fa']{} & F(A) \xrightarrow[Fa'']{} & F(A'')
\end{array}
\tag{4.32}
$$

is exact in each term, possibly excepting SA.

(b) If E *also satisfies the condition (np1) of 4.2.3, the previous sequence is exact and begins with a normal mono.*

(c) More generally and in both cases, we can replace the hypothesis that E *be homological with:* E *is right ex2 (cf. 2.3.2) and satisfies (ex3), as happens for* Gp *and* Rng.

Proof (a) We have to prove that (4.32) is exact in FA' and in SA''.

Let $P \twoheadrightarrow A$ be a normal epi. By (ex2), the composite $P \twoheadrightarrow A \twoheadrightarrow A''$ is also a normal epi; by (ex3) we can form the left commutative diagram below, with short exact rows and columns

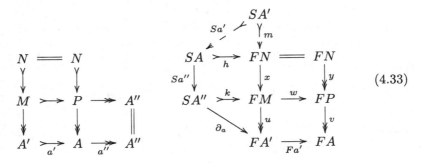

$$(4.33)$$

We use it to calculate both

$$SA = \mathrm{Ker}\,(FN \to FP), \qquad SA'' = \mathrm{Ker}\,(FM \to FP).$$

The right exact functor F produces in B the solid right diagram above, with exact columns and $h = \ker y$, $k = \ker w$. Now, the thesis follows from the modularity of B

$$\mathrm{nim}\,Sa'' = k^*k_*(Sa'')_*1 = k^*x_*h = k^*x_*y^*0 = k^*x_*x^*w^*0 = k^*x_*x^*k$$
$$= k^*(k \wedge x_*1) = (k^*k) \wedge (k^*u^*0) = k^*u^*0 = \ker \partial_a,$$

$$\mathrm{nim}\,\partial_a = u_*(k) = u_*w^*0 = u_*(w^*0 \vee u^*0) = u_*(w^*0 \vee x_*1)$$
$$= u_*w^*w_*x_*1 = u_*w^*y_*1 = u_*w^*v^*0 = u_*u^*\ker Fa' = \ker Fa'.$$

(b) By (np1), M is normally projective and SA' is the kernel of $x \colon FN \to FM$; we can complete the right diagram above with $m = \ker x \leqslant \ker wx = h$. The induced map Sa' is a normal mono, by 1.5.7(b), and

$$\mathrm{nim}\,Sa' = h^*h_*(Sa')_*1 = h^*m = h^*x^*0 = (Sa'')^*k^*0 = \ker Sa''.$$

$$\square$$

4.2.5 Theorem (The additive case)

(a) Let E *and* B *be additive semiexact categories (see 2.4.2). If* E *has sufficient normally projective objects, then every additive functor* $F \colon$ E $\to$ B *has all left satellites, that are pointwise and additive.*

(b) If moreover E *and* B *are abelian and* F *is half exact (as defined in 4.1.1), the ∂-sequence of left satellites is exact.*

Proof (a) To apply Theorem 4.2.2, let us prove that the condition 4.2.2(a) follows from the present additivity hypotheses.

Indeed, given two maps $f, g \colon a \to c$ of short exact sequences, with $P''f = P''g$ (see (4.28)), their difference is a map of short exact sequences $h = f - g \colon a \to c$, with $P''h = 0$, as in the left solid diagram below

$$
\begin{array}{ccccc}
A' & \xrightarrow{a'} & A & \xrightarrow{a''} & A'' \\
\scriptstyle h' \downarrow & \scriptstyle s \swarrow & \downarrow \bar{h} & & \downarrow 0 \\
C' & \xrightarrow[c']{} & C & \xrightarrow[c'']{} & C''
\end{array}
\qquad
\begin{array}{ccccccc}
\operatorname{Ker} Fa' & \xrightarrow{k} & FA' & \xrightarrow{Fa'} & FA & \xrightarrow{Fa''} & FA'' \\
\scriptstyle H'Fh' \downarrow & & \scriptstyle Fh' \downarrow & \scriptstyle Fs \nearrow & \downarrow F\bar{h} & & \downarrow 0 \\
\operatorname{Ker} Fc' & \xrightarrow[k']{} & FC' & \xrightarrow[Fc']{} & FC & \xrightarrow[Fc'']{} & FC''
\end{array}
\;.
$$

But $c''.\bar{h} = 0$, so that $\bar{h}$ factorises as $c's$, and the complete left diagram commutes (c' is mono). Transforming it by the functor F, we get the right commutative diagram above, in B, with $k'.H'(Fh') = Fs.(Fa'.k) = 0$, hence $H'(Fh') = 0$. By the additivity of F and $H'\hat{F}$

$$
H'\hat{F}(f) - H'\hat{F}(g) = H'\hat{F}(f - g) = H'(Fh') = 0.
$$

Therefore, by Theorem 4.2.2, F has a pointwise left satellite S, that is additive by 4.1.9. By induction, F has all left satellites, that are pointwise and additive.

(b) See [CE], Ch. III. The proof uses deeply the abelian properties of E.

$\square$

4.2.6 Satellite dimension

We say that the semiexact category E has *ls-dimension* n ($\leqslant \infty$), or left-satellite dimension n, if:

(a) every N-functor $F \colon \mathsf{E} \to \mathsf{B}$ with values in a *p-semiexact* category B has all left satellites $S_k F$,

(b) n is the least extended integer such that $S_k F$ is null, for any such F and any $k > n$.

The restriction to *p-semiexact* categories B is motivated by various facts; for instance, without this condition one can even have non-constant sequences of null satellites (see 4.4.6). See also the following lemma and its proof.

If E is additive semiexact, it can be more interesting to consider its *additive* left-satellite dimension; the definition is analogous, but uses only additive functors F with values in additive semiexact categories B. If both dimensions are defined, the additive one is lesser than or equal to the general one.

Dually, *right* satellites are used to define the *(additive) rs-dimension*.

4.2.7 Lemma (Dimension zero)

Let E *be a semiexact category.*

(a) Every N-functor $F\colon$ E $\to$ B with values in a pointed N-category, that takes normal monos to N-monos, has all left satellites given by the constant zero functor.

(b) Each of the conditions below implies the following one:

(i) every non-null normal mono of E *splits,*

(ii) every N-functor $F\colon$ E $\to$ B with values in a p-semiexact category takes all the normal monos of E *to split monos of* B,

(iii) every N-functor $F\colon$ E $\to$ B as before takes all the normal monos of E *to N-monos,*

(iv) E *has ls-dimension 0.*

(c) If E *has small hom-sets and each sequence of left satellites of a right exact functor F as before is exact, all these conditions are equivalent.*

Proof (a) It suffices to consider the first satellite. The diagram

$$
\begin{array}{ccccc}
0 & \longrightarrow & F(A') & \xrightarrow{Fa'} & F(A) \\
{\scriptstyle\varphi}\big\uparrow & & {\scriptstyle\varphi}\big\uparrow & & {\scriptstyle\varphi}\big\uparrow \\
G_1(A'') & \xrightarrow[\partial_0]{} & G_0(A') & \xrightarrow[G_0 a']{} & G_0(A)
\end{array}
$$

commutes necessarily, because Fa' is N-mono and $Fa'.\varphi.\partial_0 = \varphi.G^0a'.\partial_0 = 0$.

(b) The implications (i) $\Rightarrow$ (ii) $\Rightarrow$ (iii) are obvious, while (iii) $\Rightarrow$ (iv) follows from the previous point.

(c) Let us assume the stronger hypotheses on E, and prove that (iv) $\Rightarrow$ (i). Given a non-null normal mono $m\colon A' \rightarrowtail A$ of E (i.e. A' is non-null), we want to show that it has a retraction.

Consider the following right exact functor F, with values in the opposite of the p-homological category of pointed sets

$$F\colon \mathsf{E} \to (\mathsf{Set}_\bullet)^{\mathrm{op}},$$

$$F(C) = \mathsf{E}(C, A')/\mathrm{Nul}\mathsf{E}(C, A'), \quad F(f)[g] = f^*[g] = [g.f].$$

F can be seen as the composite of the enriched hom-functor

$$\mathrm{Hom}(-, A')\colon \mathsf{E} \to (\mathsf{Set}_2)^{\mathrm{op}}$$

(cf. 1.4.5) with the opposite of the exact functor $P\colon \mathsf{Set}_2 \to \mathsf{Set}_\bullet$ (cf. (1.114)).

By hypothesis, F has a left satellite S that is null. Then the short exact sequence $(m,p) = (A' \rightarrowtail A \twoheadrightarrow A'')$ produced by m is transformed into an exact sequence of Set.

$$FA'' \to FA \to FA' \to SA''...$$

where SA'' is null and $Fm = m^*\colon FA \to FA'$ is N-epi. In other words, m^* is a surjective pointed mapping and there is some $q \in F(A)$ such that $m^*[q] = [qm] = [1_{A'}] \in F(A')$; since A' is non-null, $1_{A'}$ is not the base point of $F(A')$ and this means indeed that $qm = 1_{A'}$. □

4.2.8 Some examples

(a) The p-homological category Set. of pointed sets has ls- and rs-dimension 0. Indeed, given a short exact sequence $(m,p) = (H \rightarrowtail A \twoheadrightarrow A/H)$, m is a section and p a retraction.

(b) The p-homological category Set_2 of pairs of sets has ls-dimension 0, as every *non-null* normal mono $(A,B) \rightarrowtail (X,B)$, where $B \subset A \subset X$ (proper inclusions) is clearly a section.

(c) The following conditions imply that E has ls-dimension $\leqslant 1$ (apply Theorem 4.2.2 and its corollary):
  - E has sufficient normally projective objects;
  - every N-functor $F\colon E \to B$ with values in a p-semiexact category satisfies the hypothesis 4.2.2(a) of the existence theorem;
  - any normal subobject of a normally projective object of E is again normally projective.
    We shall see in Section 4.4 that every category of pairs satisfies them, and that the ls-dimension of Top_2 and Gp_2 is precisely 1.

4.3 Effacements

In the general setting of semiexact categories, satellites can be identified by means of 'effacements', as in the abelian case studied by Grothendieck [Gt]. The extension is easier in the exact context, i.e. for an exact sequence of functors taking values in a g-exact category (see 4.3.1-4.3.3), more involved in the general setting (4.3.5-4.3.6).

For the sake of variety, we now deal with right satellites; the conversion to left satellites is sketched in 4.3.7. E and B are always semiexact categories and $(F^n, d^n)_{n \in \mathbb{N}}$ is a connected d-sequence of functors from E to B; the bounded case $0 \leqslant n < k$ can be treated in a similar way.

4.3.1 Sufficient effacements in the exact case

Assume first that $F = (F^n, d^n) \colon \mathsf{E} \to \mathsf{B}$ is an *exact* d-sequence of functors taking values in a *g-exact* category B. In order to prove that F is right universal, i.e. coincides with the sequence of right satellites of F^0, it is important to dispose of 'sufficient (right) effacements'; further, they must behave in a 'sufficiently natural' way, that is automatic for a sequence of additive functors between abelian categories (cf. 4.3.2), but not here.

Let us fix an integer $n > 0$. An *n-effacement* of an object A of E, with respect to F, will be a morphism m such that

$$m \colon A \rightarrowtail Q \text{ is a normal mono of E and } F^n m \text{ is null in B}, \qquad (4.34)$$

a condition that is often fulfilled with $F^n(Q)$ null in B.

The naturality problems we meet are expressed in the diagrams below

$$
\begin{array}{ccc}
A & \overset{m}{\rightarrowtail} & Q \\
\| & & \\
A & \overset{m'}{\rightarrowtail} & Q'
\end{array}
\qquad (4.35)
$$

$$
\begin{array}{ccc}
A & \overset{m}{\rightarrowtail} & Q \\
\scriptstyle f \downarrow & & \downarrow \scriptstyle f' \\
B & \overset{m'}{\rightarrowtail} & R
\end{array}
\qquad (4.36)
$$

- given two *n*-effacements of the object A as in (4.35), or more generally two *n*-effacements of A and B and a morphism $f \colon A \to B$ as in (4.36), how *to connect* them.

(a) *Functorial effacements.* The simplest way (also sufficient in various non-abelian cases, as we shall see in Chapter 6) is to assign a *functorial n-effacement* $m_A^n \colon A \rightarrowtail Q^n(A)$, i.e. a functor $Q^n \colon \mathsf{E} \to \mathsf{E}$ and a natural transformation $m^n \colon 1_\mathsf{E} \to Q^n$ whose components are *n*-effacements.

Therefore the problem expressed in (4.35) is bypassed and that of (4.36) is solved in a natural way. If this can be done for all $n > 0$, we say that F *has functorial effacements*; sometimes this goal can even be realised with *one* functorial effacement $m_A \colon A \rightarrowtail Q(A)$, independent of n.

(b) *Connected effacements.* Say that two *n*-effacements of A

$$m \colon A \rightarrowtail Q, \qquad m' \colon A \rightarrowtail Q',$$

are *directly connected* if $m' = gm$ or $m = g'm'$ (for some $g \colon Q \to Q'$ or $g' \colon Q' \to Q$). Say that they are *connected* if there is a finite sequence $m, m_1, \ldots m_r, m'$ of *n*-effacements such that all consecutive pairs $(m, m_1), \ldots (m_r, m')$ are directly connected.

We say that F *has connected effacements* if, for all $n > 0$

(i) every object A in E has some n-effacement, and all of them are connected,

(ii) every morphism $f: A \to B$ in E can be embedded in a commutative square (4.36), where m and m' are convenient n-effacements of A and B.

(c) *Sufficient effacements.* The following formulation extends the previous ones and will allow us to give a single proof for Theorem 4.3.3. Say that F *has sufficient effacements* if, for all $n > 0$

(i′) for A in E there exists (and we have chosen) a *standard* n-effacement $m_A^n: A \rightarrowtail Q$,

(ii′) every morphism $f: A \to B$ in E embeds in a commutative square (4.36), where m and m' are *quasi-standard* n-effacements of A and B (i.e. they are connected to the standard n-effacements).

4.3.2 Two simple cases

The naturality conditions are automatically satisfied in two important cases.

(a) *The abelian case.* If E and B are abelian categories and the functors F^n are additive, the d-sequence F has connected effacements if and only if it has sufficient effacements, if and only if, for each $n > 0$, *every object has some n-effacement.* Indeed, given the left diagram

$$
\begin{array}{ccc}
A & \xrightarrow{\;m\;} & Q \\
{\scriptstyle f}\downarrow & & \\
B & \xrightarrow[m']{} & R
\end{array}
\qquad\qquad
\begin{array}{ccc}
A & \xrightarrow{\;m''\;} & Q \times R \\
{\scriptstyle f}\downarrow & & \downarrow{\scriptstyle q} \\
B & \xrightarrow[m']{} & R
\end{array}
\tag{4.37}
$$

where m and m' are arbitrary n-effacements, form the right commutative diagram, where q is the second projection and $m'' = (m, m'f)$; then m'' is mono (because m is), hence a normal mono, and $F^n m'' = F^n(m, m'f) = (F^n m, F^n(m'f)) = (0, 0) = 0$ (because an additive functor preserves products); m'' is thus an effacement of A, connected to m via the first projection $Q \times R \to Q$. This proves that all the n-effacements of A are connected (take $A = B$ and $f = 1$), as well as the property 4.3.1(ii).

(b) *Normally injective effacements.* The sequence F is said to have *sufficient normally injective effacements* if every object A has a *normally injective* n-effacement $A \rightarrowtail Q$ (meaning that Q is normally injective, 4.2.1), for all $n > 0$. In such a case every left diagram (4.37) with R normally injective can be completed to a commutative square; hence, again, all the n-effacements of A are connected and 4.3.1(ii) holds: F has connected effacements.

If E has sufficient normal injectives, i.e. every object can be normally embedded in a normally injective object (cf. 4.2.1), and further all the functors F^n $(n > 0)$ annihilate on the normally injective objects of E, then the sequence F has sufficient normally injective effacements.

4.3.3 Theorem (Sufficient effacements, in the exact case)

Let an exact *sequence of functors* $F = (F^n, d^n)$: E $\to$ B *be defined on an* ex2-*category* E, *with values in a* g-exact *category* B. *If the sequence F has* sufficient effacements *(see 4.3.1(c)), then it is right universal and* $S^n F^0 = F^n$.

Proof Take a connected sequence (G^n, δ^n): E $\to$ B and a natural transformation $\varphi^0 \colon F^0 \to G^0$; assume we have already built $\varphi^p \colon F^p \to G^p$, for $0 \leqslant p < n$, satisfying the required naturality conditions and let us prove that there exists a unique natural transformation φ^n that takes on the process.

(a) *Uniqueness and construction.* The standard n-effacement of the object A yields a short exact sequence

$$(m, p) = (A \rightarrowtail Q \twoheadrightarrow A'),$$

$$m = m_A^n, \qquad p = \operatorname{cok} m, \qquad F^n m \text{ null},$$

(4.38)

and a commutative solid diagram with exact rows, in B

$$
\begin{array}{ccccccc}
F^{n-1}Q & \xrightarrow{F^{n-1}p} & F^{n-1}A' & \xrightarrow{d^{n-1}} & F^n A & \xrightarrow{null} & F^n Q \\
{\scriptstyle \varphi^{n-1}Q}\Big\downarrow & & {\scriptstyle \varphi^{n-1}A'}\Big\downarrow & & \Big\downarrow{\scriptstyle \varphi^n A} & & \\
G^{n-1}Q & \xrightarrow[G^{n-1}p]{} & G^{n-1}A' & \xrightarrow[\delta^{n-1}]{} & G^n A & &
\end{array}
$$

(4.39)

Since B is g-exact, d^{n-1} is a cokernel of $F^{n-1}p$; since

$$\delta^{n-1}.\varphi^{n-1}A'.F^{n-1}p = (\delta^{n-1}.G^{n-1}p).\varphi^{n-1}Q$$

is null, there exists precisely one morphism $\varphi^n A \colon F^n A \to G^n A$ that makes the diagram commutative. This argument defines $\varphi^n A$ and proves at the same time that it is uniquely determined by φ^{n-1}, by means of the naturality conditions.

(b) We now prove that any quasi-standard n-effacement of A (connected to the standard one) can be equally used for the definition of $\varphi^n A \colon F^n A \to G^n A$. Indeed, if $A \rightarrowtail Q'$ and $A \rightarrowtail Q''$ are quasi-standard effacements of A, let $u, v \colon F^n A \to G^n A$ be the morphisms of B that they produce by means of the previous procedure (with diagrams analogous to (4.38) and (4.39)). Plainly, we may assume that these effacements are directly connected and embed them in the following commutative diagram, with short exact rows

$$
\begin{array}{ccccc}
A & \rightarrowtail & Q' & \twoheadrightarrow & A' \\
\| & & \downarrow{\scriptstyle g} & & \downarrow{\scriptstyle g'} \\
A & \rightarrowtail & Q'' & \twoheadrightarrow & A''
\end{array}
\quad . \tag{4.40}
$$

This yields the diagram below, in B

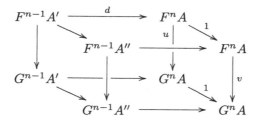

All its squares *except the right one* are already known to commute, by the naturality of φ^{n-1} (left square), of d^{n-1} upper square) and δ^{n-1} (lower square) or by the definition of u and v (back and front square). Since $d^{n-1}\colon F^{n-1}A' \to F^n A$ is epi, the right square also commutes, because of an obvious 'sixth face lemma' for cubical diagrams, that we state below (in 4.3.4). Therefore, $u = v$.

(c) *Naturality on morphisms.* The proof of the naturality of $\varphi^n\colon F^n \to G^n$ is an easy extension of the previous argument. Given a morphism $f\colon A \to B$ in E, by hypothesis 4.3.1(ii') we can embed it in the following commutative diagram, where the horizontal short exact rows are given by quasi-standard effacements m, m' of A and B.

$$
\begin{array}{ccccc}
A & \overset{m}{\rightarrowtail} & Q & \twoheadrightarrow & A' \\
\downarrow{\scriptstyle f} & & \downarrow{\scriptstyle g} & & \downarrow{\scriptstyle g'} \\
B & \underset{m'}{\rightarrowtail} & R & \twoheadrightarrow & B'
\end{array}
\quad . \tag{4.41}
$$

This produces a cube in B, whose morphism $d^{n-1}\colon F^{n-1}A' \to F^n A$ is epi;

again by the sixth face lemma, its right square also commutes

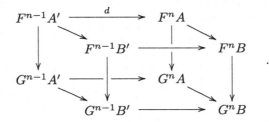

(d) *Naturality on short exact sequences.* Given a short exact sequence $A \rightarrowtail B \twoheadrightarrow C$ in E, the normal mono $f \colon A \rightarrowtail B$ can be embedded in a diagram (4.41), where $m \colon A \rightarrowtail Q$ and $m' \colon B \rightarrowtail R$ are effacements connected to the standard ones.

Now, the composition $m'' = m'f \colon A \to R$ is also a quasi-standard n-effacement of A. Indeed, it is a normal mono, by ex2; $F(m'') = Fm'.Ff$ is null because Fm' is; finally, it is directly connected to $m \colon A \rightarrowtail Q$, via the morphism g. In this way we form the following commutative diagram, with short exact rows

$$\begin{array}{ccccc} A & \overset{f}{\rightarrowtail} & B & \twoheadrightarrow & C \\ \| & & \downarrow{\scriptstyle m'} & & \downarrow \\ A & \underset{m''}{\rightarrowtail} & R & \twoheadrightarrow & X \end{array} \qquad (4.42)$$

This gives the following cube, where the back square is commutative, again by Lemma 4.3.4

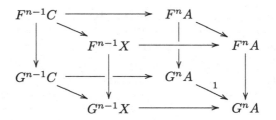

Indeed 1: $G^n A \to G^n A$ is mono, while the commutativity of the other five squares follows from the naturality of φ^{n-1} and φ^n (left and right square), from the connected sequences F and G (upper and lower square), or from the definition of φ^n (front square, because $m'' \colon A \rightarrowtail R$ is a quasi-standard effacement of A). $\qquad \square$

4.3.4 Lemma (The sixth face lemma)

Given a cubical diagram in an arbitrary category C

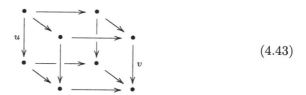

$$(4.43)$$

(a) if all its squares except possibly the upper one are commutative and v is mono, the upper face also commutes,

(a^*) if all its squares except possibly the lower one are commutative and u is epi, the lower face also commutes.

Proof Obvious. This trivial fact has been frequently used in the proof of the preceding theorem. $\qquad\square$

4.3.5 Effacements in the general case

More generally, we consider now a *connected* d-sequence $F = (F^n, d^n)_{n \in \mathbb{N}}$ between *semiexact* categories E and B.

An *n-effacement* of an object A of E, with respect to F, is now a *normal mono* $m\colon A \rightarrowtail Q$ producing a short exact sequence of E

$$(m, p) = (A \rightarrowtail Q \twoheadrightarrow A'), \qquad p = \operatorname{cok} m, \qquad (4.44)$$

such that, in the associated order two F-sequence, we have

$$\ldots F^{n-1}Q \xrightarrow{F^{n-1}p} F^{n-1}A' \xrightarrow{d^{n-1}} F^n A \xrightarrow{F^n m} F^n Q \ldots \qquad (4.45)$$

$$d^{n-1} = \operatorname{cok} F^{n-1}(p).$$

If the connected sequence F is *exact*, this condition is equivalent to

$$F^n m \text{ is null and } d^{n-1}\colon F^{n-1}A' \to F^n A \text{ is an exact morphism}, \quad (4.46)$$

so that, when the category B is also g-exact, we come back to the original formulation of (4.34): the morphism $F^n m$ is null.

The notions of *directly connected* and *connected* pairs of n-effacements of A are the same as previously, in 4.3.1(b).

We say that F has *sufficient effacements* if, for all $n > 0$:

(i) for every A in E there exists (and we have chosen) a *standard n-effacement* $m^n_A\colon A \rightarrowtail Q$,

(ii) every morphism $f: A \to B$ in E embeds in a commutative square

$$
\begin{array}{ccc}
A & \overset{m}{\rightarrowtail} & Q \\
f\downarrow & & \downarrow f' \\
B & \underset{m'}{\rightarrowtail} & R
\end{array}
\qquad (4.47)
$$

where m and m' are *quasi-standard* n-effacements of A and B (i.e. connected to the standard ones),

(iii) every normal mono $f: A \rightarrowtail B$ in E embeds in such a square (4.47), where moreover the diagonal $m'f = f'm$ is again an n-effacement of A (directly connected to m via f').

(If E is ex2, this diagonal is necessarily a normal mono. But we are not able to prove that it is necessarily an effacement, unless B is g-exact and F is exact.)

Similarly, as in 4.3.1, we can define the stronger conditions: *F has functorial, or connected, or normally injective effacements.* Clearly, if all the functors F^n (for $n > 0$) annihilate the normally injective objects of E, the last condition just means that every object A has a normal embedding in a normally injective object.

4.3.6 Theorem (Sufficient effacements in the semiexact case)

Let $F = (F^n, d^n): E \to B$ be a connected d-sequence between semiexact categories. If the sequence F has sufficient effacements (4.3.5), then it is right universal and $S^n F^0 = F^n$, for all $n \geqslant 0$.

Proof It is practically the same proof of 4.3.3, with some obvious and minor adaptations. In part (a), where $d^{n-1}: F^{n-1}A' \to F^n A$ is the cokernel of $F^{n-1}p$ because of the hypothesis (4.45); in part (d), where the condition 4.3.5(iii) is used for the construction of the diagram (4.42). $\square$

4.3.7 Effacements for ∂-sequences

Dually, we consider a connected ∂-sequence between semiexact categories $F = (F_n, \partial_n)_{n \in \mathbb{N}}: E \to B$.

An n-*effacement* of an object A of E, with respect to F, is a normal epimorphism $p: P \twoheadrightarrow A$ producing a short exact sequence of E

$$
(m, p) = (A' \rightarrowtail P \twoheadrightarrow A), \qquad m = \ker p, \qquad (4.48)
$$

such that, in the associated order two F-sequence, we have

$$\dots F_n P \xrightarrow{F_n p} F_n A \xrightarrow{\partial_n} F_{n-1} A' \xrightarrow{F_{n-1} m} F_{n-1} P \dots \qquad (4.49)$$

$$\partial_n = \ker F_{n-1}(m).$$

If the connected sequence F is *exact*, this condition is equivalent to

$$F_n p \text{ is null and } \partial_n \colon F_n A \to F_{n-1} A' \text{ is an exact morphism.} \qquad (4.50)$$

If the category B is also g-exact, it suffices to say that $F_n p$ is null.

We say that the connected ∂-sequence $F = (F_n, \partial_n)$ *has sufficient efface-ments* if, for all $n > 0$:

(i) for every A in E there exists (and we have chosen) a standard n-effacement $p_A^n \colon P \twoheadrightarrow A$,

(ii) every morphism $f \colon A \to B$ in E embeds in a commutative square as below, where p and p' are quasi-standard n-effacements of A and B (i.e. connected to the standard ones).

$$\begin{array}{ccc} P & \xrightarrow{\;p\;} & A \\ {\scriptstyle f'}\big\downarrow & & \big\downarrow{\scriptstyle f} \\ P' & \xrightarrow[\;p'\;]{} & B \end{array} \qquad (4.51)$$

(iii) every normal epi $f \colon A \twoheadrightarrow B$ in E embeds in such a square, where moreover the diagonal $fp = p'f'$ is again an n-effacement of B (directly connected to p' via f'). (This condition is superfluous if F is exact and B is g-exact.)

If this is the case, the ∂-sequence F is right universal and $S_n F_0 = F_n$, for all $n \geqslant 0$.

More particularly, one can define *functorial* or *connected* or *normally projective* effacements for ∂-sequences.

The functor $P \colon R\mathsf{Mod} \to R\mathsf{Mod}$ that associates to an R-module A the free R-module $P(A)$ on its underlying set, together with the usual trans-formation $pA \colon P(A) \twoheadrightarrow A$ is plainly a functorial projective effacement (in-dependent of degree) for any sequence of left derived functors on $R\mathsf{Mod}$.

4.4 Satellites for categories of pairs

For categories of pairs, a connected sequence of functors can be assigned in the usual reduced form, concerning the canonical short exact sequences of type $A \rightarrowtail X \twoheadrightarrow (X, A)$, as in Eilenberg-Steenrod's axioms for homology theories.

We also show that the categories of pairs have ls-dimension 0 or 1 and that all N-functors having such a domain have a left satellite of fairly easy calculation (see (4.64)). Categories of pairs provide thus an example of a fully computable situation that - generally - is not additive, nor even pointed or quasi-pointed.

C is always a ds-category, C_2 its semiexact category of pairs (Section 2.5), while B is an N-category.

4.4.1 The topological case

Recall that in Eilenberg-Steenrod [ES] the data for a homology theory (F_n, ∂_n) over Top_2 include the differential of any pair (X, A)

$$\partial_n \colon F_n(X, A) \to F_{n-1}(A). \tag{4.52}$$

It corresponds to a short exact sequence of a *particular type*, the canonical resolution of the pair (X, A)

$$(A, \emptyset) \rightarrowtail (X, \emptyset) \twoheadrightarrow (X, A). \tag{4.53}$$

This differential is successively extended to a triple $X \supset A \supset B$

$$\partial_n \colon F_n(X, A) \to F_{n-1}(A, B)$$
$$((F_n(X, A) \to F_{n-1}A \to F_{n-1}(A, B))), \tag{4.54}$$

and the triple corresponds to an arbitrary short exact sequence of Top_2

$$(A, B) \rightarrowtail (X, B) \twoheadrightarrow (X, A). \tag{4.55}$$

Finally, the exactness of the homology sequence of the pair implies the similar property for the triple (in presence of the functoriality and naturality axioms but without using the other axioms, of homotopy, excision and dimension).

We now extend this fact to an arbitrary connected (resp. exact) sequence, defined over the semiexact category of pairs $\mathsf{E} = \mathsf{C}_2$ of a ds-category C (Section 2.5), with values in an N-category (resp. modular semiexact category) B.

4.4.2 Reduced connected sequences

Recall that C embeds in C_2 by identifying the object X with the least distinguished subobject $n_X \colon N_X \to X$ of X in C; further, any 'pair' $a \colon A \to$

X of C_2 can be presented as a cokernel $X/A = n_X/n_A$, by a canonical short exact sequence $\mathrm{Sh}(a)$ (cf. 2.5.4)

$$(A \xrightarrow{n_a} X \xrightarrow{p_a} X/A) = \mathrm{Sh}(a), \tag{4.56}$$

$$\begin{array}{ccccc}
A & \xrightarrow{a} & X & \xrightarrow{1} & X \\
\uparrow{\scriptstyle n_A} & & \uparrow{\scriptstyle n_X} & & \uparrow{\scriptstyle a} \\
N_A & \xrightarrow{N_a} & N_X & \xrightarrow{v} & A
\end{array} \,. \tag{4.57}$$

A *reduced* connected (resp. exact) ∂-sequence (F_n, ∂_n) of functors $C_2 \to B$ will be defined by the following data (to be compared with the general definition in 4.1.1):

(a) a sequence of functors $F_n \colon C_2 \to B$,

(b) for every pair $a \colon A \to X$, a sequence of morphisms $\partial_n = \partial_n^a \colon F_n(a) \to F_{n-1}A$, natural on C_2, that forms a sequence of order two (resp. exact) in B

$$\dots F_n(A) \longrightarrow F_n(X) \longrightarrow F_n(a) \xrightarrow{\partial_n} F_{n-1}(A) \dots \tag{4.58}$$

4.4.3 Theorem

A reduced connected sequence $(F_n, \partial_n) \colon C_2 \to B$ can be extended to a complete connected sequence, in a unique way. If the reduced sequence is exact and the N-category B is modular semiexact, the complete sequence is also exact.

Proof An arbitrary short exact sequence (2.33) of C_2 is determined by two distinguished subobjects $a \colon A \to X$ and $b \colon B \to X$, with $a \geqslant b$ in $\mathbf{d}_X$

$$A/B \xrightarrow{m} X/B \xrightarrow{p} X/A. \tag{4.59}$$

We define the differential of this sequence by means of the differential of $\mathrm{Sh}(a)$ and the normal quotient $q \colon A \to A/B$ (i.e. the cokernel of the map $c \colon B \rightarrowtail A$ such that $ac = b$, in C_2)

$$\partial_n = F_{n-1}(q).\partial_n^a = (F_n(X/A) \to F_{n-1}(A) \to F_{n-1}(A/B)). \tag{4.60}$$

Because of the naturality condition, (4.60) is the only possible extension of the reduced differential ∂_n^a. It is thus sufficient to check that the F-sequence associated to (4.59)

$$\dots F_n(A/B) \to F_n(X/B) \to F_n(X/A) \xrightarrow{\partial_n} F_{n-1}(A/B) \dots \tag{4.61}$$

is of order two (and exact when the stronger hypotheses are satisfied). This we do in one point (out of three), $F_n(X/A)$.

In fact, the composition $\partial_n.F_n(p)$ is null, because of the following commutative diagram, whose lower row is of order two, since it comes from the canonical resolution of A/B

$$\begin{array}{ccccc}
F_n(X/B) & \xrightarrow{F_n(p)} & F_n(X/A) & \xrightarrow{\partial_n} & F_{n-1}(A/B) \\
\Big\downarrow{\partial} & & \Big\downarrow{\partial} & & \Big\| \\
F_{n-1}(B) & \longrightarrow & F_{n-1}(A) & \longrightarrow & F_{n-1}(A/B)
\end{array} \quad .$$

We now assume that the reduced sequence is exact and B is modular. Form the following diagram, where $f = F_n p$ and $g = \partial_n$

$$\begin{array}{ccccccc}
F_n(X) & \longrightarrow & F_n(X/B) & \xrightarrow{x} & F_{n-1}(B) & \xrightarrow{w} & F_{n-1}(X) \\
\Big\| & & \Big\downarrow{f} & & \Big\downarrow{u} & & \Big\| \\
F_n(X) & \xrightarrow{y} & F_n(X/A) & \xrightarrow{k} & F_{n-1}(A) & \xrightarrow{v} & F_{n-1}(X) \\
\Big\downarrow{z} & & \Big\| & & \Big\downarrow{h} & & \\
F_n(X/B) & \xrightarrow{f} & F_n(X/A) & \xrightarrow{g} & F_{n-1}(A/B) & &
\end{array} \quad .$$

By hypothesis, the first and second rows are exact, as well as the third column. It follows that the third row is also exact

$$g^*0 = k^*h^*0 = k^*u_*1 = k^*(k_*k^*u_*1) = k^*(u_*1 \wedge k_*1) = k^*(u_*1 \wedge v^*0)$$

$$= k^*(u_*u^*(v^*0)) = k^*(u_*w^*0) = k^*(u_*x_*1) = k^*(k_*f_*1) = f_*1 \vee k^*0$$

$$= f_*1 \vee y_*1 = f_*1 \vee f_*z_*1 = f_*1.$$

$\square$

4.4.4 *Projective pairs*

We calculate here the left satellite for categories of pairs, showing that these have ls-dimension $\leqslant 1$ (cf. 4.2.6) and peculiar left-homological properties, even more elementary than those of Ab, the classical abelian category of projective dimension 1.

For the sake of simplicity, we work first in a *concrete* homological category C_2 of pairs, where C is Top or Set (see 1.6.7). The results can be easily extended to a general category of pairs, as we show below (in 4.4.7); we also show in 4.4.6 that the ls-dimension of Top_2 is indeed 1 (and not 0) and that the present calculations do not become completely trivial in Set_2, even if the latter has ls-dimension 0 (by 4.2.8).

The crucial fact is that the normally projective objects of C_2 are precisely the pairs $X = (X, \emptyset)$, as shown by the following diagrams (respectively for sufficiency and necessity)

$$
\begin{array}{ccc}
& (X, \emptyset) & \\
f' \nearrow & \downarrow f & \\
(Y, C) \rightarrowtail (Y, B) &
\end{array}
\qquad
\begin{array}{ccc}
& (X, A) & \\
\nearrow & \downarrow p & \\
(X, \emptyset) \rightarrowtail (X, X) &
\end{array}
\qquad (4.62)
$$

It follows that, in C_2

(a) every normal subobject of a normally projective pair $(X, \emptyset)$ is normally projective,

(b) the lifting f' of f in the left diagram (4.62) is *uniquely* determined.

Every pair (X, A) is thus a quotient of precisely *one* normally projective object, namely $(X, \emptyset)$, and has precisely one normally projective resolution (up to isomorphism), the canonical one (as in (4.56)), that is of length 1 and functorial in (X, A)

$$
A \xrightarrow{\ n_{(X,A)}\ } X \xrightarrow{\ p_{(X,A)}\ } (X, A) \qquad (4.63)
$$

4.4.5 Left satellites for pairs

Let $F \colon C_2 \to B$ be an N-functor with values in a semiexact category.

We now show that F has all left satellites $(S_n F, \partial_n)$, with $S_n F$ null for $n > 1$. Further, if B is modular semiexact and F preserves the cokernels of the normal monos of type $A \rightarrowtail X$, for all pairs (X, A), the connected sequence is exact.

Indeed, applying 4.1.8, the first satellite $S = S_1 F$ can be computed on the (*unique*) normally projective resolution of a pair, in (4.63)

$$
S \colon C_2 \to B,
$$
$$
S(X, A) = \mathrm{Ker}_B F(n_{(X,A)}) = \mathrm{Ker}_B (F(A, \emptyset) \to F(X, \emptyset)),
$$
$$
(4.64)
$$

and really exists (by 4.1.8(ii)), since this functor S annihilates every null pair (X, X).

The differential $\partial = \partial_1$ of the general short exact sequence (m, p) produced by a triple $X \supset A \supset B$ is expressed as in (4.60), for $k = \ker F(n_{(X,A)})$ and $q \colon A \twoheadrightarrow (A, B)$

$$
\begin{array}{ccc}
(A, B) \xrightarrow{\ m\ } & (X, B) \xrightarrow{\ p\ } & (X, A) \\
q \uparrow & \uparrow & \| \\
A \rightarrowtail & X \longrightarrow & (X, A)
\end{array}
$$

$$\partial(X, A, B) = F(q).k = (S(X, A) \rightarrowtail F(A) \rightarrow F(A, B)). \qquad (4.65)$$

Further, the satellite S annihilates every normally projective pair $(X, \emptyset)$

$$S(X, \emptyset) = \mathrm{Ker}\,(F(\emptyset, \emptyset) \rightarrow F(A, \emptyset)) = 0_{F(\emptyset, \emptyset)}, \qquad (4.66)$$

so that formula (4.64) shows that its satellite is null.

Finally, assume that the more particular hypotheses above are satisfied: B is modular semiexact and F preserves the cokernels of the normal monos $A \rightarrowtail X$. By Theorem 4.4.3 it suffices to check the exactness of (F, S, ∂) over the canonical resolution of the pair (X, A). But this resolution produces a sequence of B

$$S(X, A) \twoheadrightarrow F(A) \twoheadrightarrow F(X) \twoheadrightarrow F(X, A) \qquad (4.67)$$

that is exact in FA by construction of S, and in $F(X)$ by hypothesis.

4.4.6 Some examples

The homological category Top_2 has ls-dimension 1.

Actually, the first left satellite of the singular-homology functor H_0: $\mathsf{Top}_2 \rightarrow \mathsf{Ab}$ is

$$\begin{aligned} SH_0(X, A) &= \mathrm{Ker}\,(H_0(A) \rightarrow H_0(X)) \\ &= \mathrm{Im}\,(H_1(X, A) \rightarrow H_0(A)), \end{aligned} \qquad (4.68)$$

i.e. the subgroup of $H_0(A)$ spanned by the differences $[a] - [a']$, where a and a' are points of A lying in the same pathwise connected component of X. It is non-null and of course the unique non-null satellite of H_0.

Analogously one can show that Gp_2 has ls-dimension 1, by means of the relative homology theory for groups recalled in 6.1.2(d).

Other non-trivial categories of pairs happen to have ls-dimension zero, e.g. Set_2 (Section 4.2.8). Note, however, that a sequence of left satellites defined over Set_2 need not stabilise after the first step, if the codomain is not pointed. For instance, the null endofunctor

$$N^0 : \mathsf{Set}_2 \rightarrow \mathsf{Set}_2, \quad N^0(X, A) = (X, X), \qquad (4.69)$$

has, by 4.4.5, a sequence of left satellites that only stabilises at the second

$$SN^0(X, A) = \mathrm{Ker}\,((A, A) \rightarrow (X, X)) = (A, A) = N_0(X, A), \qquad (4.70)$$

$$SN_0(X, A) = \mathrm{Ker}\,((\emptyset, \emptyset) \rightarrow (\emptyset, \emptyset)) = (\emptyset, \emptyset) = \perp(X, A). \qquad (4.71)$$

The sequence is trivially exact, since it consists of null functors.

4.4.7 Satellites for general categories of pairs

Finally, we sketch the generalisation of the above results (Sections 4.4.4 and 4.4.5) to an arbitrary category of pairs.

The normally projective objects of C_2 are precisely the objects $X = (n_X \colon N_X \to X)$ of nC (same proof as in (4.62)).

Every object $a \colon A \to X$ of C_2 is a quotient of precisely one projective object, X itself, and has precisely one normally projective resolution (up to isomorphism), namely $\mathrm{Sh}(a)$ (cf. 4.4.2), that is of length $\leqslant 1$ and functorial in a.

Every N-functor $F \colon C_2 \to B$ with values in a semiexact category extends to a universal connected ∂-sequence $(S_n F, \partial_n)$, with $S_n F$ null for $n > 1$. The first satellite $S = S_1 F$ is given by

$$S \colon C_2 \to B,$$
$$S(a) = \mathrm{Ker}\,(Fn_a) = \mathrm{Ker}\,{}_{\mathsf E}(F(n_A) \to F(n_X)), \tag{4.72}$$

so that S annihilates on every normally projective pair $X = n_X$.

4.5 Chain homology functors as satellites

As an application of the previous results on effacements, we give sufficient conditions in order that the sequence of chain homology functors of positive chain complexes (cf. 3.3.2) be universal. In particular this holds for various homological categories E, modular or non-modular, including all the additive homological categories.

Factorisation systems have been recalled in Section 2.6. Here we are interested in a more general notion, namely *weak factorisation systems* [AHRT], together with their *functorial realisations* [RT] and *natural weak factorisation systems* [GrT].

4.5.1 Graph factorisation

Let E be a homological category.

We say that E has a *functorial g-factorisation* (or *graph-factorisation*, the term being motivated by the examples below, in 4.5.3) if it comes equipped with:

(i) for every $f \colon A \to B$ in E, a *g-factorisation* $f = f''.f'$, where f' is a *normal mono* and f'' a *normal epi*,

(ii) for every commutative solid diagram, where the rows are graph factorisations

$$A \xrightarrow{\;f'\;} \bullet \xrightarrow{\;f''\;} B$$

(4.73)

with vertical morphisms a, c, b and bottom row $A' \xrightarrow{\;g'\;} \bullet \xrightarrow{\;g''\;} B'$

a morphism c making the whole diagram commutative, *consistently with vertical composition and vertical identities.*

Plainly, the last condition means that, if $a = 1$ and $b = 1$ (and $f = g$), then c is the identity.

4.5.2 Theorem (Homology and graph factorisation)

If E *is a homological category with functorial g-factorisation, the order two d-sequence* $(H^n, d^n) \colon \mathrm{Ch}^+\mathsf{E} \to \mathsf{E}$ *of cochain-homology is right universal* $(H^n = S^n H^0)$, *while the ∂-sequence* $(H_n, \partial_n) \colon \mathrm{Ch}_+\mathsf{E} \to \mathsf{E}$ *of chain-homology is left universal* $(H_n = S_n H_0)$.

Proof Since the hypothesis is self-dual, it suffices to consider one case. Given the (positive) cochain complex A in $\mathrm{Ch}^+\mathsf{E}$ and $n > 0$, we are going to construct a functorial embedding $m = m_A^n \colon A \rightarrowtail Q$ such that

$$H^n(Q) \text{ is null and } d_Q^{n-1} \text{ is an exact morphism.} \qquad (4.74)$$

This is necessarily an n-effacement of A. Indeed, the normal mono m produces a short exact sequence $(m, p) = (A \rightarrowtail Q \twoheadrightarrow C)$ and a homology sequence

$$\ldots H^{n-1}Q \to H^{n-1}C \xrightarrow{\;d^{n-1}\;} H^n A \to H^n Q \ldots \qquad (4.75)$$

Since d_Q^{n-1} is an exact morphism, by the 'central exactness' part of the homology sequence theorem (see 3.3.5(b)), this sequence is exact in $H^{n-1}C$ and in $H^n A$; its differential d^{n-1} is an exact morphism. Last, since $H^n(Q)$ is null, d^{n-1} is the cokernel of $H^{n-1}(p)$.

Now, to construct $m = m_A^n \colon A \rightarrowtail Q$, let $d_A^{n-1} = z\delta$ be the factorisation of the differential of A through the normal subobject $z \colon Z^n \to A^n$ of n-cycles of A; let $\delta = \delta''\delta'$ be the (chosen) g-factorisation of δ. We form the complex Q (with $Q^r = A^r$ for $r \neq n-1$) and the (vertical) morphism

$m \colon A \to Q$ as in the following commutative diagram

$$
\begin{array}{ccccccc}
A^{n-2} & \xrightarrow{d^{n-2}} & A^{n-1} & \xrightarrow{\quad d^{n-1}\quad} & A^n & \xrightarrow{d^n} & A^{n+1} \\
\| & & \downarrow{\scriptstyle\delta'} & \searrow^{\delta}\;\; Z^n \;\;\nearrow_{z} & \| & & \| \\
Q^{n-2} & \xrightarrow[d^{n-2}]{} & Q^{n-1} & \xrightarrow[d^{n-1}]{} & Q^n & \xrightarrow[d^n]{} & Q^{n+1}
\end{array}
\qquad (4.76)
$$

It is easy to check that Q is indeed a complex; m is a normal mono since all its components are. The condition (4.74) is satisfied, because $d_Q^{n-1} = z\delta''$ is trivially exact and $H^n(Q)$ is null

$$\mathrm{nim}\, d_Q^{n-1} = \mathrm{nim}\, z\delta'' = \mathrm{nim}\, z = \ker d_A^n = \ker d_Q^n. \qquad (4.77)$$

Therefore the functoriality of g-factorisations produces a functorial effacement $m_A^n \colon A \rightarrowtail Q(A)$ of the d-sequence H^n, provided we verify the 'diagonal' condition 4.3.5(iii). A normal monomorphism of complexes $f \colon A \rightarrowtail B$ produces a commutative diagram

$$
\begin{array}{ccc}
A & \xrightarrow{\;m\;} & Q(A) \\
{\scriptstyle f}\downarrow & & \downarrow{\scriptstyle Q(f)} \\
B & \xrightarrow[m']{} & Q(B)
\end{array}
\qquad (4.78)
$$

and we have to show that the normal mono $m'' = m'f \colon A \rightarrowtail Q(B)$ is an n-effacement of A. This can be verified as before, since the complex $Q(B)$ satisfies the hypotheses in (4.74): it is n-acyclic and its differential of degree $n-1$ is exact. □

4.5.3 Categories with functorial g-factorisation

We show, here and in the next section, that various homological categories E have a functorial g-factorisation, whence their sequence of chain or cochain homology functors are universal.

(a) *Additive homological categories.* This framework includes all abelian categories, together with the non-abelian categories of topological (or Hausdorff) K-vector spaces or abelian groups, Banach spaces, Hilbert spaces (see 1.6.3).

There is a functorial g-factorisation through the biproduct $A \oplus B$

$$
\begin{array}{ccc}
A & \xrightarrow{\quad f\quad} & B \\
{\scriptstyle g}\searrow & & \nearrow{\scriptstyle q} \\
& A \oplus B &
\end{array}
\qquad (4.79)
$$

where (writing i, j for the injections of the biproduct, and p, q for the projections, as in 2.4.1) $g = i + jf$ is the *graph* of f; and it is a normal mono since $g \sim \ker(fp - q)$. This procedure is at the origin of our general term 'graph-factorisation'.

Dually, there is a second g-factorisation, via the *co-graph* $h = fp + q \sim$ cok $(i - jf)$

$$A \xrightarrow{\quad f \quad} B \qquad . \qquad (4.80)$$
$$i \searrow \qquad \nearrow h$$
$$A \oplus B$$

Both factorisations can be enriched to form a *natural* weak factorisation system, as proved in [GrT], 4.1-4.2.

(b) The category $\mathcal{I}$ of sets and partial bijections. This non-abelian, p-exact category has a functorial g-factorisation; indeed, a partial bijection $f : A \to B$ defines an amalgamated sum

$$A +_f B = (A + B)/R, \qquad (4.81)$$

where the equivalence relation $R \subset (A + B) \times (A + B)$ is generated by the subset $f \subset A \times B$ (identifying $x \in \mathrm{Def}\, f$ with $f(x) \in \mathrm{Val}\, f$). The factorisation $A \rightarrowtail A +_f B \twoheadrightarrow B$, consisting of an embedding and a coinclusion, is now obvious.

It can be noted that the morphism f is determined by two injections $A \hookleftarrow X \rightarrowtail B$ (where X is $\mathrm{Def}\, f$) and our amalgamated sum $A +_f B$ is the pushout of this diagram *in the category of sets*, or in the category of injections; but it is not a pushout in $\mathcal{I}$, that lacks general pushouts and pullbacks.

(c) The homological categories Inj_2 and Inc_2 (1.7.4, 2.3.2) have a functorial g-factorisation, by decomposing the morphism $f : (X, A) \to (Y, B)$ in

$$(X, A) \rightarrowtail (Y, fA) \twoheadrightarrow (Y, B). \qquad (4.82)$$

(d) Plainly, a homological category E that is not g-exact but whose exact morphisms are stable under composition can*not* have g-factorisations; this is the case of $\mathsf{Set}_\bullet$, $\mathcal{S}$, $\mathcal{T}$ and $\mathcal{C}$ (1.6.4-1.6.5).

More generally, the same negative conclusion holds whenever the exact morphisms of E generate a proper (i.e. non-total) subcategory. This is the case of Set_2 and Top_2, since their exact morphisms are injective mappings (cf. 1.4.2). It is also the case of the p-homological category Ltc, because its exact morphisms $f : X \to Y$, characterised in 1.2.5 by the conditions (for all $x \in X$, $y \in Y$)

$$f^\bullet f_\bullet(x) = x \vee f^\bullet 0, \qquad f_\bullet f^\bullet(y) = y \wedge f_\bullet 1,$$

satisfy the following properties, that are closed under composition

$$f^\bullet(f_\bullet x \vee y) = x \vee f^\bullet y, \qquad f_\bullet(f^\bullet y \wedge x) = y \wedge f_\bullet x.$$

4.5.4 The amalgam of a modular connection

On the other hand, the p-exact categories Mlc and Dlc of modular connections between modular or distributive lattices (introduced in 1.2.8) have (consistent) functorial g-factorisations. We just sketch the construction, based on a sort of *amalgam* of a connection, that is similar to the amalgam of a partial bijection considered above (in (4.81)).

A modular connection $f \colon X \to Y$ has a canonical epi-mono factorisation, that coincides with the normal factorisation of general lattice connections (of 1.2.4), $f = (\operatorname{nim} f)g(\operatorname{ncm} f)$, where g is an iso

$$\begin{array}{ccc}
X & \xrightarrow{\ f\ } & Y \\
{\scriptstyle p}\downarrow & & \uparrow{\scriptstyle m} \\
\uparrow f^\bullet 0 & \xrightarrow{\ g\ } & \downarrow f_\bullet 1
\end{array} \tag{4.83}$$

$$\mathrm{cl}X = f^\bullet Y = \{x \in X \mid f^\bullet f_\bullet(x) = x\} = \uparrow f^\bullet 0,$$
$$\mathrm{cl}Y = f_\bullet X = \{y \in Y \mid f_\bullet f^\bullet(y) = y\} = \downarrow f_\bullet 1,$$

$$\begin{array}{lll}
p = \operatorname{cok} \ker f, & p_\bullet(x) = x \vee f^\bullet 0, & p^\bullet(x) = x, \\
g \colon \uparrow f^\bullet 0 \to \downarrow f_\bullet 1, & g_\bullet(x) = f_\bullet(x), & g^\bullet(y) = f^\bullet(y), \\
m = \ker \operatorname{cok} f, & m_\bullet(y) = y, & m^\bullet(y) = y \wedge f_\bullet 1.
\end{array} \tag{4.84}$$

This factorisation yields a cospan (a, b) in the category of *modular lattices and injective quasi homomorphisms*

$$\begin{array}{ccc}
X & \overset{c}{\dashrightarrow} & Z \\
{\scriptstyle a}\uparrow & & \uparrow{\scriptstyle d} \\
\mathrm{cl}X & \xrightarrow{\ b\ } & Y
\end{array} \qquad\qquad
\begin{array}{l}
a(x) = p^\bullet(x) = x, \\[1em]
b(x) = m_\bullet g_\bullet(x) = f_\bullet(x).
\end{array} \tag{4.85}$$

Its pushout Z is the 'amalgamated sum' that we want to use, as in the following example, consisting of quasi sublattices of the ordered plane $\mathbb{R}\times\mathbb{R}$

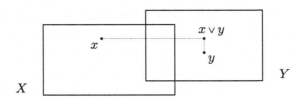

Here, each lattice X, Y is a product of two compact real intervals, with $0_X \leqslant 0_Y \leqslant 1_X \leqslant 1_Y$ in $\mathbb{R} \times \mathbb{R}$; moreover, $\mathrm{cl} X = \mathrm{cl} Y = X \cap Y$, a and b are inclusions and $Z = X \cup Y$ is again a quasi sublattice of $\mathbb{R} \times \mathbb{R}$.

In the general case, Z is the set-pushout (where we identify each $x \in \mathrm{cl} X$ with $bx = f_\bullet x \in \mathrm{cl} Y$), equipped with the order relation $\leqslant$ that extends as follows the orderings of X and Y (for $x \in X$ and $y \in Y$)

$$\begin{aligned}
x \leqslant y \qquad &\text{if } f_\bullet x \leqslant y &(\Leftrightarrow x \leqslant f^\bullet y), \\
y \leqslant x \qquad &\text{if } x \in \mathrm{cl} X \text{ and } y \leqslant f_\bullet x &(\Leftrightarrow y \in \mathrm{cl} Y,\ f^\bullet y \leqslant x).
\end{aligned}$$
(4.86)

This is indeed a modular lattice (distributive if X and Y are), with $0_Z = 0_X$, $1_Z = 1_Y$ and

$$x \vee y = f_\bullet x \vee y \text{ (in } Y), \qquad x \wedge y = x \wedge f^\bullet y \text{ (in } X). \tag{4.87}$$

Now, c and d give two modular connections $X \rightarrowtail Z \twoheadrightarrow Y$, whose composition is f

$$\begin{aligned}
(c, c^\bullet)\colon X \rightarrowtail Z, \quad c^\bullet x = x, \quad c^\bullet y = f^\bullet y, \\
c^\bullet cx = x = x \vee c^\bullet 0_X, \\
cc^\bullet x = x = x \wedge c(1_X), \\
cc^\bullet y = f^\bullet y = 1_X \wedge y = y \wedge c(1_X),
\end{aligned}$$
(4.88)

$$\begin{aligned}
(d_\bullet, d)\colon Z \twoheadrightarrow Y, \quad d_\bullet x = f_\bullet x, \quad d_\bullet y = y, \\
dd_\bullet x = f_\bullet x = x \vee 0_Y = x \vee d^\bullet 0_Y, \\
dd_\bullet y = y = y \vee d^\bullet 0_Y, \\
d_\bullet dy = y = 1_Y \wedge y = d_\bullet 1_Z \wedge y,
\end{aligned}$$
(4.89)

The construction is functorial.

5

Universal constructions

This chapter deals with universal constructions of various kinds. It begins by the diagram 5.1.1, showing the concrete 2-categories EX_1, EX_2,..., EX_5, EX that have been introduced in 1.7.0 and 2.2.8, together with other full sub-2-categories of EX_1 determined by the modular and distributive properties.

The first section deals with universal constructions, such as the exact centre or the modular expansion, that are *coreflectors* (or right adjoints) of embeddings between these 2-categories (cf. A3.4). The second and third are concerned with categories of fractions of semiexact categories, together with *reflectors* (or left adjoints) of such embeddings.

Section 5.4 is a review of categories of *relations* for g-exact categories, an easy extension of the p-exact case described in Part I.

Finally, Sections 5.5 and 5.6 study categories of fractions for g-exact categories and prove that EX_4, EX_5, EX and AB *have a structure of (unrestricted) homological category*, involving thick subcategories (the normal subobjects) and Serre quotients (the normal quotients).

The projective category associated to an ex2-category forms a reflector (for a subcategory of projective ex2-categories), that has already been considered (see 2.1.4).

5.1 Coreflectors, exact centres and expansions

This section contains universal 'coreflective' constructions, as the exact centre, the modular centre, the modular expansion or the exact expansion of an ex2-category. E is always an ex2-category.

5.1.1 A semilattice of 2-categories

The following diagram shows the meet-semilattice of *full* sub-2-categories of EX_1 that we have considered

$$
\begin{array}{ccccccccccc}
EX_6 & \to & EX_5 & \to & EX_4 & \to & EX_3 & \to & EX_2 & \to & EX_1 \\
\| & & \| & & \| & & \uparrow & & \uparrow & & \uparrow \\
EX_{6m} & \to & EX_{5m} & \to & EX_{4m} & \to & EX_{3m} & \to & EX_{2m} & = & EX_{1m} \\
\uparrow & & \uparrow & & \uparrow & & \uparrow & & \uparrow & & \uparrow \\
EX_{6d} & \to & EX_{5d} & \to & EX_{4d} & \to & EX_{3d} & \to & EX_{2d} & = & EX_{1d}
\end{array}
\qquad (5.1)
$$

where ex6 means p-exact (and $EX_6 = EX$), ex3m-category means a *modular homological* category, ex3d-category means a *distributive homological* category, and so on.

The three vertical equalities at the left come from the fact that every g-exact category is modular; the two horizontal equalities at the right come from the fact that every modular semiexact category is ex2 (Sections 2.3.1 - 2.3.3). Let us also recall that the distinction between the left column and the next one is merely a question of connected components (cf. 2.2.7).

It will be useful to use an isomorphic (semi)lattice

$$
\Lambda = \{1, 2, 3, 4, 5, 6, 2m, 3m, 2d, 3d, 4d, 5d, 6d\}, \qquad (5.2)
$$

consisting of the 'indices' of (5.1), modulo the identifications above.

Moreover, for $i \in \Lambda$, EX'_i will denote the sub-2-category of EX_i formed by the nsb-full functors (and all natural transformations between them); these 2-categories produce a second semilattice of the same form.

The universal constructions that we now consider are coreflectors of some of the inclusions above.

5.1.2 Homological and exact objects

We say that the object A is *homological* in the ex2-category E if the axiom (ex3) 'holds locally' in A, i.e. the following equivalent conditions are satisfied:

(a) given $m \in NsbA$ and $q \in NqtA$ with $m \geqslant \ker q$, the morphism qm is exact,

(a') the pullback of a diagram $A \twoheadrightarrow \bullet \leftarrowtail S$ is of type $A \leftarrowtail \bullet \twoheadrightarrow S$,

(a'') the pushout of a diagram $A \leftarrowtail \bullet \twoheadrightarrow S$ is of type $A \twoheadrightarrow \bullet \leftarrowtail S$.

The proof of the equivalence is the same as in the global case, in 2.1.2.

More particularly, we have already defined A to be *exact* in E if the axiom (ex4) 'holds locally' in A (Section 3.2.4), that is:

(b) for every $m \in \mathrm{Nsb}A$ and $q \in \mathrm{Nqt}A$, the morphism qm is exact.

Homological and exact objects are preserved by every nsb-full ex2-functor and reflected by every conservative ex2-functor.

The *homological centre* $\mathrm{Hmc}\,\mathsf{E}$ of the ex2-category E will be the full subcategory formed by its homological objects. The *exact centre* $\mathrm{Exc}\,\mathsf{E}$ will be the subcategory formed by its exact objects and by the exact morphisms between them.

Each of these categories is equipped with the null maps of E which belong to it.

5.1.3 Theorem (The homological and the exact centre)

Let E be an ex2-category. Then $\mathrm{Exc}\,\mathsf{E}$ is an nsb-full ex2-subcategory (cf. 1.3.7) of E and a g-exact category in its own right. Furthermore, every nsb-full exact functor $F \colon \mathsf{E}' \to \mathsf{E}$, where E' is a g-exact category, factorises uniquely through the inclusion $\mathrm{Exc}\,\mathsf{E} \to \mathsf{E}$, which is thus the universal arrow from the inclusion functor $\mathsf{EX}'_4 \subset \mathsf{EX}'_2$ to the 'object' E (cf. A2.1).

Analogously, $\mathrm{Hmc}\,\mathsf{E}$ is an nsb-full ex2-subcategory of E and a homological category. Every nsb-full exact functor $F \colon \mathsf{H} \to \mathsf{E}$ defined on a homological category factorises uniquely through the inclusion $\mathrm{Hmc}\,\mathsf{E} \to \mathsf{E}$, which is thus the universal arrow from the inclusion $\mathsf{EX}'_3 \subset \mathsf{EX}'_2$ to E.

In other words, the categories EX'_3 (of homological categories and nsb-full ex3-functors) and EX'_4 (of g-exact categories and nsb-full ex4-functors) are both (strictly) coreflective in EX'_2.

Proof (A) First, $\mathsf{C} = \mathrm{Exc}\,\mathsf{E}$ is a subcategory of E. In fact, if $f \colon A \to B$ and $g \colon B \to C$ are exact morphisms between exact objects of E and $f = mp$, $g = nq$ are normal factorisations in E

$$A \xrightarrow{\ p\ } \bullet \xrightarrowtail{\ m\ } B \xrightarrow{\ q\ } \bullet \xrightarrowtail{\ n\ } C \tag{5.3}$$

the composite is exact, since $gf = n(qm)p = (nn').(p'p)$, where $qm = n'p'$ is a normal factorisation of the morphism qm, that is exact because the object B is.

(B) We now verify that C is a (conservative) nsb-full semiexact subcategory, by means of 1.7.3(c). Condition 1.7.3(iii) is easy to check. Let $mf = g$, where m is a normal mono in E and m and g belong to C; then all the three objects are exact in E and the morphism f is also exact, by the detection property 1.5.7(c).

As to 1.7.3(ii'), we must prove that, if $k \colon K \rightarrowtail A$ is a normal mono of E with values in an exact object A, then the object K is also exact, so that k is in C. (The dual property on the normal quotients of A follows by duality.) Therefore, let m and p be a normal subobject and a normal quotient of K, as below, and let us form the pushout of p and k

$$
\begin{array}{ccc}
\bullet \xrightarrowtail{m} & K \xrightarrow{p} & \bullet \\
{\scriptstyle k}\downarrow & & \downarrow {\scriptstyle k'} \\
A & \xdashrightarrow{p'} & \bullet
\end{array}
\qquad (5.4)
$$

Since A is exact, this produces a normal mono k' (cf. 5.1.2); now the morphism $k'(pm) = p'(km)$ is exact (by ex2 and because A is exact), hence pm is also (by 1.5.7(c)).

(C) Last, C is g-exact because the embedding preserves normal factorisations and reflects isomorphisms. The universal property holds trivially, because every nsb-full ex2-functor preserves the exact objects and the exact morphisms.

(D) The proof for the homological centre follows the same outline, but is simpler: point (A) and the first part of (B) can be omitted, because Hmc E is a full subcategory of E. $\qquad\square$

5.1.4 Examples

(a) The exact centre of the homological category Ltc of lattices and connections is the p-exact category Mlc of modular lattices and modular connections (Section 1.2.8).

Indeed, we already know that the exact morphisms of Ltc are the exact connections (see 1.2.5). In order to determine when a lattice X is an exact object, let us take a normal subobject $m\colon \downarrow a \to X$ and a normal quotient $p\colon X \to \uparrow b$. Then the connection $f = mp\colon \downarrow a \to \uparrow b$ satisfies

$$
f^\bullet f_\bullet(x) = m^\bullet p^\bullet p_\bullet m_\bullet(x) = (x \vee b) \wedge a
$$
$$
\geqslant x \vee (b \wedge a) = x \vee f^\bullet b \quad (\text{for } x \leqslant a),
$$

$$
f_\bullet f^\bullet(y) = p_\bullet m_\bullet m^\bullet p^\bullet(y) = (y \wedge a) \vee b
$$
$$
\leqslant y \wedge (a \vee b) = y \wedge f_\bullet a \quad (\text{for } y \geqslant b).
$$

Therefore $f = mp$ is an exact connection (for all m and p, i.e. for all $a, b \in X$) if and only if X is a modular lattice.

(This proves again that every g-exact category E is modular, since its transfer functor Nsb$\colon$ E $\to$ Ltc has to factorise through the exact centre of Ltc.)

(b) The exact centre of the category Ban of Banach spaces is the full sub-category of *finite dimensional* Banach spaces, equivalent to the (abelian) category of finite dimensional (real or complex) vector spaces.

Indeed, the exact morphisms are the morphisms with closed image (by 1.6.3). If $m\colon A \to X$ and $p\colon X \to X/B$ are a normal subobject and quotient of the space X, the image of $f = mp$ is $(A + B)/B$ (where $A + B$ is the algebraic join) and is closed if and only if $A + B$ is closed in X; therefore the object X is exact if and only if the algebraic sum of two closed linear subspaces is always closed. Clearly this holds if the dimension of X is finite, but the latter condition is also necessary as it can be proved by an argument concerning the Markuševič bases in a separable Banach space (see [Si], Section 8; the author is grateful to Laura Burlando for this reference). Hlb has a similar exact centre, since the exact forgetful functor Hlb $\to$ Ban is nsb-full.

(c) We know (Section 1.6.4) that the exact morphisms of the category Set. of pointed sets are the (pointed) mappings which are injective outside of their kernel; since they are stable under composition, each object is exact and Exc Set. coincides with the subcategory $\mathcal{I}'$ of its exact morphisms (equivalent to the p-exact category $\mathcal{I}$ of sets and partial bijections, see 1.4.3).

(d) In Set_2 the exact objects reduce to the null ones, their exact morphisms to the isomorphisms $(X, X) \to (Y, Y)$. The exact centre Exc Set_2 is thus isomorphic to the groupoid of sets and bijections; all maps are null, kernels and cokernels are identities (and the category is trivially g-exact).

5.1.5 The modular and distributive centre of an ex2-category

Every ex2-category E contains an invariant, nsb-full ex2-subcategory Mdc E, its *modular centre*, such that every nsb-full ex2-functor with values in E, whose domain is some modular ex2-category, factorises uniquely through Mdc E $\to$ E (the universal arrow from the inclusion $\mathsf{EX}'_{2m} \subset \mathsf{EX}'_2$ to E, cf. A2.1).

The objects and morphisms of Mdc E consists of the *modular* objects and morphisms of E, i.e. those that are taken by the ex2-functor Nsb: E $\to$ Ltc to modular lattices and modular connections, respectively.

Analogously we define the *distributive centre* Dsc E. The distributive and the exact centre are contained in the modular centre; in particular, every exact object is modular.

For example, consider the category Set_2 of pairs of sets. All its lattices of normal subobjects are distributive, but a map $f\colon (X, A) \to (Y, B)$ is

modular if and only if it is injective out of $f^{-1}(B)$ (Section 2.3.2), whence
$\mathsf{Mdc\,Set_2} = \mathsf{Dsc\,Set_2}$ is the subcategory of all pairs of sets and such mappings. It contains $\mathsf{Inj_2}$ and $\mathsf{Inc_2}$ (pairs of sets and injections or respectively
inclusions, 1.7.4), that are distributive homological categories but not g-
exact. Similar facts hold for $\mathsf{Top_2}$.

Again, $\mathsf{Set_\bullet}$ is not modular and $\mathsf{Mdc\,Set_\bullet} = \mathsf{Dsc\,Set_\bullet} = \mathsf{Exc\,Set_\bullet} = \mathcal{I}'$, the
distributive p-exact subcategory of those mappings f which are injective
outside of their kernel (Section 1.6.4).

5.1.6 The general expansion of an ex2-category

The objects of the *general expansion* $\mathsf{Lat\,E}$ are the pairs (A, X), where A
is an object of E and X is a sublattice of $\mathrm{Nsb}A$ (hence with the same 0
and 1). A morphism $(f, X, Y) \colon (A, X) \to (B, Y)$ is given by any morphism
$f \colon A \to B$ in E such that $f_*(X) \subset Y$ and $f^*(Y) \subset X$; this morphism
(f, X, Y) will also be written as f, when the lattices X and Y are clear
from the context. The composition is given by that of E.

We now show that the conservative forgetful functor

$$U \colon \mathsf{Lat\,E} \to \mathsf{E}, \qquad (A, X) \mapsto A, \quad f = (f, X, Y) \mapsto f, \qquad (5.5)$$

creates an ex2-structure on $\mathsf{Lat\,E}$. Since U is conservative and faithful, it
will follow that, if E is homological, or g-exact, or componentwise p-exact,
or modular or distributive, so is $\mathsf{Lat\,E}$.

The null morphisms f are those for which $U(f)$ is null. The kernel and
cokernel of f are

$$m \colon (\mathrm{Ker}\, f, m^*X) \rightarrowtail (A, X), \qquad p \colon (B, Y) \twoheadrightarrow (\mathrm{Cok}\, f, p_*(Y)), \qquad (5.6)$$

where $m = \ker f = f^*0 \in X$ and $p = \mathrm{cok}\, f$.

Let us verify the description of the kernel. For $x \in X$, $m_*m^*(x) = x \wedge m \in X$. Moreover the mappings m_* and m^* establish an isomorphism
between $\mathrm{Nsb}(\mathrm{Ker}\, f)$ and the *quasi* sublattice $\downarrow m$ of $\mathrm{Nsb}A$; in this isomor-
phism, the subset m^*X of $\mathrm{Nsb}(\mathrm{Ker}\, f)$ corresponds to

$$m_*m^*(X) = \{x \in X \mid x \leqslant m\} = \downarrow m \cap X,$$

a sublattice of $\downarrow m$; it follows that m^*X is also a sublattice of $\mathrm{Nsb}(\mathrm{Ker}\, f)$.
Thus, m is well defined. Last, if $g \colon (C, Z) \to (A, X)$ is annihilated by f,
the morphism g factorises uniquely as mh in E and h can be lifted to $\mathsf{Lat\,E}$,
since

$$h_*(Z) = m^*m_*h_*(Z) = m^*g_*(Z) \subset m^*X,$$
$$h^*(m^*X) = g^*(X) \subset Z. \tag{5.7}$$

Now, the expansion $\operatorname{Lat} \mathsf{E}$ is semiexact by 1.5.4(b); it is plainly ex2, as well as the functor U. The latter is not nsb-full, since there are isomorphisms

$$X \cong \operatorname{Nsb}(A, X), \qquad m \mapsto m = (m, m^*X, X), \qquad (5.8)$$

that allow us to identify the functor $\operatorname{Nsb}: \operatorname{Lat} \mathsf{E} \to \operatorname{Ltc}$ with the forgetful functor

$$V: \operatorname{Lat} \mathsf{E} \to \operatorname{Ltc},$$
$$V(A, X) = X, \quad V(f, X, Y) = (f_*: X \to Y, f^*: Y \to X). \qquad (5.9)$$

The (obvious) universal property of the general expansion says that, for every ex2-category X and every ex2-functor $F: \mathsf{X} \to \mathsf{E}$, there is a *lifting* $F^\sharp: \mathsf{X} \to \operatorname{Lat} \mathsf{E}$ (with $UF^\sharp = F$) that is nsb-full (Section 1.7.1), and determined up to isomorphism.

In other words, $U: \operatorname{Lat} \mathsf{E} \to \mathsf{E}$ is a *biuniversal* arrow from the inclusion $\mathsf{EX}'_2 \subset \mathsf{EX}_2$ to E (cf. A4.6), and the category EX'_2 of ex2-categories and nsb-full ex2-functors is bi-coreflective in EX_2; analogously EX'_i is bi-coreflective in EX_i, for $i \geqslant 2$ in Λ (cf. (5.2)).

5.1.7 The modular and distributive expansions

The *modular expansion* $\operatorname{Mdl} \mathsf{E}$ is the modular centre of $\operatorname{Lat} \mathsf{E}$, i.e. (by (5.9)) the subcategory whose objects are the pairs (A, X) with X modular and whose morphisms are the triples $f = (f, X, Y)$ with $f_*: X \to Y$ and $f^*: Y \to X$ forming a modular connection. $\operatorname{Mdl} \mathsf{E}$ is a modular ex2-category.

Similarly, the *distributive expansion* $\operatorname{Dst} \mathsf{E}$ is the distributive centre of $\operatorname{Lat} \mathsf{E}$, i.e. the full subcategory of $\operatorname{Mdl} \mathsf{E}$ whose objects are the pairs (A, X) with X a distributive lattice.

If E is homological, or g-exact, or p-exact, so are $\operatorname{Mdl} \mathsf{E}$ and $\operatorname{Dst} \mathsf{E}$. The distributive expansion of a *p-exact* category has already been introduced in I.2.8. For $\mathcal{J} = \operatorname{Dst}(\mathcal{I})$ see below (Section 5.1.9).

The universal property of these constructions says that, for every modular (resp. distributive) ex2-category X and every ex2-functor $F: \mathsf{X} \to \mathsf{E}$, there is a *lifting* $F^\sharp: \mathsf{X} \to \operatorname{Mdl} \mathsf{E}$ (resp. $\mathsf{X} \to \operatorname{Dst} \mathsf{E}$) that is nsb-full and determined up to isomorphism. Of course, 'lifting' means that $UF^\sharp = F$, and the forgetful functor $U: \operatorname{Mdl} \mathsf{E} \to \mathsf{E}$ is a biuniversal arrow from the inclusion $\mathsf{EX}'_{2m} \subset \mathsf{EX}_2$ to E (cf. A4.6); analogously for the distributive case.

5.1.8 *The exact expansion and the distributive exact expansion*

E is always an ex2-category. Its *exact expansion* will be the exact category

$$\mathrm{Ex\,E} = \mathrm{Exc\,Lat\,E} = \mathrm{Exc\,Mdl\,E}, \qquad (5.10)$$

equipped with the obvious conservative, forgetful functor $U\colon \mathrm{Ex\,E} \to \mathrm{E}$.

Indeed, recall that the forgetful functor $\mathrm{Lat\,E} \to \mathrm{E}$ is conservative. Therefore the object (A, X) of $\mathrm{Lat\,E}$ is exact if and only if X is an E-*exact* sublattice of $\mathrm{Nsb}A$, in the sense that:

$$\text{if } m, n \in X, \text{ then the morphism } (\mathrm{cok}\,n).m \text{ is exact in E.} \qquad (5.11)$$

In this case $X \cong \mathrm{Nsb}(A, X)$ is an exact object of Ltc, i.e. a modular lattice. By the same reason, the morphism $f\colon (A, X) \to (A, Y)$ of $\mathrm{Lat\,E}$ is exact if and only if f is an exact morphism in E, and in this case $\mathrm{Nsb}(f)$ is an exact connection.

The *distributive exact expansion* of E will be the distributive exact category

$$\mathrm{Dx\,E} = \mathrm{Exc\,Dst\,E}, \qquad (5.12)$$

i.e. the full subcategory of $\mathrm{Ex\,E}$ whose objects (A, X) are the objects of $\mathrm{Lat\,E}$ where X is an E-exact sublattice of $\mathrm{Nsb}A$, as defined in (5.11), and a distributive lattice in its own right.

The universal property says that, for every exact (resp. distributive exact) category X and every ex2-functor $F\colon \mathrm{X} \to \mathrm{E}$, there is a lifting $F^\sharp\colon \mathrm{X} \to \mathrm{Ex\,E}$ (resp. $\mathrm{X} \to \mathrm{Dx\,E}$) that is nsb-full and determined up to isomorphism. In other words, EX'_2 and EX'_{4d} are bi-coreflective (A4.6) in EX_2.

5.1.9 *The p-exact category of semitopological spaces*

Even though $\mathcal{I}$ is already distributive, we have used its distributive expansion $\mathcal{J} = \mathrm{Dst}\,(\mathcal{I})$ (see I.2.8.8), because it allows for a better description of the universal models of distributive exact theories (Part I). It is a distributive p-exact category and is *not boolean*, whence it is not categorically equivalent to $\mathcal{I}$.

An object of $\mathcal{J}$ is called a *semitopological space*. It is a (small) set X equipped with a sublattice $\mathrm{Cls}X$ of $\mathcal{P}X$ (containing $\emptyset$ and X), whose elements are called *closed* subsets. We use the usual terminology concerning *open* subset (the complement of a closed one), *locally closed* subset (an intersection of a closed and an open subset), *continuous mapping, homeomorphism* and *induced semitopology* on some subset of X.

A $\mathcal{J}$-morphism $f = (U, K; f_0): X \to Y$, called a *partial homeomorphism* of semitopological spaces, is a partial bijection between the underlying sets which preserves closed parts, by direct and inverse images (computed in $\mathcal{I}$, by (1.83)). In particular:

$$\text{Ker}\, f = f^*(\emptyset) = X \setminus U \text{ is closed in } X,$$

$$\text{Cok}\, f = Y \setminus K \text{ is open in } Y.$$

$$\text{Def}\, f = U \text{ is open in } X, \tag{5.13}$$

$$\text{Im}\, f = f_*(X) = K \text{ is closed in } Y,$$

Therefore f amounts to a *homeomorphism* $f_0: \text{Def}\, f \to \text{Im}\, f$ *from an open subspace* ($\text{Def}\, f$) *of* X *to a closed subspace* ($\text{Im}\, f$) *of* Y.

The zero-object of $\mathcal{J}$ is the empty space. Subobjects can be identified with the inclusions of closed subspaces H, and quotients with the 'co-inclusions' of open subspaces U

$$H: H \rightarrowtail S, \qquad U: S \twoheadrightarrow U. \tag{5.14}$$

The category $\mathcal{J}$ is self-dual, via the anti-automorphism:

$$X \mapsto X^{\text{op}}, \quad (f: X \to Y) \mapsto (f^{\text{op}}: Y^{\text{op}} \to X^{\text{op}}), \tag{5.15}$$

where X^{op} is the *opposite* semitopological space (interchange closed and open subsets) and $(f^{\text{op}})_0 = f_0^{-1}$. (Essentially, this is a consequence of $\mathcal{I}$ being boolean.)

We have also used the p-exact subcategory $\mathcal{J}_0$ of *semitopological spaces and open-closed parts*, introduced in I.4.6.3. A morphism $L: S \to T$ is given by any *common subspace* L of S and T (i.e. $L \subset S \cap T$ has the same induced semitopology) which is open in S and closed in T. (Notice that all spaces are subsets of the universe $\mathcal{U}$.) Composition is the intersection (cf. I.4.6.3).

$\mathcal{J}_0$ may also be viewed as the distributive expansion of the p-exact category $\mathcal{I}_0$ of sets and common parts. Also $\mathcal{J}_0$ is self-dual, via the restriction of (5.15).

5.2 Fractions for semiexact categories

Categories of fractions informally arose in the *langage modulo C* of Serre [Ser]; they were formalised for abelian categories by Grothendieck [Gt] and Gabriel [Ga] and extended to general category theory in Gabriel-Zisman [GaZ]. Later, Bénabou [Be2] gave an extended study of this topic for various categorical frameworks, including regular categories.

The main goal of this section is to set up fractions for semiexact and homological categories. It should be noted that these fractions *lack a bounded calculus* and we manage to get a solution in some EX_i by restricting the universal problem *within* EX_1, i.e. within exact functors. More effective solutions will be given in Sections 5.3 and 5.5; see also Sections 6.3, 6.5.

E is always a semiexact category.

5.2.1 *Categories of fractions*

Let us recall that, given a set Σ of morphisms of the category C, the *category of fractions* of C produced by Σ is the universal solution of making the arrows of Σ invertible. In other words, we look for a functor $P: C \to C'$ that carries all the morphisms of Σ to isomorphisms and such that every functor defined on C which does the same (with values in an arbitrary category), factorises uniquely through P.

The solution exists and is determined up to isomorphism. The general construction, by finite 'linear' diagrams made up of arrows of C and 'reversed' arrows of Σ, is formally easy (and similar to the construction which we shall give later for ex1-fractions, in 5.2.2), but not very practical. Under suitable hypotheses, a well known and much more useful construction based on a two-arrow calculus can be given [GaZ].

However, this general solution need not respect the semiexact structure; therefore, we restrict the problem to ex1-categories and ex1-functors. Given a set Σ_0 of morphisms of the semiexact category E, we shall call Σ_0-*functor* any exact functor F from E to some *unrestricted* ex1-category (see 1.5.1) that takes all the morphisms of Σ_0 to isomorphisms.

Then the category of *ex1-fractions of* E *produced by* Σ_0 is defined as a universal Σ_0-functor, i.e. a Σ_0-functor $P: E \to \Sigma_0^{-1}E$ such that every such functor factorises uniquely through P and an exact functor. We show below that the solution exists; moreover $\Sigma_0^{-1}E$ is an ex1-category (not just an unrestricted one) that need not have small hom-sets, even if E does.

Unrestricted codomains are used to prove that the solution does not depend on the universe $\mathcal{U}$. Our argument below establishes a contravariant Galois connection between sets of maps of E and 'illegitimate' sets of functors.

Given an ex1-functor $F: E \to E'$, its *isokernel* $\mathrm{Ikr}F$ will be the subset of MorE consisting of the F-*relative isomorphisms*, i.e. the maps that are taken by F to isomorphisms of E'.

On the other hand, its (annihilation) *kernel* $\mathrm{Ker}\,F$ will be the set of objects that are annihilated by F (i.e. carried to null objects of E'). Equivalently, one can use the full subcategory $\mathrm{Ker}\,F \subset E$ formed by these objects;

it is the kernel of F in the category EX_1, with respect to the closed ideal of null ex1-functors (annihilating the whole domain).

In a different context, Bénabou [Be2] uses the term 'kernel' for what is called here an isokernel.

5.2.2 Theorem (Ex1-fractions)

Let E be an ex1-category, let Σ_0 be a set of morphisms of E and let Σ be the set of the maps of E that are made invertible by every Σ_0-functor.

The category of ex1-fractions of E with respect to Σ_0 (as defined in Section 5.2.1) exists and will be written as $P \colon \mathsf{E} \to \Sigma_0^{-1}\mathsf{E}$, $f \mapsto [f]$. $\Sigma_0^{-1}\mathsf{E}$ is an ex1-category (in the restricted sense).

Every morphism in $\Sigma_0^{-1}\mathsf{E}$ factorises as

$$[h_r]^{-1}[f_r]\ldots[h_1]^{-1}[f_1] \qquad (f_i \text{ in } \mathsf{E},\ h_i \text{ in } \Sigma). \tag{5.16}$$

The functor P is nsb-full: a normal mono (resp. epi) of $\Sigma_0^{-1}\mathsf{E}$ is always of the form $Pf = [f]$, for some normal mono (resp. epi) f of E. The set Σ coincides with $\operatorname{Ikr}P$ and is the least ex1-isokernel of E containing Σ_0.

Proof (A) We begin by forming a category $\mathsf{E}^\sharp$. The objects are those of E and the morphisms are the finite, linear diagrams consisting of arrows of E and 'reversed' (dotted) arrows of Σ

$$A \xrightarrow{\ f_1\ } \bullet \xleftarrow{\ \ h_1\ \ } \bullet \ \ldots\ \bullet \xrightarrow{\ f_r\ } \bullet \xleftarrow{\ \ h_r\ \ } B \tag{5.17}$$

up to composition of consecutive arrows of E or Σ and deleting identities. The composition is obvious.

$\mathsf{E}^\sharp$ is a $\mathcal{U}$-category. E embeds in $\mathsf{E}^\sharp$ and a morphism φ of the latter can always be factorised as

$$\varphi = h_r^\sharp f_r \ldots h_1^\sharp f_1, \quad (f_i \text{ in } \mathsf{E},\ h_i \text{ in } \Sigma), \tag{5.18}$$

where $h^\sharp$ denotes the reversed arrow associated to h. The morphism φ defines two transfer mappings (independently of the factorisation)

$$\begin{aligned} \varphi_* &= h_r^* f_{r*} \ldots h_1^* f_{1*} \colon \mathrm{Nsb}A \to \mathrm{Nsb}B, \\ \varphi^* &= f_1^* h_{1*} \ldots f_r^* h_{r*} \colon \mathrm{Nsb}B \to \mathrm{Nsb}A. \end{aligned} \tag{5.19}$$

Every Σ_0-functor F extends uniquely to $\mathsf{E}^\sharp$, preserving these mappings

$$F'(h_r^\sharp f_r...h_1^\sharp f_1) = (Fh_r)^{-1}.(Ff_r)...(Fh_1)^{-1}.(Ff_1),$$

$$F(\varphi_* x) = F(h_r^*...f_{1*}(x)) \sim (Fh_r)^*...(Ff_{1*})(Fx)$$
$$= (F'\varphi)_*(Fx),$$

(5.20)

$$F(\varphi^* y) \sim (F'\varphi)^*(Fy).$$

Consider now the congruence $\varphi R \psi$ of $\mathsf{E}^\sharp$ defined by $F'\varphi = F'\psi$ for all Σ_0-functors F; let $\mathsf{C} = \mathsf{E}^\sharp/R$, with projection $P \colon \mathsf{E} \to \mathsf{C}$. Plainly, every Σ_0-functor F factorises uniquely as $F = F''P$ with $F''[\varphi] = F'\varphi$.

We say that a morphism $[\varphi]$ of C is *null* if and only if $F''[\varphi] = F'\varphi$ is null for all Σ_0-functors F; this happens if and only if $[\varphi_* 1]$ is null in C, if and only if the normal subobject $\varphi^* 0$ belongs to Σ.

We want to prove that this functor $P \colon \mathsf{E} \to \mathsf{C}$ is the solution.

(B) P takes each commutative square (5.21) of E, with arrows h, k in Σ

$$
\begin{array}{ccc}
\bullet & \overset{h}{\dashrightarrow} & \bullet \\
{\scriptstyle g}\big\uparrow & & \big\uparrow{\scriptstyle f} \\
\bullet & \underset{k}{\dashrightarrow} & \bullet
\end{array}
$$

(5.21)

to a 'bicommutative' square of C, in the sense that $[h^\sharp f] = [gk^\sharp]$. Indeed, each Σ_0-functor F transforms (5.21) into a commutative square with two horizontal isomorphisms.

(C) Moreover, P takes every $h \in \Sigma$ to an isomorphism $[h]$ of C, with inverse $[h^\sharp]$, as it follows applying point (B) to the following squares of E

$$
\begin{array}{ccc}
\bullet & \overset{h}{\dashrightarrow} & \bullet \\
{\scriptstyle 1}\big\uparrow & & \big\uparrow{\scriptstyle h} \\
\bullet & \underset{1}{\dashrightarrow} & \bullet
\end{array}
\qquad
\begin{array}{ccc}
\bullet & \overset{1}{\dashrightarrow} & \bullet \\
{\scriptstyle h}\big\uparrow & & \big\uparrow{\scriptstyle 1} \\
\bullet & \underset{h}{\dashrightarrow} & \bullet
\end{array} .
$$

(5.22)

(D) We now prove that C has kernels (and dually cokernels), showing that the kernel of $[\varphi]$ in C is $[m]$, for $m = \varphi^* 0$. First, $[\varphi m]$ is null, as $(\varphi m)^* 0 = m^*(\varphi^* 0) = m^*(m) = 1$.

Let us assume that $[\varphi\psi]$ is null, whence $h = (\varphi\psi)^* 0 = \psi^*(m) \in \Sigma$. Let $\psi = h_1^\sharp f_1...h_r^\sharp f_r$ and form the following diagram in E, from right to left, by direct and inverse images of normal subobjects

$$
\begin{array}{ccccccccccccc}
A' & \overset{f_r}{\longrightarrow} & \bullet & \overset{h_r}{\dashleftarrow} & \bullet & \cdots & \bullet & \overset{f_1}{\longrightarrow} & \bullet & \overset{h_1}{\dashleftarrow} & A & \overset{\varphi}{-\!\!\!\to} & B \\
{\scriptstyle h}\big\uparrow & & {\scriptstyle m_r}\big\uparrow & & \big\uparrow & & & {\scriptstyle n_1}\big\uparrow & & {\scriptstyle m_1}\big\uparrow & {\scriptstyle m}\big\uparrow & & \\
\bullet' & \underset{g_r}{\longrightarrow} & \bullet & \underset{k_r}{\dashleftarrow} & \bullet & \cdots & \bullet & \underset{g_1}{\longrightarrow} & \bullet & \underset{k_1}{\dashleftarrow} & M & &
\end{array}
$$

(5.23)

Therefore, $m_1 = h_{1*}(m)$, $n_1 = f_1^*(m_1)$ and so on, ending with

$$f_r^*(m_r) = \psi^*(m) = h \in \Sigma.$$

Each Σ_0-functor $F \colon \mathsf{E} \to \mathsf{X}$ carries the (leftward) 'direct image' squares of (5.23) to analogous squares of X, with Fh_i an isomorphism; it follows that Fk_i is also invertible, and all k_i are in Σ. Furthermore, these squares are 'bicommutative' in C by point (B), which means that the $\mathsf{E}^\sharp$-morphism $\lambda = k_1^\sharp g_1 ... k_r^\sharp g_r h^\sharp \colon A' \to M$ solves our problem: $[m][\lambda] = [\psi]$.

Now it suffices to check that $[m]$ is mono in C; indeed, if $[m][\lambda] = [m][\lambda']$, then $Fm.F'\lambda = Fm.F'\lambda'$ (for all F); since Fm is a normal mono in the codomain of F, $F'\lambda = F'\lambda'$ (for all F) and $[\lambda] = [\lambda']$.

(E) As a consequence, a normal mono of C is always of the form $Pm = [m]$, for some normal mono m of E; dually, a normal epi of C can be written as an equivalence class $[p]$, for some normal epi p of E. In particular, the sets $\mathrm{Nsb}_\mathsf{C}(A)$ are small.

Furthermore, every normal mono $[m]$ of C is N-mono. Indeed, if $[m][\varphi]$ is null, then $Fm.F'\varphi$ is null (for all Σ_0-functors F), whence $F'\varphi$ is null and $[\varphi]$ is also. Therefore C is a semiexact $\mathcal{U}$-category (by 1.5.4, Criterion B), P is exact (by point (D)) and nsb-full. Moreover, for all Σ_0-functors F, the induced functor F'' is exact, by (5.20) and (D)).

(F) Finally, we already know that $\Sigma \subset \mathrm{Ikr}P$. The other inclusion comes from the fact that all the functors F factorise through P; this also proves the last assertion of the statement.

(G) *Remark.* The congruence R above can also be characterised as the least congruence of categories S of $\mathsf{E}^\sharp$ such that

(i) for each commutative square (5.21), $(h^\sharp f)\, S\, (gk^\sharp)$,

(ii) if m is a normal mono, p is a normal epi (of E) and $(m\varphi p)\, S\, (m\psi p)$, then $\varphi\, S\, \psi$.

Indeed it is easy to see that the intersection R of all such congruences S satisfies our problem as well, by an argument analogous to the previous one. □

5.2.3 Remarks

(a) Every ex1-functor $F \colon \mathsf{E} \to \mathsf{E}'$ factorises uniquely, up to isomorphism, through the category of fractions $P \colon \mathsf{E} \to \Sigma^{-1}\mathsf{E}$ of its isokernel $\Sigma = \mathrm{Ikr}F$

$$\mathsf{E} \to \Sigma^{-1}\mathsf{E} \to \mathsf{E}', \tag{5.24}$$

and $\mathrm{Ikr}F = \mathrm{Ikr}P$ (by 5.2.2).

The second functor, likely, need not be conservative and we do not get a good factorisation structure for ex1-functors. However, the categories of fractions are useful for universal constructions, as well as for making new homological categories, as we shall see.

(b) The ex1-isokernels of E are stable under arbitrary intersection. In fact, given a family $\Sigma_F = \mathrm{Ikr}F$ of isokernels, with $F \in \mathcal{F}$, let $\Sigma = \cap \Sigma_F$ and form the category of fractions $P \colon \mathsf{E} \to \Sigma^{-1}\mathsf{E}$; then $\mathrm{Ikr}P \supset \Sigma$; the other inclusion follows from the fact that all the functors $F \in \mathcal{F}$ factorise through P.

5.2.4 *Theorem (Ex1-fractions continued)*

Let Σ be an ex1-isokernel of the semiexact category E. *If* E *is ex2 (or ex3, ex4, exm, exd), the ex1-category of fractions $\Sigma^{-1}\mathsf{E}$ is also ex2 (and so on), and solves the problem of 'making the maps of Σ invertible' for ex2-categories and ex2-functors (and so on).*

In particular, in the ex4 case, every morphism in the category of fractions factorises as follows, with p a normal epi, m a normal mono of E *and $h \in \Sigma$*

$$[\varphi] = [m][h^\sharp][p], \tag{5.25}$$

$$\ker[\varphi] = [\ker p], \quad \mathrm{cok}[\varphi] = [\mathrm{cok}\, m]. \tag{5.26}$$

Moreover, always in the ex4-case, the isokernel $\Sigma = \mathrm{Ikr}P$ and the annihilation kernel $\mathrm{Ker}\, P$ determine each other

$$\Sigma = \{f \in \mathsf{E} \mid \mathrm{Ker}\, f, \mathrm{Cok}\, f \text{ belong to } \mathrm{Ker}\, P\},$$
$$\mathrm{Ker}\, P = \{A \in \mathrm{Ob}\mathsf{E} \mid 0_A \in \Sigma\}. \tag{5.27}$$

Note. Categories of fractions for g-exact categories are studied more deeply in Sections 5.5, 5.6.

Proof If E is ex2, two consecutive normal monos $[n], [m]$ of $\Sigma^{-1}\mathsf{E}$ have a composite $[m].[n] = [mn]$, that is normal. If E is ex3, let $[m], [n]$ be normal subobjects of A in $\Sigma^{-1}\mathsf{E}$, with $[n] \leqslant [m]$; according to a general property of ex1-functors (already observed in 1.7.1), we can assume that $n \leqslant m$ in E (otherwise, we replace n with $n' = m \wedge n$, so that $[n'] = [m] \wedge [n] = [n]$); now, if $p = \mathrm{cok}\, n$, then pm is exact in E and $[pm]$ is exact in $\Sigma^{-1}\mathsf{E}$.

Let us assume that E is ex4, and prove (5.25). Take a morphism $\varphi = h_r^\sharp f_r ... h_1^\sharp f_1$ in the category $\mathsf{E}^\sharp$ (Section 5.2.2), with f_i in E and h_i in Σ; let $f_i = m_i p_i$ be normal factorisations. Each m_i can be 'pushed to the left', by direct and inverse images (preserving the morphisms of Σ); we thus get a composite of normal monos in the category of fractions, that can be expressed by a single $[m]$, because of our first argument here. Similarly, all p_i can be 'pushed to the right'. Last, we compose all the morphisms of Σ, and get the factorisation (5.25).

Now (5.26) follows from (5.25), because the central morphism $[h^\sharp]$ of the factorisation (5.25) is an isomorphism in the category of fractions. Property (5.27) is obvious: for each Σ-functor F, Ff is invertible if and only if $\ker f$ and $\operatorname{cok} f$ are annihilated by F, whereas FA is null if and only if $F(0_A)$ is an isomorphism. Last, $\Sigma^{-1}\mathsf{E}$ is g-exact: if $\ker[\varphi] = [\ker p]$ and $cok[\varphi] = [\operatorname{cok} m]$ are null, p and m are in Σ, by (5.27), and $[\varphi]$ is an isomorphism.

Finally, for the cases exm and exd, it suffices to note that the functor P is nsb-full and surjective on objects; therefore, every lattice of normal subobjects of A in $\Sigma^{-1}\mathsf{E}$ is a quotient of the lattice $\mathrm{Nsb}_\mathsf{E} A$. $\qquad\square$

5.2.5 Theorem and Definition (Exj-fractions)

More generally, given an ex1-category E, a set Σ_0 of morphisms of E and an index $j \in \Lambda$ (Section 5.2.1), one can find an ex1-functor $P\colon \mathsf{E} \to \mathsf{C}$ with values in an exj-category that makes every morphism in Σ_0 invertible and is universal for this property. The solution will be called the exj-category of fractions of the ex1-category E, produced by Σ_0.

In particular, in the ex4-case, every morphism $[\varphi]$ in the category of fractions factorises as follows, with p a normal epi, m a normal mono of E and $h_i \in \Sigma = \mathrm{Ikr} P$

$$[\varphi] = [m][h_r^\sharp h_{r-1}...h_2^\sharp h_1][p], \tag{5.28}$$

$$\ker[\varphi] = [\ker p], \quad \operatorname{cok}[\varphi] = [\operatorname{cok} m], \tag{5.29}$$

Moreover, the isokernel and the annihilation kernel $\mathrm{Ker}\, P$ *determine each other*

$$\Sigma = \{f \in \mathsf{E} \mid \mathrm{Ker}\, f, \mathrm{Cok}\, f \ \text{belong to} \ \mathrm{Ker}\, P\},$$
$$\mathrm{Ker}\, P = \{A \in \mathrm{Ob}\mathsf{E} \mid \dot{0}_A \in \Sigma\}. \tag{5.30}$$

Proof The construction is analogous to that of 5.2.2, and we use the notation of its proof.

Let the expression Σ_{0j}-*functor* mean here any exact functor from E to an *unrestricted exj-category* that makes the maps of Σ_0 invertible; let Σ be the set of morphisms of E which are made invertible by all such functors. Then the ex1-functor $P\colon \mathsf{E} \to \mathsf{C} = \mathsf{E}^\sharp/R$ constructed as in the proof of 5.2.2 (from Σ) is such that all Σ_{0j}-functors factorise through it in EX_1, and we only need to show that C is exj.

In case ex2, let $f = nm$ be the composite of two normal monos of E; then $f = n'h$, with $n' = \operatorname{nim}(nm)$ and Fh is an isomorphism for all Σ_{0j}-functors F (because F takes values in an ex2-category). Therefore $[f] \sim [n']$ is a normal mono in our category of fractions.

For ex3, let $[m], [n]$ be normal subobjects of E in Σ^{-1}E, with $[n] \leqslant [m]$. As before (in 5.2.4), we can assume that $n \leqslant m$ in E and take $p = \operatorname{cok} n$. The normal factorisation $pm = k.g.q$ (in E) is preserved by every Σ_{0j}-functor F; since the codomain of F is homological, Fg is exact. In other words, $g \in \Sigma$ and $[pm]$ is exact in Σ^{-1}E.

In case ex4, let us begin by proving (5.28). Take a morphism $\varphi = h_r^\sharp f_r...h_1^\sharp f_1$ in $\mathsf{E}^\sharp$, with f_i in E and h_i in Σ; then the normal factorisation $f_i = m_i k_i p_i$ has central morphism k_i in Σ (by our assumption on the Σ_{0j}-functors F); as in the proof of 5.2.4, each m_i can be 'pushed to the left', by direct and inverse images, and all p_i can be 'pushed to the right'.

Now (5.29) follows from (5.28), since the central bracket of the factorisation (5.28)) is an isomorphism in the category of fractions. For (5.30), the object A is annihilated by P if and only if $P0_A$ is an isomorphism, if and only if $0_A \in \Sigma$ (if and only if A is annihilated by every Σ_{0j}-functor F); conversely, $f \in \Sigma$ if and only if Ff is an isomorphism (for every F), if and only if $\ker f$ and $\operatorname{cok} f$ are annihilated by all F, if and only if they are annihilated by P.

Last, C is g-exact: if $\ker[\varphi] = [\ker p] = 0$ and $\operatorname{cok}[\varphi] = [\operatorname{cok} m] = 0$, then p and m are in Σ, by (5.30), and $[\varphi]$ is an isomorphism.

For ex5, two parallel morphisms of $\mathsf{E}^\sharp$ between null objects of C are identified by the functors F' (for each Σ_{0j}-functor F), hence coincide.

Finally, the modular and distributive cases are easily proved, as above.

$\square$

5.3 Complements and reflectors

As a by-product of the previous study of categories of fractions, we get (strict) reflectors for the inclusions $\mathsf{EX}_i \to \mathsf{EX}_j$ (Section 5.3.1). We also show how full reflective subcategories can be seen as categories of fractions (Sections 5.3.3 - 5.3.5).

E is always a semiexact category.

5.3.1 The exact quotient of a semiexact category

Every ex1-category E gives an ex1-functor from E to a g-exact category, called the *exact quotient* of E

$$P \colon \mathsf{E} \to \mathrm{Exq}\, \mathsf{E}, \qquad (5.31)$$

which is universal for this property: every exact functor $F \colon \mathsf{E} \to \mathsf{E}'$ with values in an (unrestricted) g-exact category factorises uniquely through P.

The solution consists plainly in the ex4-category of fractions of E produced by the empty set $\Sigma_0 = \emptyset$ of mappings (cf. 5.2.5), or equivalently by the set Σ of all the mappings that are made invertible by every exact functor with values in an unrestricted exact category. Every map in $\mathrm{Exq}\, \mathsf{E}$ factorises as in (5.28).

This construction will be used in 7.1.2, for the exact quotient of Inj_2.

We have seen that EX_4 is (strictly) reflective in EX_1. In the same way, one gets the reflector of EX_j in EX_i, for every inclusion of the semilattice (5.1); and in particular the *homological quotient* of a semiexact category.

5.3.2 Factorisation of ex4-functors

Let $F \colon \mathsf{E} \to \mathsf{E}'$ be an ex1-functor *with values in a g-exact category* E'. Then, in its factorisation $F = GP$ through the ex4-category of fractions determined by $\Sigma = \mathrm{Ikr}F$, the functor G is conservative.

Indeed, using the factorisation (5.28) of a morphism $[\varphi]$ of $\Sigma^{-1}\mathsf{E}$, the map

$$G[\varphi] = G([m].[h_r^\sharp h_{r-1}...h_2^\sharp h_1].[p]) = Fm.(Fh_r)^{-1}.....(Fh_1).Fp, \qquad (5.32)$$

is an isomorphism in E' if and only if Fm and Fp are, if and only if m and p belong to Σ, if and only if $[\varphi]$ itself is an isomorphism.

Our factorisation $F = GP$ can be characterised by saying that P is an ex4-fraction functor (of an ex1-category) and G a conservative ex4-functor, as these properties plainly imply that $\mathrm{Ikr}P = \mathrm{Ikr}F$.

In particular, if F (i.e. E) is ex4, we find a factorisation structure of EX_4 by fraction-functors and conservative functors, which will be further analysed in Section 5.5.

5.3.3 Full reflective subcategories

Full reflective subcategories can be seen as categories of fractions, up to equivalence; this is also of interest in the semiexact case.

First, let an adjunction $P \dashv J$ be given, between arbitrary categories

$$P \colon \mathsf{C} \rightleftarrows \mathsf{X} \colon J,$$

$$r \colon 1_\mathsf{C} \to JP, \quad s \colon PJ \cong 1_\mathsf{X},$$

(5.33)

where the counit s is an isomorphism, or equivalently J is full and faithful ([M5], IV.3, Theorem 1).

This plainly amounts to saying that X is equivalent to a *full reflective subcategory* of C, namely its image $J\mathsf{X}$; and is isomorphic to the latter whenever J is injective on objects. It is a very common situation as we remarked when introducing N-adjunctions (in 1.7.5). The main examples we are interested in are treated below (in 5.3.4 and 5.3.5).

On the other hand, our situation - observed from the left adjoint side - involves *categories of fractions*. Let $\Sigma_P = \operatorname{Ikr}P$ be the set of P-relative isomorphisms, i.e. the maps of C that P makes invertible. Let Σ_R be the set of components $r_C \colon C \to JPC$ of the unit. Clearly, $\Sigma_R \subset \Sigma_P$, as $Pr_C = (s_{PC})^{-1}$; moreover, the problem of inverting the maps of Σ_P or Σ_R is the same, because the naturality diagram of $r \colon 1 \to JP$ on any map f of C

$$
\begin{array}{ccc}
A & \overset{f}{\longrightarrow} & B \\
{\scriptstyle r}\downarrow & & \downarrow{\scriptstyle r} \\
JPA & \underset{JPf}{\longrightarrow} & JPB
\end{array}
$$

(5.34)

shows that if $f \in \Sigma_P$, each functor F that makes the components of r invertible must do the same on f.

It is easy to see that the functor P solves the problem of inverting its own isokernel Σ_P, in the following weak sense: each functor $F \colon \mathsf{C} \to \mathsf{D}$ that makes these maps invertible factorises *up to isomorphism* through P ($F \cong TP$), by means of a functor $T \colon \mathsf{X} \to \mathsf{D}$ determined *up to isomorphism*. Indeed, we can always take $T = FJ$ and $Fr \colon F \cong FJP = TP$; conversely, if T' is any solution, i.e. $T'P \cong F$, then $T = FJ \cong T'PJ \cong T'$.

This means that X is equivalent to the category of fractions $\Sigma_P^{-1}\mathsf{C} = \Sigma_R^{-1}\mathsf{C}$, in the sense of general fractions, i.e. within CAT (see 5.2.1).

We assume now that C and X are semiexact categories and P, J are N-functors, forming an N-adjunction as before, with counit $s \colon PJ \cong 1$. Therefore P is right exact and J is left exact (Section 1.7.5).

J, as the composite of the equivalence $\mathsf{X} \to J\mathsf{X}$ with the embedding $J\mathsf{X} \to \mathsf{C}$, is exact if and only if its image $J\mathsf{X}$ is a semiexact subcategory of C.

On the other hand, it is easy to show that:

- P is the solution of the problem of inverting Σ_P (or Σ_R) within semiexact categories and *right exact* functors;

- if it is exact, it is also the solution within ex1-functors (see 5.2.1).

To see this, take $F: \mathsf{C} \to \mathsf{D}$ right exact with $\mathrm{Ikr}F \supset \Sigma_P$, and let us prove that $T = FJ$ preserves cokernels. Given $f: A \to B$ in C, $J(\mathrm{cok}\, f)$ factorises uniquely as $\mathrm{cok}\,(Jf)$ followed by a map x in X

$$J(\mathrm{cok}\, f) = x.\mathrm{cok}\,(Jf): \mathrm{Cok}\, Jf \to J(\mathrm{Cok}\, f),$$

$$PJ(\mathrm{cok}\, f) = Px.P(\mathrm{cok}\, Jf) = Px.\mathrm{cok}\,(PJf), \tag{5.35}$$

$$FJ(\mathrm{cok}\, f) = Fx.\mathrm{cok}\,(FJf).$$

But $PJ \cong 1$ is exact, whence Px is an isomorphism, $x \in \Sigma_P$ and Fx is an isomorphism. The last assertion is trivial: if F is exact, $T = FJ$ preserves kernels by composition.

5.3.4 Pairs and pointed objects

We have already seen, in 1.7.5(b), that the obvious (full) embedding $J: \mathsf{Top}_\bullet \to \mathsf{Top}_2$ is reflective, with a left adjoint P (the *reflector*) that is exact

$$J: \mathsf{Top}_\bullet \to \mathsf{Top}_2, \quad X \mapsto (X, \{0_X\}),$$

$$P: \mathsf{Top}_2 \to \mathsf{Top}_\bullet, \quad (X, A) \mapsto X/A,$$

$$r: 1 \to JP, \quad r_{XA}: (X, A) \to JP(X, A) = (X/A, \{0\}), \tag{5.36}$$

$$s: PJ \to 1, \quad s_X: PJX = X/\{0\} \cong X.$$

(Recall that $P(X, A) = X/A$ denotes the pushout of the inclusion $A \to X$ along the map $A \to \top = \{*\}$, so that $X/\emptyset = X^+$ is the space X with an isolated base-point added).

The adjunction is N-consistent, i.e. J and P are N-functors (Section 1.7.5), J is left exact but P is even exact. Therefore $\mathsf{Top}_\bullet$ is equivalent to the category of fractions $\Sigma_P^{-1}\mathsf{Top}_2 = \Sigma_R^{-1}\mathsf{Top}_2$, with $\Sigma_P = \mathrm{Ikr}P$; the maps of Σ_P are precisely the relative homeomorphisms, linking the present situation with the relative-homeomorphism axiom of homology theories (cf. 6.1).

Analogous facts hold for $\mathsf{Set}_\bullet$ and Set_2. More generally, the relations between pairs and pointed objects can be formalised as follows.

Let C be a category with distinguished subobjects (Section 2.5) and assume that:

(a) C has a terminal object $\top$, and every map $\top \to A$ is a distinguished mono,

(b) every distinguished subobject $a\colon A \to X$ has a pushout - say X/A - along the map $A \to \top$.

Then $\mathsf{C}_\bullet$ is the full subcategory of C_2 of 'pairs' $\top \to A$. It is pointed, with $0 = (\top \to \top)$, and p-semiexact, with kernels constructed as in C_2 and cokernel of $f\colon (\top \to A) \to (\top \to B)$ given by B/fA, where fA denotes the distinguished subobject $f_*(1_A)$ of B. Again, if $\top$ is a zero object $\mathsf{C}_\bullet$ is isomorphic to C.

The (full) inclusion $J\colon \mathsf{C}_\bullet \to \mathsf{C}_2$ has left adjoint $P\colon (X, A) \mapsto X/A$, constructed with the pushout (b); the adjunction is N-consistent, J is left exact and P is right exact. By 5.3.3, $P\colon \mathsf{C}_2 \to \mathsf{C}_\bullet$ is equivalent to the category of fractions $\Sigma_P^{-1}\mathsf{C}_2 = \Sigma_R^{-1}\mathsf{C}_2$, in the general sense as well as within semiexact categories and *right exact* functors.

5.3.5 *Groups as pairs*

The category Gp ($\cong \mathsf{Gp}_\bullet$) has a reflective full embedding I in Gp_2 (already considered in 1.6.8), with a reflector K

$$K\colon \mathsf{Gp}_2 \rightleftarrows \mathsf{Gp} : I \qquad\qquad K \dashv I,$$
$$I(G) = (G, 0), \qquad\qquad K(X, A) = X/\overline{A}, \qquad (5.37)$$
$$r\colon (X, A) \to (X/\overline{A}, 0), \qquad s\colon G/0 \cong G,$$

where $\overline{A}$ is the invariant closure of A in X. I is left exact and K is right exact; both are not exact.

5.4 Relations for g-exact categories

The use of categories of relations in homological algebra, as a tool for formalising subquotients and induced morphisms, goes back at least to Mac Lane's 'Homology' [M3], for categories of modules.

These categories of relations have been extended to the abelian framework [M2, P2, Hi], and later to the p-exact one [T1, T2, BP, G1], as expounded in Part I. In the last case, a two-map calculus is no longer available, because of the lack of general pullbacks and general pushouts, and one has to set up a more complicated construction by four-map diagrams (see I.2.5).

The further extension to generalised exact categories, as defined here, is fairly easy; actually, it remains in the setting of 'quaternary' categories, already treated by the author in [G1]. In this way, the methods of Part I can be extended to g-exact categories. It might be interesting to study their extension to a homological category: but this would give a non-associative composition of relations, likely forming a lax or colax 2-category; we shall not attempt this, here.

E is always a g-exact category.

5.4.0 Involutive ordered categories

We begin by recalling some terminology, already introduced in I.2.4.

An *involutive category* is a category A equipped with an involution, i.e. a contravariant involutive endofunctor, that is the identity on objects and, acting on a map $a \colon A \to B$, yields a map $a^\sharp \colon B \to A$. As in Part I, we are only interested in *regular involutions*, in von Neumann sense: $a = aa^\sharp a$, for every morphism a. (This has nothing to do with 'regular categories'.)

An *involutive ordered* category $A = (A, \sharp, \leqslant)$ is a category equipped with a regular involution $(-)^\sharp$ and a consistent (categorical) order relation $\leqslant$. Our axioms, omitting the obvious ones for the order relation, are thus as follows

$$\operatorname{Dom} a^\sharp = \operatorname{Cod} a, \qquad\qquad 1^\sharp = 1,$$
$$(ba)^\sharp = a^\sharp b^\sharp, \qquad\qquad a^{\sharp\sharp} = a, \qquad\qquad (5.38)$$
$$a = aa^\sharp a, \qquad\qquad a \leqslant b \ \Rightarrow\ a^\sharp \leqslant b^\sharp.$$

Because of the involution and of its regularity, our category is selfdual and balanced

(a) a is mono $\Leftrightarrow a^\sharp a = 1 \Leftrightarrow a$ is a split mono $\Leftrightarrow a^\sharp$ is epi,

(b) a is epi $\Leftrightarrow aa^\sharp = 1 \Leftrightarrow a$ is a split epi $\Leftrightarrow a^\sharp$ is mono,

(c) a is iso $\Leftrightarrow (a^\sharp a = 1$ and $aa^\sharp = 1) \Leftrightarrow a$ is mono and epi.

Therefore, an epi-mono factorisation of a morphism must be essentially unique.

The *proper* morphisms u are defined by the following conditions:

$$u^\sharp u \geqslant 1, \qquad uu^\sharp \leqslant 1. \qquad\qquad (5.39)$$

They form a subcategory $\operatorname{Prp} A$, obviously non-closed under involution. The order, restricted to proper maps, becomes trivial

$$u \leqslant v \ \Rightarrow\ u = v. \qquad\qquad (5.40)$$

A *projection* of the object A is a symmetric idempotent endomap $e \colon A \to A$, characterised by the following equivalent properties:

$$e = e^\sharp = ee \ \Leftrightarrow\ e = e^\sharp e \ \Leftrightarrow\ e = ee^\sharp. \qquad\qquad (5.41)$$

They form the set $\operatorname{Prj} A$ of projections of A. This set is not closed under composition (the product of two projections being just idempotent, generally); it has two canonical orders: the restriction of $\leqslant$ and the order $e \prec f$ produced by factorisation, and characterised as follows

$$e = ef \iff e = fe$$
$$\iff (\exists e' \in \mathrm{Prj}A : e = e'f) \iff (\exists e' \in \mathrm{Prj}A : e = fe'). \tag{5.42}$$

A *RO-category* (I.2.4.5, or [G5]) is an involutive ordered category that is a $\mathcal{U}$-category and such that all sets $\mathrm{Prj}A$ of projections are small.

A *RO-functor* is a functor between such categories that preserves the involution and the order; then it also preserves proper maps, projections and their two order-relations. A *RO-transformation* $\varphi \colon F \to G \colon \mathsf{A} \to \mathsf{B}$ (between RO-functors) assigns to each object A in A a *proper* morphism $\varphi A \colon FA \to GA$ of B, so that: for every map $a \colon A \to A'$ in A, $\varphi A'.Fa \leqslant Ga.\varphi A$ (*lax-naturality*); note that, because of (5.40), φ *is natural on proper maps*.

All this forms the 2-category RO of RO-categories. It is *strictly* complete (I.3.7.4) and the obvious 2-functor Prp : RO $\to$ CAT preserves 2-limits.

5.4.1 Relations for generalised exact categories

A g-exact category E is *biquaternary* (I.2.5.1, or [G1]), i.e. satisfies the following axioms (Q1, Q2, Q3, Q3*, Q6):

(Q1) every morphism has a unique epi-mono factorisation,

(Q2) inverse images of monos and direct images of epis exist,

(Q3) the pullback axiom holds (Section 2.2.2),

(Q3*) the pushout axiom holds (Section 2.2.2),

(Q6) a square formed of two parallel monos and two parallel epis is a pullback if and only if it is a pushout.

Indeed, (Q1) follows from 2.2.6(b) (the epi-mono factorisation coincides with the normal one); (Q2) from 1.5.6; (Q3), (Q3*) from 2.2.2; (Q6) from 2.2.1.

Therefore, as proved in I.2.5 (or [G1]) and sketched below, the construction of relations for E can be performed via four-map factorisations, in the same way as for p-exact categories [T1, T2, BP], and yields an involutive ordered category (Section 5.4.0) Rel E with the same objects.

5.4.2 Construction of relations

A *relation* $a\colon A \to B$ is an equivalence-class of four-map diagrams (or *w-diagrams*) of monomorphisms and epimorphisms of E

$$A \overset{m}{\nwarrow} \bullet \overset{p}{\nearrow} \bullet \overset{q}{\nwarrow} \bullet \overset{n}{\nearrow} B \qquad a = [m, p, q, n], \tag{5.43}$$

two such diagrams being identified when there exist three isomorphisms u, v, w (uniquely determined) that make the following diagram commutative

$$\begin{array}{ccccccc}
A & \xleftarrow{m} & \bullet & \xrightarrow{p} & \bullet & \xleftarrow{q} & \bullet & \xrightarrow{n} & B \\
\| & & \downarrow u & & \downarrow w & & \downarrow v & & \| \\
A & \xleftarrow{m'} & \bullet & \xrightarrow{p'} & \bullet & \xleftarrow{q'} & \bullet & \xrightarrow{n'} & B
\end{array} \,. \tag{5.44}$$

An identity is defined by four identities of E; the involution by reversing the diagrams; the order $a \leqslant a'$ by a commutative diagram as above, where u, v and w are just supposed to be morphisms; these are again uniquely determined, so that the conditions $a \leqslant a'$ and $a' \leqslant a$ do imply $a = a'$.

Last, the composition of relations is obtained by using epi-mono factorisations, together with limits and colimits which exist in E, namely inverse images (pullbacks) of monos and direct images (pushouts) of epis. More precisely, given the relations

$$a = [m, p, q, n]\colon A \to B, \qquad b = [m', p', q', n']\colon B \to C,$$

the composite $ba\colon A \to C$ is the class of the diagonal, dashed four-tuple in the following commutative diagram of E

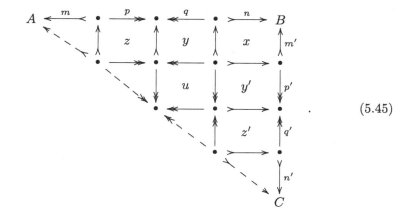

$$(5.45)$$

Here, the square (x) is a pullback, (y) and (y') are commutative (by epi-mono factorisation), (z) and (z') are pullbacks and (u) is a pushout.

The only not obvious part of the construction is the associativity of this composition; it can be proved as in I.2.5.8 (where we follow the same line as in the first proof, by Tsalenko [T2], for the p-exact case).

We have thus constructed the category of relations Rel E. E is canonically embedded in its category of relations, by identifying the morphism $u = mp$ (epi-mono factorisation) with the relation $[1, p, 1, m]$, plainly independent of the chosen factorisation of u. This embedding is consistent with composition and identities.

By construction, every relation has a *w-factorisation*

$$a = nq^{\sharp}pm^{\sharp} \quad (p, q \text{ epis of E; } m, n \text{ monos of E}). \tag{5.46}$$

The pullback of a mono along itself is the identity, and dually; it follows that, for m mono, p epi and u invertible in E:

$$m^{\sharp}m = 1, \qquad pp^{\sharp} = 1, \qquad u^{\sharp} = u^{-1}. \tag{5.47}$$

This implies the regularity of the involution

$$aa^{\sharp}a = (nq^{\sharp}pm^{\sharp}).(mp^{\sharp}qn^{\sharp}).(nq^{\sharp}pm^{\sharp}) = nq^{\sharp}pm^{\sharp} = a. \tag{5.48}$$

Notice that, *if E has small hom-sets, the same holds for* Rel E. Indeed, we can rewrite the w-factorisation (5.46) as $a = nq^{\sharp}.i.pm^{\sharp}$, where m and n are subobjects, p and q are quotients and i is an isomorphism; then the thesis follows from the fact that E is (already assumed to be) well-powered and well-copowered.

5.4.3 Functors and transformations

We know that an exact functor $F: E \to E'$ between g-exact categories preserves monos and their inverse images, epis and their direct images; therefore it extends uniquely to an involution-preserving functor

$$\mathrm{Rel}\, F: \mathrm{Rel}\, E \to \mathrm{Rel}\, E'. \tag{5.49}$$

(Conversely, if such an extension exists and F preserves the null objects, then F is exact, with the same proof as in I.2.6.2 for the p-exact case). Rel F also preserves the order $\leqslant$ and the *null* relations, i.e. those which factorise through a null object.

A natural transformation $\varphi: F \to G: E \to E'$ between exact functors gives a RO-transformation Rel $\varphi: \mathrm{Rel}\, F \to \mathrm{Rel}\, G$, that has the *same* (proper) components as φ (but is lax-natural on relations!)

$$(\mathrm{Rel}\, \varphi)A = \varphi A: FA \to GA. \tag{5.50}$$

All this defines a 2-functor Rel : $EX_4 \to RO$, once we verify the smallness property of projection-sets (in 5.4.4(d)). We list below some of the main properties of this construction.

5.4.4 Proposition (Monorelations and isomorphisms)

(a) For a relation $s\colon S \to A$, the following conditions are equivalent:

(i) s is a monorelation *(i.e. a monomorphism of* Rel E*),*

(ii) $s^\sharp s = 1$,

(iii) $s = nq^\sharp$, with n mono and q epi in E *(reduced w-factorisation).*

(b) The isomorphisms of Rel E *coincide with those of* E.

(c) A morphism u of E *is mono (resp. epi) in* Rel E *if and only if it is in* E.

(d) An endorelation $e\colon A \to A$ is a projection if and only if it has a w-factorisation of type $e = mp^\sharp pm^\sharp$; all sets Prj$A$ *are small.*

(e) Every relation a has an essentially unique epi-mono factorisation in Rel E, *that can be deduced from the w-factorisation (5.46)*

$$a = (nq^\sharp).(pm^\sharp). \tag{5.51}$$

Proof (a) (i) $\Leftrightarrow$ (ii) By the characterisation of monos in a category with regular involution, in 5.4.0(a). (ii) $\Leftrightarrow$ (iii) If $s = nq^\sharp pm^\sharp$ is a monorelation, it follows that 1 has the following w-factorisation

$$1 = s^\sharp s = (mp^\sharp qn^\sharp).(nq^\sharp pm^\sharp) = mp^\sharp pm^\sharp. \tag{5.52}$$

From the uniqueness of the latter, m and p are isomorphisms (and can be chosen to be identities). The converse follows trivially from (5.47).

(b) An isomorphism a of Rel E is mono and epi, whence its w-factorisation is composed of isomorphisms of E and $a = mp^\sharp qn^\sharp = mp^{-1}qn^{-1}$ is an isomorphism in E.

The rest of the proof follows easily, along the same lines. The smallness of the sets Prj$A$ follows from the factorisation of projections and the smallness of the sets of subobjects and quotients of E. $\qquad\square$

5.4.5 Other factorisations

A relation $a\colon A \to B$ determines up to isomorphism a diagram of E, whose two squares are bicartesian in E (see 2.2.1); this diagram contains the following 'main' factorisations of the relation a (same terminology as in I.2.5.6):

$$A \xrightarrow{\quad r \quad} X \xrightarrow{\quad s \quad} B \qquad f = n'p, \quad g = m'q, \qquad (5.53)$$

$$a = nq^\sharp pm^\sharp \qquad\qquad (\textit{w-factorisation}, \text{ along the lower path}),$$

$$a = q'^\sharp n'm'^\sharp p' \qquad\qquad (\textit{w*-factorisation}, \text{ along the upper path}),$$

$$a = q'^\sharp.f.m^\sharp \qquad\qquad (\textit{ternary factorisation}, \text{ along the solid path}),$$

$$a = n.g^\sharp.p' \qquad\qquad (\textit{coternary factorisation}, \text{ along the dashed path}),$$

$$a = s.r = (nq^\sharp).(pm^\sharp) = (q'^\sharp n').(m'^\sharp p')$$
$$(\textit{epi-mono factorisation}, \text{ along the straight path}).$$

The central object X of the diagram is the coimage-image of a in $\mathrm{Rel}\,\mathsf{E}$.

The composition by w*-factorisations is dual to the procedure (5.45); this shows that $\mathrm{Rel}\,\mathsf{E}^{\mathrm{op}}$ is isomorphic to $\mathrm{Rel}\,\mathsf{E}$. The composition by ternary or coternary factorisations, described below in (5.75), is easy to guess and can be followed for an alternative construction of $\mathrm{Rel}\,\mathsf{E}$, that is simpler but has the drawback of being asymmetric, hence not stable under involution.

The w- and w*-factorisations of a relation, both *invariant under involution*, are more appropriate for the study of relations, of subquotients and induced relations. On the other hand, the ternary and coternary factorisations (interchanged by the involution) are more suitable for studying categories of fractions (see Section 5.5), and we shall occasionally use them.

The direct and inverse images of subobjects extend in the obvious way to relations (as in I.4.2.5): if $a = p^\sharp fm^\sharp = ng^\sharp q : A \to B$, we get two increasing mappings:

$$a_* : \mathrm{Sub}A \to \mathrm{Sub}B, \quad a_* = p^*f_*m^* = n_*g^*q_*,$$
$$a^* : \mathrm{Sub}B \to \mathrm{Sub}A, \quad a^* = m_*f^*p_* = q^*g_*n^*. \qquad (5.54)$$

This pair forms a 'modular relation' (I.4.2.3). But here we only need the fact that this procedure is consistent with the composition of relations: $(ba)_* = b_*a_*$ and $(ba)^* = a^*b^*$.

Finally, let us notice that the transfer mappings of a subquotient $s = mq'^\sharp = q^\sharp m' : S \to A$, defined in 3.1.2 for a *homological* category E

$$s_* : \mathrm{Nsb}S \to \mathrm{Nsb}A, \qquad s_*(y) = m_*q'^*(y) = q^*m'_*(y),$$
$$s\hat{\,}, s\check{\,} : \mathrm{Nsb}A \to \mathrm{Nsb}S, \quad s\hat{\,}(x) = q'_*m^*(x) \leqslant s\check{\,}(x) = m'^*q_*(x), \qquad (5.55)$$

agree with the present ones, with $s^* = s^\wedge = s^\vee$. (The coincidence $s^\wedge = s^\vee$ follows from the modularity of lattices of subobjects, in the present g-exact case, cf. 3.1.2.)

5.4.6 Subquotients and projections

In particular, a monorelation s is characterised by having two (unique) factorisations, that form a bicartesian square in E

$$
\begin{array}{ccc}
M & \overset{m}{\rightarrowtail} & A \\
q\downarrow & \nearrow_{s} & \downarrow p \\
S & \underset{n}{\rightarrowtail} & A/N
\end{array}
\qquad (5.56)
$$

$s = mq^\sharp$, with m mono, q epi   (*reduced w-factorisation*),

$s = p^\sharp n$, with n mono, p epi   (*reduced w*-factorisation*).

Therefore $S \cong M/N$ and the subobjects of A in $\mathrm{Rel}\,\mathsf{E}$ can be identified with the *subquotients* of A with respect to E (1.3.6; 2.2.1).

A subquotient s of A can also be identified with a projection $e \in \mathrm{Prj}\,A$, via the bijection

$$
s \mapsto ss^\sharp, \qquad e \mapsto \mathrm{im}\,e. \qquad (5.57)
$$

This formulation bypasses any problem of choice of representatives, since s and t are equivalent monorelations with codomain A if and only if $ss^\sharp = tt^\sharp$.

$\mathrm{Rel}\,\mathsf{E}$ makes possible to consider a much more general notion of induction than the regular one (studied in Sections 3.1, 3.2 for homological categories). In this sense, as considered in Mac Lane's [M3] for categories of modules, given a relation $a\colon A \to B$, a subquotient $s\colon S \to A$ of its domain and a subquotient $t\colon T \to B$ of its codomain, a *induces from s to t* the relation

$$
t^\sharp as\colon S \to T. \qquad (5.58)
$$

This notion extends regular induction on subquotients. Indeed, if a is in E, s and t are normal subobjects (in E) and $a_*(s) \leqslant t$, the induced morphism a' is determined by the condition $tm' = as$, so that, in $\mathrm{Rel}\,\mathsf{E}$, $m' = t^\sharp tm' = t^\sharp as$; analogously, the two notions agree when s and t are reversed relations of normal quotients. By the transitivity of regular induction (Section 3.2.3) and of the present generalised notion (obvious), the equality also holds in the general case.

As already remarked in [M3], general induction need *not* be consistent with composition.

We have seen in Part I that, for a p-exact category E, general induction has a good behaviour if and only if E is *distributive*, if and only if Rel E is *orthodox* (its idempotent endomaps are stable under composition). Indeed, by the 'coherence theorem for homological algebra' (Theorem I.2.7.6), these conditions are necessary and sufficient in order that the induced isomorphisms between subquotients be preserved by composition. This fact has a real interest, since various theories of homological algebra have a *distributive* classifying p-exact category: for instance, as proved in I.6, this is the case of the filtered complex, the double complex, the exact couple, etc.

5.4.7 *RE'-categories*

The notion of RE-category was studied in [G5] and I.3, in order to have an embedding Rel : EX → RE in a *strictly* complete 2-category, where universal problems can be solved by means of Freyd's theorems; in particular, the existence of universal models for theories.

The idea is to simulate kernels and cokernels, that are only preserved *up to isomorphism* by an exact functor F, by means of projection-operators, the numerator and denominator of a projection, that are *strictly* preserved by the associated functor Rel F.

All this extends to generalised exact categories, by means of a slight extension of the previous notion. A *RE'-category* $A = (A, \natural, \leqslant)$ is defined by the following axioms (where the axiom (RE.2) is replaced with a weaker form (RE'.2)):

(RE.0) A is a RO-category (Section 5.4.0);

(RE.1) every projection e has precisely one pair of projections, the *numerator* $n(e)$ and the *c-denominator* $d^c(e)$, that satisfy the following conditions (mutually dual with respect to the order $\leqslant$):

$$e.ne = e, \quad n(e) \leqslant 1, \quad n(e) \leqslant e, \tag{5.59}$$

$$e.d^c(e) = e, \quad d^c(e) \geqslant 1, \quad d^c(e) \geqslant e; \tag{5.60}$$

(RE'.2) for every object A, there exist two relations $\omega_A = \min(\mathrm{Prj}A)$ and $\Omega_A = \max(\mathrm{Prj}A)$ so that

$$\omega_A \Omega_A \omega_A = \omega_A, \quad \Omega_A \omega_A \Omega_A = \Omega_A. \tag{5.61}$$

Actually, if $A = \mathrm{Rel}\,E$ is the category of relations of a g-exact category, given a projection e with w- and w*-factorisations

$$e = (A \leftarrowtail M \twoheadrightarrow M/N \leftarrow M \rightarrowtail A)$$
$$= (A \twoheadrightarrow A/N \leftarrowtail M/N \rightarrowtail A/N \leftarrow A), \tag{5.62}$$

its numerator and co-denominator are

$$n(e) = (A \leftarrowtail M \rightarrowtail A), \quad d^c(e) = (A \twoheadrightarrow A/N \leftarrowtail A). \tag{5.63}$$

Further, the least and the greatest projection of A with respect to $\leqslant$ are

$$\begin{aligned}
\omega_A &= (A \leftarrowtail A_0 \rightarrowtail A) = \min(\mathrm{Prj}A), \\
\Omega_A &= (A \twoheadrightarrow A_0 \leftarrowtail A) = \max(\mathrm{Prj}A),
\end{aligned} \tag{5.64}$$

where A_0 is any null (quasi-zero) object in the connected component of A and $A_0 \rightarrowtail A \twoheadrightarrow A_0$ are respectively the least subobject and the least quotient of A (Section 2.2.3).

If E is componentwise p-exact, it is easy to see that these relations are also the minimum and maximum of $\mathrm{Rel}\,E(A, A)$, so that the original axiom (RE.2) is satisfied.

In this way, the categories of relations of g-exact categories can be characterised as RE'-categories with epi-mono factorisations, extending a similar result concerning p-exact categories (I.4.1.1 or [G5]). The precise statement is as follows.

5.4.8 Theorem (Characterisation of categories of relations)

Let $A = (A, \sharp, \leqslant)$ *be a RO-category (Section 5.4.0) and* $E = \mathrm{Prp}\,A$ *its subcategory of proper morphisms. The following conditions are equivalent:*

(a) A *is a RE'-category with (unique) epi-mono factorisations,*

(b) E *is a g-exact category and* A *is RO-isomorphic to* $\mathrm{Rel}\,E$,

(c) E *is a g-exact category and there is a (unique) involution-preserving isomorphism* $\mathrm{Rel}\,E \to A$ *consistent with the embeddings* $E \to \mathrm{Rel}\,E$ *and* $E \to A$.

Proof As for I.4.1.1, with minor adaptations. $\qquad\square$

5.4.9* The projection-completion of a RO-category

Finally, we show that every RE'-category A has a canonical full embedding in the category of relations of an associated g-exact category, $E = \mathrm{Prp}\,\mathrm{Fct}\,A$. Because of the previous theorem, this is equivalent to embed A in a RE'-category with epi-mono factorisations.

First, let us recall the construction of the projection-completion $\mathrm{Fct}\,A$ of a RO-category A (I.3.3.4). Because of the regular involution, every morphism

$a\colon A \to B$ has two associated projections that, respectively, simulate its coimage and its image

$$c(a) = a^\sharp a \in \mathrm{Prj}A, \qquad i(a) = aa^\sharp \in \mathrm{Prj}B, \qquad (5.65)$$

$$a = a.c(a) = i(a).a = i(a).a.c(a). \qquad (5.66)$$

It is easy to see that A has epi-mono factorisations if and only if its projections *split* (have such a factorisation). Indeed, if the projection e splits, one has

$$e = s.r, \qquad rr^\sharp = 1, \qquad s^\sharp s = 1, \qquad (5.67)$$

$$ss^\sharp = srr^\sharp s^\sharp = ee^\sharp = e, \qquad r^\sharp r = e, \qquad (5.68)$$

Thus, splitting $c(a) = s.r$, we get an epi-mono factorisation of a (essentially unique, by 5.4.0)

$$a = (as).r, \quad (as)^\sharp(as) = s^\sharp a^\sharp a s = s^\sharp ss^\sharp s = 1. \qquad (5.69)$$

We now define the *projection-completion* Fct A of the RO-category A (a full subcategory of the well-known 'idempotent-completion' of a category). An object is an A-projection, a morphism

$$a = (a; e, f)\colon e \to f \quad (e \in \mathrm{Prj}A,\ f \in \mathrm{Prj}B), \qquad (5.70)$$

is given by any A-morphism $a\colon A \to B$ with $a = fae$. Fct A is a RO-category with composition, involution and order induced by those of A; the identity of the projection e is $e\colon e \to e$.

A projection $e_0\colon e \to e$ is given by any A-projection $e_0 \prec e$ and splits epi-mono as follows

$$e_0 = (e_0; e_0, e).(e_0; e, e_0)\colon e \to e_0 \to e. \qquad (5.71)$$

Therefore Fct A has (essentially unique) epi-mono factorisations. A has a full RO-embedding

$$U\colon A \to \mathrm{Fct}\,A, \qquad A \mapsto 1_A, \quad a \mapsto (a; 1, 1), \qquad (5.72)$$

and every RO-functor $F\colon A \to B$ with values in a RO-category with epi-mono factorisations can be extended to Fct A, uniquely up to isomorphism.

Now, if A is a RE′-category, so is Fct A (same proof as in I.4.1.4, for RE-categories). It follows, by 5.4.8, that $E = \mathrm{Prp\,Fct}\,A$ is a g-exact category and A is a full subcategory of Fct A $\cong$ Rel E.

Since every full subcategory of a RE′-category is plainly a RE′-category, we have also proved that *RE′-categories coincide (up to isomorphism) with the full subcategories of the categories of relations over g-exact categories.*

5.5 Fractions for g-exact categories

Categories of fractions for p-exact categories have been studied in [G12]. We extend here these results to g-exact categories, showing that their categories of fractions can be given an explicit construction, with a three-arrow calculus inherited from relations. They solve the universal problem of inverting a set of morphisms, closed under some conditions, at the same time in CAT and in EX$_4$, i.e. for general functors as well as for exact functors.

All this works similarly for p-exact and abelian categories.

The next section will show that such categories of fractions are equivalent to (a generalisation of) Serre's quotients modulo thick subcategories.

5.5.1 Exact isokernels

Let E be a g-exact category. We now consider *general* categories of fractions of E, that solve the universal problem, of inverting a set of morphisms, for all categories and functors - not within EX$_1$ as in Section 5.2.

A set Σ of morphisms of E will be said to be an *exact isokernel* if:

(ik.0) Σ contains all the isomorphisms of E,

(ik.1) *two-out-of three property*: given $h = gf$ in E, if two maps out of f, g, h are in Σ, so is the third,

(ik.2) *factorisation*: if $f = mp$ is an epi-mono factorisation, f belongs to Σ if and only if p and m do; furthermore, in the following commutative square, $m \in \Sigma$ implies $n \in \Sigma$, while $p \in \Sigma$ implies $q \in \Sigma$

$$\begin{array}{ccc} \bullet & \overset{m}{\rightarrowtail} & \bullet \\ {\scriptstyle q}\downarrow & & \downarrow{\scriptstyle p} \\ \bullet & \underset{n}{\rightarrowtail} & \bullet \end{array} \qquad (5.73)$$

(ik.3)

  - if the left diagram below is a *pullback*, $f \in \Sigma \Rightarrow g \in \Sigma$ and $n \in \Sigma \Rightarrow m \in \Sigma$,

  - if the right diagram below is a *pushout*, $f \in \Sigma \Rightarrow g \in \Sigma$ and $p \in \Sigma \Rightarrow q \in \Sigma$,

$$\begin{array}{ccc} \bullet & \overset{f}{\rightarrow} & \bullet \\ {\scriptstyle m}\uparrow & & \uparrow{\scriptstyle n} \\ \bullet & \underset{g}{\rightarrow} & \bullet \end{array} \qquad \begin{array}{ccc} \bullet & \overset{f}{\rightarrow} & \bullet \\ {\scriptstyle p}\downarrow & & \downarrow{\scriptstyle q} \\ \bullet & \underset{g}{\rightarrow} & \bullet \end{array} \quad . \qquad (5.74)$$

Clearly, *the isokernel* Ikr$F$ *of an ex4-functor* F *is an exact isokernel.* We shall prove that the converse is also true (Theorem 5.5.8).

5.5.2 Lemma (Properties of exact isokernels)

Let E *be a g-exact category and* Σ *a set of morphisms of* E *that satisfies (ik.0, 2, 3). The remaining axiom (ik.1) is equivalent to the conjunction of:*

(ik.1a) a composite of two monos (resp. epis) belongs to Σ *if and only if each of them does,*

(ik.1b) in the commutative square (5.73): m and q belong to Σ *if and only if n and p do.*

If these hold (and Σ *is an exact isokernel of* E*), then:*

(a) if $n \prec m$ *within monomorphisms and* $n \in \Sigma$*, then* $m \in \Sigma$*,*

(a*) if $p \prec q$ *within epimorphisms and* $p \in \Sigma$*, then* $q \in \Sigma$*.*

Proof Let Σ be an exact isokernel. The left square below is a pullback:

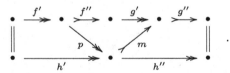

Thus, if $n \in \Sigma$, also $h \in \Sigma$, by (ik.3), and m too, by (ik.1). Consider now the commutative square (5.73), rewritten above, at the right: if m and q are in Σ, also n is, by (ik.2), and p too, by (ik.1); the other implication is dual to the former.

Conversely, let us assume (ik.1a), (ik.1b) and prove (ik.1). We consider a composite $h = gf$ and factor epi-mono these three morphisms: $f = f'' f'$ and so on:

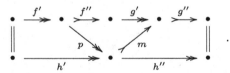

Now, if f and g are in Σ, so are f', f'', g', g'', by (ik.2), whence also p and m (by (ik.2)), h' and h'' (by (ik.1a)) and h (by (ik.2)). On the other hand, if h and f are in Σ, so are h', h'', f', f''; then $p, mg'' \in \Sigma$, by (ik.1a); then $g' \in \Sigma$, because of (ik.1b), and $g \in \Sigma$, by (ik.2). The last case follows by duality. $\qquad\square$

5.5.3 Lemma

Let Σ be an exact isokernel of an abelian category E. Then the set Σ is closed under binary biproducts: if the morphisms $f_i \colon A_i \to B_i$ $(i = 1, 2)$ belong to Σ, also $f_1 \oplus f_2 \colon A_1 \oplus A_2 \to B_1 \oplus B_2$ does.

Proof We begin by noting that, if a mono $m\colon A \rightarrowtail B$ is in Σ, also $m \oplus C\colon A \oplus C \rightarrowtail B \oplus C$ is (for every object C): we apply (ik.3) to the left pullback below, where p_1 and q_1 are projections

$$
\begin{array}{ccc}
B \oplus C & \xrightarrow{\ q_1\ } & B \\
{\scriptstyle m\oplus C}\Big\uparrow & & \Big\uparrow{\scriptstyle m} \\
A \oplus C & \xrightarrow[\ p_1\ ]{} & A
\end{array}
\qquad
\begin{array}{ccc}
A & \xrightarrow{\ u_1\ } & A \oplus C \\
{\scriptstyle p}\Big\downarrow & & \Big\downarrow{\scriptstyle p\oplus C} \\
\bullet & \xrightarrowtail[\ v_1\ ]{} & B \oplus C
\end{array}\ .
$$

Dually, if $p\colon A \twoheadrightarrow B$ is in Σ, also $p \oplus C\colon A \oplus C \rightarrowtail B \oplus C$ is: we apply (ik.3) to the right pushout above, where u_1 and v_1 are injections.

Now, if the two morphisms f_i are in Σ, their epi-mono factorisations $m_i p_i\colon A_i \to X_i \to B_i$ are also, by (ik.2). By the previous remark, $m_1 \oplus m_2 = (X_1 \oplus m_2)(m_1 \oplus A_2)$ is in Σ, as well as $p_1 \oplus p_2$. Finally, also $f_1 \oplus f_2 = (m_1 \oplus m_2)(p_1 \oplus p_2)$ is in Σ. $\qquad\square$

5.5.4 A three-arrow calculus

Let E be a g-exact category and Σ an exact isokernel of E.

The construction of the category of fractions $\Sigma^{-1}\mathsf{E}$ can be realised as a quotient $\mathsf{E}^\sharp/R$, of the subcategory $\mathsf{E}^\sharp$ of Rel E obtained by reversing only the morphisms of Σ. This provides an explicit three-arrow calculus.

More precisely, $\mathsf{E}^\sharp$ is the subcategory of Rel E generated by E and by the reversed arrows $h^\sharp$ of Σ. Each morphism φ of $\mathsf{E}^\sharp$ has a ternary factorisation (Section 5.4.5) of the following kind, where the dashed arrows are supposed to belong to Σ

$$
A \xleftarrowtail{\ m\ } \bullet \xrightarrow{\ f\ } \bullet \xleftarrow{\ p\ } B \qquad \varphi = p^\sharp f m^\sharp \quad (m, p \in \Sigma),
$$

Indeed such relations are closed under composition: this is computed by the following diagram, where the square (x) is commutative, (y) is a pullback and (z) a pushout

$$\tag{5.75}$$

The congruence R in $\mathsf{E}^\sharp$ is the following relation between parallel maps: $\varphi R\psi$ if there is a commutative diagram in E

$$A \xleftarrow{\ m\ } \bullet \xrightarrow{\ f\ } \bullet \xleftarrow{\ p\ } B \qquad \varphi = p^\sharp f m^\sharp$$

$$A \xleftarrow{\ \ } \bullet \longrightarrow \bullet \xleftarrow{\ \ } B \qquad (\eta)$$

$$A \xleftarrow{\ n\ } \bullet \xrightarrow{\ g\ } \bullet \xleftarrow{\ q\ } B \qquad \psi = q^\sharp g n^\sharp$$

i.e. if there is some $\eta \prec \varphi, \psi$ (the central morphism of $\mathsf{E}^\sharp$), where the relation $\eta \prec \varphi$ is described in the upper half of the diagram.

We now prove that R is indeed a congruence of categories and that $\mathsf{E}^\sharp / R = \Sigma^{-1}\mathsf{E}$.

5.5.5 Theorem

Let E be a g-exact category and Σ an exact isokernel of E (Section 5.5.1). Then the relation R defined above is a congruence of categories in $\mathsf{E}^\sharp$, and the natural functor

$$P\colon \mathsf{E} \to \mathsf{E}^\sharp / R, \quad P(f) = [f] = [1^\sharp . f . 1^\sharp], \tag{5.76}$$

is the category of fractions $\Sigma^{-1}\mathsf{E}$ produced by the set of morphisms Σ.

Proof (A) First, the relation $\eta \prec \varphi$ is a obviously a preorder. It is generated by the relations $\eta \prec_1 \zeta$ and $\zeta \prec_2 \varphi$ represented below (because, given $\eta \prec \varphi$, we can always interpose ζ as below)

$$A \xleftarrow{\ m\ } \bullet \xrightarrow{\ f\ } \bullet \xleftarrow{\ p\ } B \qquad \varphi = p^\sharp f m^\sharp$$

$$A \xleftarrow{\ n\ } \bullet \xrightarrow{\ h\ } \bullet \xleftarrow{\ p\ } B \qquad \zeta = p^\sharp h n^\sharp$$

$$A \xleftarrow{\ n\ } \bullet \xrightarrow{\ g\ } \bullet \xleftarrow{\ q\ } B \qquad \eta = q^\sharp g n^\sharp .$$

(B) It will also be useful to remark that, given a *commutative* diagram that links φ and ψ

$$A \xleftarrow{\ m\ } \bullet \xrightarrow{\ f\ } \bullet \xleftarrow{\ p\ } B \qquad \varphi = p^\sharp f m^\sharp$$

$$A \xleftarrow{\ n\ } \bullet \xrightarrow{\ g\ } \bullet \xleftarrow{\ q\ } B \qquad \psi = q^\sharp g n^\sharp$$

then $\varphi R\psi$; indeed the intermediate morphism $p^\sharp un^\sharp$ precedes both φ and ψ, with respect to $\prec$.

(C) The relation R is transitive. Since $\prec$ is transitive, it suffices to prove that, if $\varphi \prec \zeta$ and $\psi \prec \zeta$ in $\mathsf{E}^\sharp$

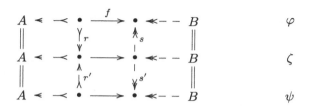

there is some η that precedes both φ and ψ in $\mathsf{E}^\sharp$: in fact, it can be constructed with the pullback of r, r' and the pushout of s, s'.

(D) R is consistent with composition; this is not trivial, since $\prec$ is not consistent. We shall prove that, if $\varphi \prec \psi$ and $\eta \prec \zeta$, then $\eta\varphi \, R \, \zeta\psi$. Because of (A) and (C), we can break the proof into four cases.

Case 1. $\varphi \prec_1 \psi$ and $\eta = \zeta$:

$$
\begin{array}{ccccccccccc}
A & \twoheadleftarrow\!\!\prec & \bullet & \longrightarrow & \bullet & \twoheadleftarrow\!- & B & \twoheadleftarrow\!\!\prec & \bullet & \longrightarrow & \bullet & \twoheadleftarrow\!- & C & \quad \eta\varphi \\
\| & & \| & & \downarrow & & \| & & \| & & \| & & \| \\
A & \twoheadleftarrow\!\!\prec & \bullet & \longrightarrow & \bullet & \twoheadleftarrow\!- & B & \twoheadleftarrow\!\!\prec & \bullet & \longrightarrow & \bullet & \twoheadleftarrow\!- & C & \quad \zeta\psi .
\end{array}
$$

The composition of the rows will be obtained in two steps. First, by canonical factorisation of the central part, around the object B

$$
\begin{array}{ccccccccccccc}
A & \twoheadleftarrow\!\!\prec & \bullet & \longrightarrow & \bullet & \prec\!\!-\!\!\prec & I & \twoheadleftarrow\!- & \bullet & \longrightarrow & \bullet & \twoheadleftarrow\!- & C & \quad \eta\varphi \\
\| & & \| & & \downarrow & & \downarrow & & \| & & \| & & \| \\
A & \twoheadleftarrow\!\!\prec & \bullet & \longrightarrow & \bullet & \prec\!\!-\!\!\prec & J & \twoheadleftarrow\!- & \bullet & \longrightarrow & \bullet & \twoheadleftarrow\!- & C & \quad \zeta\psi .
\end{array}
$$

Second, by two pullbacks of monos (X, Y) on the left side and two pushouts of epis (U, V) on the right one

$$
\begin{array}{ccccccccccccc}
A & \twoheadleftarrow\!\!\prec & \bullet & \prec\!\!-\!\!\prec & X & \longrightarrow & I & \longrightarrow & U & \twoheadleftarrow\!- & \bullet & \twoheadleftarrow\!- & C & \quad \eta\varphi \\
\| & & \| & & \downarrow & & \downarrow & & \downarrow & & \| & & \| \\
A & \twoheadleftarrow\!\!\prec & \bullet & \prec\!\!-\!\!\prec & Y & \longrightarrow & J & \longrightarrow & V & \twoheadleftarrow\!- & \bullet & \twoheadleftarrow\!- & C & \quad \zeta\psi .
\end{array}
$$

The thesis follows now from point (B).

Case 2. $\varphi \prec_2 \psi$ and $\eta = \zeta$:

$$
\begin{array}{ccccccccccc}
A & \leftarrowtail & \bullet & \longrightarrow & \bullet & \leftarrow\!\!- & B & \leftarrowtail & \bullet & \longrightarrow & \bullet & \leftarrow\!\!- & C & \qquad \eta\varphi \\
\| & & \uparrow\lambda & & \| & & \| & & \| & & \| & & \| & \\
A & \leftarrowtail & \bullet & \longrightarrow & \bullet & \leftarrow\!\!- & B & \leftarrowtail & \bullet & \longrightarrow & \bullet & \leftarrow\!\!- & C & \qquad \zeta\psi \, .
\end{array}
$$

We proceed as above. First, by canonical factorisation

$$
\begin{array}{ccccccccccccc}
A & \leftarrowtail & \bullet & \longrightarrow & \bullet & \leftarrowtail & I & \leftarrow\!\!- & \bullet & \longrightarrow & \bullet & \leftarrow\!\!- & C & \quad \eta\varphi \\
\| & & \uparrow\lambda & & \| & & \| & & \| & & \| & & \| & \\
A & \leftarrowtail & \bullet & \longrightarrow & \bullet & \leftarrowtail & I & \leftarrow\!\!- & \bullet & \longrightarrow & \bullet & \leftarrow\!\!- & C & \quad \zeta\psi
\end{array}
$$

then with two pullbacks of monos (X, Y) and one pushout of an epi (U)

$$
\begin{array}{ccccccccccccc}
A & \leftarrowtail & \bullet & \leftarrowtail & X & \longrightarrow & I & \longrightarrow & U & \leftarrow\!\!- & \bullet & \leftarrow\!\!- & C & \quad \eta\varphi \\
\| & & \| & & \uparrow\lambda & & \downarrow & & \| & & \| & & \| & \\
A & \leftarrowtail & \bullet & \leftarrowtail & Y & \longrightarrow & I & \longrightarrow & U & \leftarrow\!\!- & \bullet & \leftarrow\!\!- & C & \quad \zeta\psi \, .
\end{array}
$$

Case 3. $\varphi = \psi$ and $\eta \prec_1 \zeta$:

$$
\begin{array}{ccccccccccc}
A & \leftarrowtail & \bullet & \longrightarrow & \bullet & \leftarrow\!\!- & B & \leftarrowtail & \bullet & \longrightarrow & \bullet & \leftarrow\!\!- & C & \quad \eta\varphi \\
\| & & \| & & \| & & \| & & \| & & \downarrow & & \| & \\
A & \leftarrowtail & \bullet & \longrightarrow & \bullet & \leftarrow\!\!- & B & \leftarrowtail & \bullet & \longrightarrow & \bullet & \leftarrow\!\!- & C & \quad \zeta\psi \, .
\end{array}
$$

Again, we proceed as in Case 1. First, by canonical factorisation

$$
\begin{array}{ccccccccccccc}
A & \leftarrowtail & \bullet & \longrightarrow & \bullet & \leftarrowtail & I & \leftarrow\!\!- & \bullet & \longrightarrow & \bullet & \leftarrow\!\!- & C & \quad \eta\varphi \\
\| & & \| & & \| & & \| & & \| & & \downarrow & & \| & \\
A & \leftarrowtail & \bullet & \longrightarrow & \bullet & \leftarrowtail & I & \leftarrow\!\!- & \bullet & \longrightarrow & \bullet & \leftarrow\!\!- & C & \quad \zeta\psi
\end{array}
$$

second, with a pullback of a mono (X) and two pushouts of epis (U, V)

$$
\begin{array}{ccccccccccccc}
A & \leftarrowtail & \bullet & \leftarrowtail & X & \longrightarrow & I & \longrightarrow & U & \leftarrow\!\!- & \bullet & \leftarrow\!\!- & C & \quad \eta\varphi \\
\| & & \| & & \| & & \| & & \downarrow & & \downarrow & & \| & \\
A & \leftarrowtail & \bullet & \leftarrowtail & X & \longrightarrow & I & \longrightarrow & V & \leftarrow\!\!- & \bullet & \leftarrow\!\!- & C & \quad \zeta\psi \, .
\end{array}
$$

Case 4. $\varphi = \psi$ and $\eta \prec_2 \zeta$:

$$
\begin{array}{ccccccccccc}
A & \leftarrowtail & \bullet & \longrightarrow & \bullet & \leftarrow\!\!- & B & \leftarrowtail & \bullet & \longrightarrow & \bullet & \leftarrow\!\!- & C & \quad \eta\varphi \\
\| & & \| & & \| & & \| & & \uparrow\lambda & & \| & & \| & \\
A & \leftarrowtail & \bullet & \longrightarrow & \bullet & \leftarrow\!\!- & B & \leftarrowtail & \bullet & \longrightarrow & \bullet & \leftarrow\!\!- & C & \quad \zeta\psi \, .
\end{array}
$$

We proceed as above, by canonical factorisation

$$
\begin{array}{ccccccccccc}
A & \twoheadleftarrow & \bullet & \longrightarrow & \bullet & \twoheadleftarrow & I & \dashleftarrow & \bullet & \longrightarrow & \bullet & \twoheadleftarrow & C & & \eta\varphi \\
\| & & \| & & \| & & \uparrow & & \uparrow & & \| & & \| & & \\
A & \twoheadleftarrow & \bullet & \longrightarrow & \bullet & \twoheadleftarrow & J & \dashleftarrow & \bullet & \longrightarrow & \bullet & \twoheadleftarrow & C & & \zeta\psi
\end{array}
$$

and then with two pullbacks of monos (X, Y) and two pushouts of epis (U, V)

$$
\begin{array}{ccccccccccc}
A & \twoheadleftarrow & \bullet & \twoheadleftarrow & X & \longrightarrow & I & \longrightarrow & U & \dashleftarrow & \bullet & \twoheadleftarrow & C & & \eta\varphi \\
\| & & \| & & \uparrow & & \uparrow & & \uparrow & & \| & & \| & & \\
A & \twoheadleftarrow & \bullet & \twoheadleftarrow & Y & \longrightarrow & J & \longrightarrow & V & \dashleftarrow & \bullet & \twoheadleftarrow & C & & \zeta\psi
\end{array}
$$

(E) It is now easy to verify that the functor (5.76) is indeed the category of fractions of E produced by Σ.

First, P takes every morphism of Σ to an isomorphism; by (ik.2), it suffices to prove this fact for monos (and dually for epis). Now, if $m \in \Sigma$ is a mono, $m^\sharp m = 1$ already holds in $E^\sharp \subset \operatorname{Rel} E$, while the relation $[m].[m^\sharp] = [m.m^\sharp] = 1$ follows by the relation $m.m^\sharp \prec 1_A$

$$
\begin{array}{ccccc}
A & \overset{m}{\twoheadleftarrow} & M & \overset{m}{\succ\!\!-} & A \\
\| & & \downarrow{\scriptstyle m} & & \| \\
A & =\!\!=\!\!= & A & =\!\!=\!\!= & A
\end{array}
\quad .
$$

Furthermore, let $F \colon E \to C$ be a functor, with values in an arbitrary category, that takes every map of Σ to an isomorphism. Then F factorises uniquely through P, via the functor

$$
G \colon E^\sharp/R \to C, \quad G[p^\sharp f m^\sharp] = (Fp)^{-1}.(Ff).(Fm)^{-1}.
$$

Of course one has to show that this definition is consistent (transforming the diagram (5.5.4), that defines R, by the functor F) and preserves composition (transforming the diagram (5.75) that expresses composition of our ternary diagrams). $\qquad\square$

5.5.6 Factorisations and null morphisms

We have proved that every morphism α in $\Sigma^{-1}E$ has the following *ternary* factorisation

$$
\alpha = [p]^{-1}.[f].[m]^{-1}, \tag{5.77}
$$

with f in E and p, m in Σ (resp. epi and mono).

Factorising $f = n'q'$, we can form two bicartesian squares

$$
A \;\; I \;\; B \tag{5.78}
$$

This yields a second three-arrow factorisation, the *coternary* one:

$$
\alpha = [n].[h]^{-1}.[q], \tag{5.79}
$$

with $h = m'p'$ in Σ and q, n in E (resp. epi and mono).

The diagram (5.78) contains also a quaternary factorisation (the lower path) and a coquaternary one (the upper path), which we shall not use here.

By definition, a morphism of $\Sigma^{-1}\mathsf{E}$ is *null* if it factorises through a null object of E, or equivalently trough the image of a null morphism of E. Such morphisms thus form the closed ideal of $\Sigma^{-1}\mathsf{E}$ associated to the null objects of E (Section 1.3.2). Other characterisations are given below.

5.5.7 Theorem (Exactness properties)

Let E be a g-exact category and Σ an exact isokernel of E (Section 5.5.1). Let us fix a morphism $\alpha = [\varphi] = [p^\sharp f m^\sharp] = [n g^\sharp q] \colon A \to B$ of $\Sigma^{-1}\mathsf{E}$ (in ternary and coternary factorisations).

(a) The morphism $\alpha = [\varphi]$ is null in $\Sigma^{-1}\mathsf{E}$ (as defined above) if and only if $\varphi^*(0) \in \Sigma$. The latter is computed as follows (cf. (5.54)), for $\varphi = p^\sharp f m^\sharp = n g^\sharp q$

$$
\begin{aligned}
\varphi^*(0) &= m_* f^* p_*(0) = m_* f^*(0), \\
\varphi^*(0) &= q^* h_* n^*(0) = q^*(0) = \ker q.
\end{aligned} \tag{5.80}
$$

The null objects of $\Sigma^{-1}\mathsf{E}$ are precisely those which are isomorphic in $\Sigma^{-1}\mathsf{E}$ to some object null in E; they are also characterised by each of the following equivalent properties

$$
(0_A \colon A_0 \rightarrowtail A) \in \Sigma, \qquad (0^A \colon A \twoheadrightarrow A^0) \in \Sigma.
$$

(b) The category $\Sigma^{-1}\mathsf{E}$ is g-exact (with respect to such null morphisms). If E has small hom-sets, the same holds in $\Sigma^{-1}\mathsf{E}$.

(c) The functor P is exact and Sub-full. If E' is a g-exact category and the functor $F \colon \mathsf{E} \to \mathsf{E}'$ takes every map of Σ to an isomorphism, then F is exact if and only if the unique functor $G \colon \Sigma^{-1}\mathsf{E} \to \mathsf{E}'$ such that $F = GP$ is exact.

(d) Kernels and cokernels, images and coimages of $\alpha = [\varphi]$, in $\Sigma^{-1}E$, are computed as follows

$$\ker \alpha = [\varphi^*(0)] = [\ker q], \qquad \operatorname{coim} \alpha = [q],$$

$$\operatorname{cok} \alpha = [\operatorname{cok} n], \qquad \operatorname{im} \alpha = [\varphi_*(1)] = [n]. \qquad (5.81)$$

(e) The morphism $\alpha = [\varphi]$ is an isomorphism if and only if n and q are in Σ, if and only if f is in Σ; in this case $\alpha^{-1} = [mf^\sharp p] = [q^\sharp gn^\sharp]$. Moreover, $\operatorname{Ikr} P = \Sigma$.

(f) If E is componentwise p-exact, or p-exact, so is $\Sigma^{-1}E$.

(g) If E is abelian, so is $\Sigma^{-1}E$.

Proof (a) Suppose that $\alpha = [\varphi]\colon A \to B$ has $k = \varphi^*(0) = \ker q \in \Sigma$. Then

$$\alpha = \alpha[k][k]^{-1} = [ng^\sharp].[qk].[k]^{-1},$$

factorises through the image of qk, which is null in E. Conversely, suppose that φ can be factorised in $E^\sharp$ as $\chi\psi\colon A \to Z \to B$, where Z is a null object of E. Then $\chi^*(0) = 1_Z$ (the unique subobject of Z, cf. 1.5.3(e)), and $\varphi^*(0) = \psi^*\chi^*(0) = \psi^*(1_Z)$ can be easily computed on a ternary factorisation $\psi = p'^\sharp g'm'^\sharp$, obtaining $\varphi^*(0) = m'_* g'^* p'_*(1_Z) \sim m' \in \Sigma$

Let us now consider an arbitrary object A, its identity $\alpha = \operatorname{id}A = [11^\sharp 1]$ in $\Sigma^{-1}E$ and $k = \ker(1_A) = 0_A\colon A_0 \rightarrowtail A$, the null subobject of A in E. If A is null *in* $\Sigma^{-1}E$, then α is a null morphism and $k \in \Sigma$. If $k \in \Sigma$, then $[k]\colon A_0 \to A$ is an isomorphism of $\Sigma^{-1}E$ between A and a null object of E. But the existence of such an isomorphism implies that A is null in $\Sigma^{-1}E$. By duality, also the condition $(0^A\colon A \twoheadrightarrow A^0) \in \Sigma$ is equivalent to A being null in $\Sigma^{-1}E$.

(d) Let us prove the first formula of (5.81). First, the composite $\alpha[\ker q] = [n].[g]^{-1}.[q.\ker q]$ is null. Second, let $[\varphi].[\psi]$ be null, with $\psi = p'^\sharp g'm'^\sharp\colon X \to A$. We form the following diagram, from right to left by direct and inverse images of monomorphisms, so that $k = (\varphi\psi)^*(0) = \psi^*\varphi^*(0) \in \Sigma$

$$X \xleftarrow{\;m'\;} \bullet \xrightarrow{\;g'\;} \bullet \xleftarrow{\;p'\;} A \cdots\overset{\varphi}{\cdots}\!\!\!> B$$

All the horizontal leftward arrows are in Σ, and become isomorphisms in $\Sigma^{-1}E$; therefore, reversing such isomorphisms, their squares still commute, which means that $[\psi] = [\psi.k][k]^{-1}$ factorises through $[\varphi^*(0)]$.

Dually, cokernels exist and are calculated as in (5.81).

(b) This proves that $\Sigma^{-1}E$ is g-exact, as every morphism factorises $[\varphi] =$

$[n].[g]^{-1}.[q]$, through a normal epi, an isomorphism and a normal mono. We have also shown that its sets of normal subobjects are small (because they are in E). If E has small hom-sets, the same holds for Rel E (Section 5.4.2), and also for $E^\sharp$ and $\Sigma^{-1}E$.

(c) The functor P is exact and Sub-full, by the previous characterisation of kernels and cokernels in $\Sigma^{-1}E$. Moreover, if $F = GP$ is exact, so is G: take $\varphi = ng^\sharp q$ in $E^\sharp$

$$\ker G[\varphi] = \ker (Fn.(Fg)^{-1}.Fq) = \ker (Fq)$$
$$= F(\ker q) = G[\ker q] = G[\ker \varphi].$$

(e) The condition $n, q \in \Sigma$ (or $f \in \Sigma$) plainly implies that α is an isomorphism, since all the morphisms of Σ become iso in $\Sigma^{-1}E$.

Conversely, let us begin by noting that, if the *endomap* $\varphi = p^\sharp f m^\sharp$: $A \to A$ is R-equivalent to 1_A in $E^\sharp$, then $f \in \Sigma$

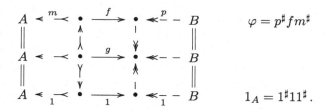

In fact, g is in Σ by composition, and f too by 'decomposition' (ik.1). By diagram (5.78), it follows easily that, if an endomorphism of $E^\sharp$ with coternary factorisation $\varphi = ng^\sharp q$ is R-equivalent to 1, then q and n belong to Σ.

Consider now two morphisms $\varphi: A \to B$ and $\psi: B \to A$ of $E^\sharp$, such that $\psi\varphi R 1_A$ and $\varphi\psi R 1_B$. Using the composition of their *coternary factorisations* (dual to (5.75)) and the previous result, it is easy to deduce that φ and ψ satisfy the conditions of the thesis. It follows that $\operatorname{Ikr} P = \Sigma$.

(f) Let E be componentwise p-exact, which amounts to saying that (it is g-exact and) the ideal NulE is strict (as defined in 2.2.4): if Z, Z' are null objects in the same connected component of E, there is precisely one morphism $Z \to Z'$ (which is then an isomorphism).

Given two arbitrary arrows $[\varphi], [\psi]: A \to B$ between two objects null *in*

E,

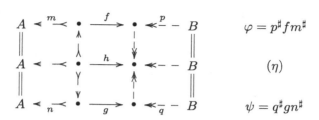

$$\varphi = p^\sharp f m^\sharp$$

$$(\eta)$$

$$\psi = q^\sharp g n^\sharp$$

the normal monos m, n and the normal epis p, q are necessarily *isomorphisms of* E and we can complete the diagram above. Note that the two 'candidates' for h, given by the upper and lower 'paths', are parallel maps between null objects and *coincide*, because NulE is strict.

This shows that $[\varphi] = [\psi]$. But the property we have proved holds more generally for the null objects of Σ^{-1}E, because each of them is isomorphic to some null object of E. Therefore Σ^{-1}E is componentwise p-exact (and p-exact when E is).

(g) Let E be an abelian category. We already know that Σ^{-1}E is p-exact, and we only need to show that a binary biproduct $A = A_1 \oplus A_2$ in E is also a product in Σ^{-1}E.

Let two arrows $\varphi_i \colon X \to A_i$ be given in $E^\sharp$, in coternary factorisation. We form the following commutative diagram in E

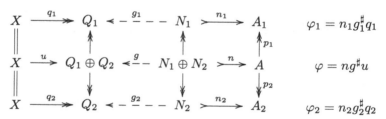

$$\varphi_1 = n_1 g_1^\sharp q_1$$

$$\varphi = n g^\sharp u$$

$$\varphi_2 = n_2 g_2^\sharp q_2$$

where $u = (q_1, q_2) \colon X \to Q_1 \oplus Q_2$ (*not* an epi, generally), $g = g_1 \oplus g_2 \in \Sigma$ (by Lemma 5.5.3) and $n = n_1 \oplus n_2$. Note that $\varphi = n g^\sharp u$ is a map of $E^\sharp$, even though it is *not* written in a coternary factorisation. This diagram projects to a commutative diagram of Σ^{-1}E where the leftward horizontal arrows are invertible. Therefore $[p_i].[\varphi] = [\varphi_i]$.

Finally, to prove that the pair of projections $[p_i]$ is jointly monic in Σ^{-1}E it suffices to note that the exact functor $P \colon E \to \Sigma^{-1}$E preserves kernels and their meets

$$\text{Ker}\,[p_1] \cap \text{Ker}\,[p_2] = [\text{Ker}\,(p_1) \cap \text{Ker}\,(p_2)] = 0.$$

$\square$

5.5.8 Theorem (General fractions for g-exact categories)

Let E *be a g-exact category. The following conditions on a set Σ of morphisms of* E *are equivalent:*

(a) Σ *is an exact isokernel,*

(b) there exists an ex4-functor $F \colon E \to E'$ *whose isokernel is* Σ,

(c) the (general) category of fractions $\Sigma^{-1}E$ *is g-exact, the functor* $P \colon$ $E \to \Sigma^{-1}E$ *is exact and* $\mathrm{Ikr}P = \Sigma$.

In this case, P *satisfies the universal problem (of inverting the morphisms of* Σ) *in* CAT, *as already stated, and also in* EX4. *In other words, if* $G \colon \Sigma^{-1}E \to E'$ *is a functor with values in a g-exact category and* GP *is exact, so is* G.

The same holds in the ex5 *and* abelian *cases (always with respect to exact isokernels, as defined in 5.5.1).*

Proof The statement is a synopsis of the results of this section.   □

5.6 The homological structure of EX and AB

The category EX_4 of g-exact $\mathcal{U}$-categories and exact functors has a natural ideal of null arrows: the functors that send each object to a null one.

With respect to this ideal, EX_4 *itself is an unrestricted homological category.* Its normal subobjects coincide with the thick subcategories of E (defined in 5.6.1, as in the abelian case), and are in bijective correspondence with the exact isokernels of E. Its normal quotients coincide with the Serre quotients and with the categories of fractions of the previous section.

EX_5, EX and AB are homological subcategories of EX_4, and have similar properties.

These results first appeared in [G12], for the p-exact case.

5.6.1 Thick subcategories and Serre quotients

Let E be a g-exact category. A (generally large) *thick set of objects* K of E is defined as in the abelian case:

(tk.1) K contains all the null objects of E; given a short exact sequence $B \rightarrowtail A \twoheadrightarrow C$ of E, the object A is in K if and only if both B and C are.

This is equivalent to considering a *thick subcategory* K of the g-exact category E, i.e. a full subcategory whose objects satisfy (tk.1).

As a consequence, K (equipped with the null objects of E) is a g-exact, full conservative subcategory of E, i.e. it is g-exact in its own right, it embeds fully and exactly in E, and every object of E isomorphic to some object of K belongs to the latter.

As in the abelian case we write E/K for the *Serre quotient* of E modulo K, i.e. the solution of the universal problem of annihilating K, among all the exact functors from E to some g-exact category; we prove below that the solution exists, and is given by a suitable category of fractions Σ^{-1}E.

Plainly, every thick subcategory of a componentwise p-exact, or p-exact, or abelian category is of the same type. It is also well-known that the Serre quotient of an abelian category modulo a thick subcategory is abelian [Gt, Ga]; a fact that will result, here, from 5.5.7(g).

5.6.2 Lemma

Given a g-exact category E, *the two obvious mappings between sets of morphisms Σ and sets of objects K of* E

$$K(\Sigma) = \{A \mid (0_A : A_0 \rightarrowtail A) \in \Sigma\} = \{A \mid (0^A : A \twoheadrightarrow A^0) \in \Sigma\}$$
$$= \{A \mid A \text{ is null in } \Sigma^{-1}\text{E}\}, \tag{5.82}$$
$$\Sigma(K) = \{f \mid \operatorname{Ker} f, \operatorname{Cok} f \text{ belong to } K\},$$

establish a bijective correspondence between exact isokernels and thick sets of objects (or thick subcategories).

Proof (A) Let Σ be an exact isokernel. The equality between the three formulas for $K(\Sigma)$ is already known (see 5.5.7(a)). We have to prove that $K = K(\Sigma)$ is a thick set of objects in E.

First, K contains all the null objects, because Σ contains all the isos. Second, let a short exact sequence $B \rightarrowtail A \twoheadrightarrow C$ be given in E and consider the commutative diagram below

$$\begin{array}{ccc}
B & \overset{r}{\rightarrowtail} A & \overset{s}{\twoheadrightarrow} C \\
{\scriptstyle p'}\downarrow & {\scriptstyle p}\downarrow & \downarrow{\scriptstyle p''} \\
B^0 & \underset{r'}{\rightarrowtail} C^0 & = C^0
\end{array} \tag{5.83}$$

Notice that the null quotient of C yields the null quotient of A, while the epi-mono factorisation of the null morphism $pr: B \to C^0$ yields the null quotient of B.

If A is in K, then $p = 0^A$ is in Σ, and so is p' (ik.2); B thus belongs to K; dually, also C does.

Conversely, assume that B and C are in K, so that $p', p'' \in \Sigma$. The diagram (5.83) has short exact rows (because $r' \colon B^0 \rightarrowtail C^0$ is null, and its cokernel is the identity); it projects to a similar diagram in the g-exact category $\Sigma^{-1}\mathsf{E}$, where $[p']$ and $[p'']$ are invertible; therefore also $[p]$ is, and $p \in \mathrm{Ikr}P = \Sigma$.

(B) We now assume that K is a thick set of objects and prove that $\Sigma = \{f \mid \mathrm{Ker}\, f, \mathrm{Cok}\, f \in K\}$ is an exact isokernel of E; we make use of Lemma 5.5.2, replacing the condition (ik.1) with (ik.1a, ik.1b).

The conditions (ik.0, 2) hold trivially. For (ik.1a), consider the composition of two monomorphisms, as in the left diagram below, and the associated short exact sequence of E, at the right

$$A'' \rightarrowtail A' \rightarrowtail A \qquad A'/A'' \rightarrowtail A/A'' \twoheadrightarrow A/A'.$$

Thus (tk.1) proves that these two monomorphisms are in Σ (i.e. A/A' and A'/A'' are in K) if and only if their composition is (i.e. A/A'' is in K). Dually for epimorphisms.

To prove (ik.1b), let us start from the commutative square occupying the lower left position of the following diagram

$$
\begin{array}{ccccc}
H \wedge L & \rightarrowtail & L & \twoheadrightarrow & L/(H \wedge L) \\
\downarrow & & \downarrow & & \downarrow \\
H & \rightarrowtail & A & \twoheadrightarrow & A/H \\
\downarrow & & \downarrow & & \downarrow \\
(H \vee L)/L & \rightarrowtail & A/L & \twoheadrightarrow & A/(H \vee L)
\end{array}
\qquad (5.84)
$$

We complete the diagram by two cokernels (A/H and $A/(H \vee L)$) and the three kernels of the upper row: then the latter is short exact, by the 3×3-Lemma 2.3.7. Now, if the two arrows starting at H are in Σ, the objects $H \wedge L$ and A/H are in K, whence $L/(H \wedge L)$ and $A/(H \vee L)$ are also in K, and L too by (tk.1). It follows that the two arrows ending at A/L are in Σ.

For (ik.3), we prove the pullback case, which we split in two. A (general) pullback of a mono along a mono intervenes in the left upper square of the previous diagram (5.84), which again we can complete with short exact rows and columns. Now, if $L \rightarrowtail A$ is in Σ, A/L is in K, whence also $(H \vee L)/L$ is, and the thesis follows.

Last, a pullback of a mono along an epi intervenes in the lower left square

of the following diagram

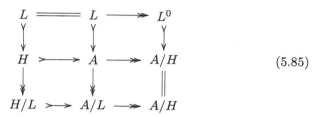

$$(5.85)$$

which again is commutative with short exact rows and columns. It is now easy to check that if $H/L \rightarrowtail A/L$ (resp. $A \twoheadrightarrow A/L$) is in Σ, so is $H \rightarrowtail A$ (resp. $H \twoheadrightarrow H/L$). $\square$

5.6.3 Theorem

Given a g-exact category E, *a set* K *of objects is thick if and only if it is the kernel of some exact functor* $F: E \to E'$ *with values in a g-exact category (with respect to the ideal of null functors).*

Furthermore, if K *is a thick subcategory and* Σ *the associated exact isok-ernel (cf. 5.6.2), the functor* $P: E \to \Sigma^{-1}E$ *satisfies the universal problem of annihilating* K, *so that* $E/K = \Sigma^{-1}E$ *and* $\operatorname{Ker} P = K$.

Since P is the identity on the objects, $\operatorname{Ker} P$ *is the full subcategory of* E *on the null objects of* $E/K = \Sigma^{-1}E$ *(characterised in 5.5.7(a)).*

Proof After the previous lemma, the thesis follows trivially from the fact that, for every exact functor $F: E \to E'$, the isokernel and the annihilation kernel are related by the correspondence (5.82). $\square$

5.6.4 * The structure of EX$_4$

These kernels and quotients belong to a natural *homological* structure of EX$_4$ (in the unrestricted sense, of course - this is not a $\mathcal{U}$-category, in the sense of 1.5.1).

EX$_4$ has no zero-object, as its terminal object 1 is not even weakly initial. But it has a natural closed ideal given by the null functors, i.e. those which annihilate all the objects, or equivalently which factorise through a g-exact category whose objects are all null (a 'null groupoid', in the sense of 2.2.6).

Every exact functor $F: E \to E'$ has a kernel with respect to this ideal, its annihilation-kernel $\operatorname{Ker} F$, i.e. the full subcategory of E formed by the objects annihilated by F; it also has a normal coimage $E/\operatorname{Ker} F$ (a Serre quotient) and a normal image $\mathsf{Nim}\, F$, i.e. the least thick subcategory of E' that contains the set-theoretic image $F(\mathsf{E})$.

EX_4 is thus a semiexact category (by Criterion A in 1.5.4); the normal subobjects coincide with the thick subcategories, the normal quotients with the g-exact categories of fractions, or equivalently with the Serre quotients. We verify below that EX_4 is actually homological.

EX_5, EX and AB are full (hence conservative) subcategories of $\mathsf{EX4}$, closed under normal subobjects and quotients (cf. 5.5.5, 5.6.1), hence they are homological subcategories (by 1.7.3).

The categories EX_4, EX_5, EX and AB are not g-exact with respect to the above ideal. In other words an exact functor $F\colon \mathsf{E} \to \mathsf{E}'$ between abelian categories need not be an exact morphism, i.e. need not induce an isomorphism from its normal coimage ($\mathsf{E}/\mathsf{Ker}\, F$) to its normal image ($\mathsf{Nim}\, F$); indeed, it suffices to consider the embedding of an abelian subcategory that is not thick.

5.6.5 * The homological properties

It will be useful to note that, given an ex4-functor $F\colon \mathsf{E} \to \mathsf{E}'$ and a thick subcategory L of E', the inverse image $F^*(\mathsf{L})$ coincides with the set-theoretic pre-image $F^{-1}(\mathsf{L})$; indeed, the latter is plainly a thick subcategory of E, therefore it is g-exact and yields the pullback of L along F, in EX_4. We now verify the axioms (ex2, 3).

First, if K is a thick subcategory of H which is thick in E (g-exact), then K is plainly thick in E.

Second, in order to prove that the normal quotients of EX_4 are also closed under composition, it suffices to show that every normal quotient $P\colon \mathsf{E} \to \mathsf{E}'$ is fully normal (Section 2.1.1), which means that the associated transfer connection for thick subcategories (or thick subsets of objects) satisfies $P_*P^* = 1$. Now, P is the identity on the objects. It follows that, if L is a thick subset of E', $P^*(L) = P^{-1}(L)$ coincides with L and P_*P^*L, the least thick subset of E' containing P^*L, coincides again with L. In other words, the thick subsets of E' are precisely the thick subsets of E that contain $\mathsf{Ker}\, P$, i.e. the set of null objects of E'.

Finally, we have to check the subquotient axiom. Let $\mathsf{K} \subset \mathsf{H}$ be thick subcategories of E; then K is thick in H and we can form a commutative diagram of 'natural' ex4-functors, where H/K is the normal coimage of the composite QH

$$
\begin{array}{ccc}
H & \xrightarrow{\ H\ } & E & \xrightarrow{\ P\ } & E/H \\
\downarrow & & \downarrow{\scriptstyle Q} & & \| \\
H/K & \xrightarrow[H']{} & E/K & \xrightarrow[V]{} & E/H
\end{array}
\qquad (5.86)
$$

We have to prove that the functor H' (induced by the embedding H) is the kernel of V. Now, $\mathsf{Ker}\,V$ is the full subcategory of the objects A of E/K that annihilate in E/H, i.e. those of H. Its maps are diagrams of type

$$
M' \;\prec\!\!\!-\!\!\!\prec\; A' \;\longrightarrow\; A'' \;\prec\!\!\!-\; M''
$$

(cf. (5.76)), where A' and A'' are again in H (thick in E), modulo the equivalence relation R (cf. 5.5.4) described by a diagram of E that, because of the same reason, belongs to H. All this shows that $\mathsf{Ker}\,V$ is a realisation of the category of fractions H/K.

6

Applications to algebraic topology

This chapter begins by studying homology theories defined on homological categories. The classical terminology of homology theories is reviewed in Section 6.1; transformations of their bases, allowing one to pull-back the theories, are considered. In particular the connection between relative theories satisfying the relative-homeomorphism axiom and single space theories is formalised by means of the full reflective embedding of Top. in the category of pairs Top$_2$.

A singular homology theory for fibrations and covering maps is proposed in Section 6.2.

Then we move to the main goal of this chapter: providing a framework for the spectral sequences of homotopy theory, based on convenient homological categories.

In Section 6.3 we introduce the category Ngp *of normalised groups*, a homological category of fractions of Gp$_2$, obtained by excision of the invariant subgroups, or also as a quotient of a concrete homological category. We show that a tower of fibrations *between path-connected spaces* defines a homotopy exact couple *in* Ngp, so that the theory developed in Section 3.5 can be applied, and yields the spectral sequence of our tower of fibrations.

In Sections 6.4 and 6.5 we remove the restriction to pathwise connected spaces, using the category Act of actions of groups on pointed sets, or *actions* for short, in which Set., Gp and Gp$_2$ can be embedded, in a natural way.

We prove that Act is homological and show that the homotopy spectral sequence of a tower of fibrations can be established in a suitable homological category Nac of *normalised actions*: it is a category of fractions of Act, that again can be realised as a quotient of a concrete homological category.

6.1 Homology theories and their transformations

E is always a semiexact category, B an N-category. The present definition of a homology theory, from E to B, adapts to this setting the Eilenberg-Steenrod's notion of 'h-theory on an h-category' [ES].

The relations between single-space homology theories and relative homology theories that satisfy the relative homeomorphism axiom are reviewed, in order to infer which notions of transformation of theories are relevant.

Let us recall that connected and exact ∂-sequences of functors from a semiexact category E to an N-category B are dealt with in Section 4.1.

6.1.1 Homology theories

Let E be a semiexact category, equipped with a relation R between parallel morphisms (called *homotopy* relation) and a set Σ of morphisms (called *excision* maps). Such a triple $E^\sharp = (E, R, \Sigma)$ will be called a *homological basis* (for homology theories) on E.

A (generalised) *homology theory*

$$H = ((H_n), (\partial_n))_{n \in \mathbb{Z}},$$

on the basis (E, R, Σ), with values in the N-category B, has to satisfy the following axioms:

(hl.1) (*Exactness*) H is an exact ∂-sequence of functors from E to B,

(hl.2) (*Homotopy*) if $f R g$ in E, then $H_n(f) = H_n(g)$, for all n,

(hl.3) (*Excision*) if $f \in \Sigma$, then $H_n(f)$ is an isomorphism of B, for all n.

More generally, a *weak homology theory* is defined by weakening axiom (hl.1), so that (H_n, ∂_n) is only required to be a *connected* sequence.

A homology theory is said to be *ordinary* if moreover an object P is assigned (the *standard point*) and:

(hl.4) (*Dimension*) $H_n(P)$ is null in B, for $n \neq 0$.

Then $H_0(P)$ is called the *coefficient object* of H. The interest of choosing, as a point, a *precise* object (instead of a class of isomorphic objects) will be clear in 6.1.2(d). The term 'excision', that comes from a well-known topological situation (see 6.1.2(a)), can be replaced with *pre-isomorphism* when it can be misleading (see 6.1.4).

Plainly, every exact sequence of functors is a homology theory, with respect to trivial assignments for homotopy and excision. But a major interest of the definition, as shown in [ES], resides in catching in the axioms the main features of a given family of 'theories', in order to determine them at some relevant extent; for instance on the 'finitely presentable' objects of E, in some sense.

A *morphism* (resp. isomorphism) of homology theories is a natural transformation (resp. an isomorphism) of the underlying connected sequences of functors (cf. 4.1.4). This defines the category $Th(E^\sharp, B)$ of homology theories on the basis $E^\sharp = (E, R, \Sigma)$, with values in the N-category B.

Finally, let us recall that the differential of a connected sequence (H_n, ∂_n) can be seen as a natural transformation (cf. 4.1.1)

$$\partial_n \colon H_n P'' \to H_{n-1} P' \colon ShE \to B, \qquad (6.1)$$

where ShE is the N-category of short exact sequences of E, while P', P'': ShE $\to$ E are the first and last projection.

Of course, a *cohomology theory* on the same data is a homology theory with values in B^{op}.

6.1.2 Examples

This framework contains various classical homology theories, that in part will be further examined in the sequel.

(a) Eilenberg-Steenrod's homology theories for pairs of spaces. Take for E the homological category Top_2 (Section 1.6.7) with the classical notions of homotopy, excision, and point; take B = Ab. Then, the axioms and the group of coefficients $H_0(P)$ determine the theory over finite CW-complexes, up to isomorphism. Recall that Σ is the set of inclusions $(X \setminus U, A \setminus U) \to (X, A)$, where U is an open subspace of X whose closure is contained in the interior of A; the excision axiom thus says that an open subspace U of A can be cut out of the pair, provided that it is 'not near' the boundary of A in X.

Of frequent use as a domain of homology theories are also the full homological subcategories Cph_2 of *compact Hausdorff pairs* and Cpm_2 of *compact metric pairs*, i.e. pairs (X, A), where X is a compact Hausdorff (resp. metric) space and A is a closed subspace of X.

With respect to Eilenberg-Steenrod's axioms, the 'exactness axiom' above comprises the two functorial conditions, the naturality of ∂ and the exactness sequence for a *triple* of spaces; the latter is well-known to be equivalent to the exact sequence for a *pair* of spaces (see [ES], I.10; or here, 4.4.2 - 4.4.3).

(b) Eilenberg-Steenrod's *cohomology* theories. Take E as before and B = Ab^{op}.

(c) Single space homology theories for locally compact Hausdorff spaces

([ES], X.7). Take $\mathsf{E} = \mathcal{T}_{LC}$, the category of locally compact Hausdorff spaces and partial proper maps, defined on open subspaces (cf. 1.7.4(d)), with the usual notions of homotopy and point, and no excision maps. More recently, single space homology theories for general spaces have been produced (see 6.1.9).

(Eilenberg-Steenrod's formulation of these theories ([ES], X.7) uses the subcategory $\mathcal{B}_{LC}$ of $\mathcal{T}_{LC}$ formed by its everywhere-defined mappings; a theory consists of (covariant) functors H_n on $\mathcal{B}_{LC}$, plus contravariant homomorphisms $H_n(X) \to H_n(U)$ for the open subsets U of X. Since a map $f \colon X \to Y$ of $\mathcal{T}_{LC}$ factorises uniquely through a 'coinclusion' $p \colon X \twoheadrightarrow U$ and an everywhere-defined mapping $U \to Y$, these two aspects can be combined, obtaining *one* functor $H_n \colon \mathcal{T}_{LC} \to \mathsf{B}$ in every degree. Globally, we get an *exact* sequence of functors, as the exactness axiom concerns precisely the general short exact sequence of $\mathcal{T}_{LC}$, that is of the form $H \rightarrowtail X \twoheadrightarrow X \setminus H$, for a closed subspace H of X.)

(d) The classical homology of groups can be extended to a *relative* theory for group-pairs (X, Y), credited to Massey [Mas2, Ta, Ri]; this produces an exact sequence of functors H_n from the category Gp_2 of pairs of groups with values in Ab [G13]. An excision theorem for this theory is in [Ri]. The dimension axiom (with a shift of one unit, because $H_0(X, Y)$ is always null), is satisfied for the standard point $P = (\mathbb{Z}, 0)$. Note that the fact of choosing as 'points' all the pairs isomorphic to this P would lead here to some loss of information for the coefficient object, since the pair $(\mathbb{Z}, 0)$ has *two* automorphisms.

(e) Relative homotopy can be seen as an exact sequence of functors from $\mathsf{Top}_{\bullet}2$ (pairs of pointed spaces) with values in the homological, non-exact category Act of actions of groups on pointed sets (Section 6.4); it is a homology theory, with respect to the standard homotopy relation, the standard point and pre-isomorphisms induced by fibrations [Mi1].

(f) Čech homology is a weak homology theory, that is exact over triangulable pairs ([ES], IX.9.10). Non-ordinary homology theories include K-theory, stable homotopy, cobordism theories; a general exposition can be found in [Sw].

6.1.3 Single space homology theories

Consider the p-homological category $\mathsf{Top}_{\bullet}$ of pointed spaces, with the following data for homology theories: the homotopy relation R is given by the usual pointed homotopies, the set Σ of excision maps is formed by the pointed homeomorphisms (or equivalently is empty) and the standard point is $S^0 = \{-1, 1\}$, with base-point (say) 1.

A *single space homology theory* is a theory $T = ((T_n), (\partial_n))$ on Top$_\bullet$, with values in an N-category B. The exactness axiom says that, for every pair (X, A) of pointed spaces, determining the general short exact sequence of Top$_\bullet$

$$A \rightarrowtail X \twoheadrightarrow X/A, \tag{6.2}$$

we have an exact sequence of B, natural for maps of pairs

$$\ldots T_n(A) \to T_n(X) \to T_n(X/A) \xrightarrow{\partial_n} T_{n-1}(A) \ldots \tag{6.3}$$

We are going to establish a correspondence between single-space homology theories and homology theories over Top$_2$ that satisfy the relative-homeomorphism axiom.

Single space theories should not be confused with the absolute formulation of (reduced) homology theories for spaces, based on the suspension endofunctor [DT, Ke1] - a notion that is entirely equivalent to that of Eilenberg-Steenrod's theories over Top$_2$.

6.1.4 Relative homeomorphisms

Everything follows from the N-adjunction $P \dashv J$, with counit $s \colon PJ \cong 1$, already considered in 1.7.5(b) and 5.3.4

$$J \colon \mathsf{Top}_\bullet \to \mathsf{Top}_2, \quad J(X) = (X, \{0_X\}) \qquad \text{(left exact)}, \tag{6.4}$$

$$P \colon \mathsf{Top}_2 \to \mathsf{Top}_\bullet, \quad P(X, A) = X/A \qquad \text{(exact)}, \tag{6.5}$$

where $P(X, A) = X/A$ denotes the pushout of the inclusion $A \to X$ along the map $A \to \{*\}$. (In particular $X/\emptyset = X^+$ is the space X with an isolated base-point added.)

The *relative homeomorphisms* are, by definition, those map of pairs $f \colon (X, A) \to (Y, B)$ that induce a homeomorphism $Pf \colon X/A \to Y/B$; they form the isokernel of P (Section 5.2) and contain all the components of the unit $r_{XA} \colon (X, A) \to (X/A, \{0\})$ of the adjunction. They also contain all the classical excision maps $i \colon (X \setminus U, A \setminus U) \to (X, A)$ (recalled in 6.1.2(a))

$$\Sigma_P = \mathrm{Ikr} P \supset \Sigma. \tag{6.6}$$

Indeed, if $U \neq A$, it is obvious that $Pi \colon (X \setminus U)/(A \setminus U) \to X/A$ is a homeomorphism. Otherwise, if $U = A$, the hypothesis $U \subset \overline{U} \subset \mathrm{int} A \subset A$ implies that $U = A$ is open and closed in X, so that $(X \setminus U)/(A \setminus U) = (X \setminus U)^+ \cong X/A$.

A homology theory $H = (H_n, \partial_n)$ from Top_2 to B is said to satisfy the *relative homeomorphism axiom* if the following equivalent conditions hold:

(RH) all the functors H_n carry the relative homeomorphisms into isomorphisms,

(RH') for every pair (X, A) and every n,

$$H_n(r_{XA}): H_n(X, A) \to H_n(X/A, \{0\})$$

is an isomorphism.

Since $\Sigma_P \supset \Sigma$, this amounts to consider new, finer homological data for Top_2, with pre-isomorphisms consisting of all the relative homeomorphisms (or, equivalently, all the components r_{XA})

$$\mathsf{RHTop}_2 = (\mathsf{Top}_2, R, \Sigma_P). \tag{6.7}$$

Recall now the general result we already found for full reflective subcategories (in 5.3.3, 5.3.4): $\mathsf{Top}_\bullet$ is equivalent to the category of fractions

$$\Sigma_P^{-1}\mathsf{Top}_2 = \Sigma_R^{-1}\mathsf{Top}_2,$$

both in the sense of general fractions and within EX_1 (P is exact). We want to show that this equivalence between functors on $\mathsf{Top}_\bullet$, on the one hand, and functors on Top_2 satisfying the relative homeomorphism condition, on the other hand, induces an equivalence between analogous homological theories.

To do this, we also have to transform the differential ∂_n; this is trivial for the exact functor P, less trivial for J whose lack of exactness has to be balanced with relative homeomorphisms. Therefore we abandon for the moment the concrete situation and establish some general terminology for transforming homology theories via functors.

6.1.5 Homological transformations

Let $\mathsf{E}^\sharp = (\mathsf{E}, R, \Sigma)$ and $\mathsf{X}^\sharp = (\mathsf{X}, R', \Sigma')$ be two homological bases for homology theories, as defined in 6.1.1.

A *short exact* (resp. *exact*) *transformation* (of homological bases) $F: \mathsf{X}^\sharp \to \mathsf{E}^\sharp$ will be a short exact (resp. exact) functor $F: \mathsf{X} \to \mathsf{E}$ between the underlying semiexact categories such that, if $H = (H_n, \partial_n)$ is a homology theory over E, then the (backwords) transformed sequence

$$F^*(H) = (H_n F, \partial_n^*), \qquad \partial_n^* = \partial_n \hat{F}, \tag{6.8}$$

is a homology theory over X (where $\hat{F}: \mathsf{ShX} \to \mathsf{ShE}$ is the obvious extension of F).

This certainly happens for a *strict transformation*, i.e. a short exact functor $F\colon \mathsf{X} \to \mathsf{E}$ that strictly preserves the additional structure

$$f R' g \text{ in } \mathsf{X} \;\Rightarrow\; (Ff)\,R\,(Fg) \text{ in } \mathsf{E}, \qquad\qquad F(\Sigma') \subset \Sigma, \qquad (6.9)$$

but the same also works if F takes R' and Σ' into some appropriate 'closure' of R and Σ, respectively.

Each of these three kinds of transformations (short exact, exact, strict) is closed under composition.

Now, let $F\colon \mathsf{X}^\sharp \to \mathsf{E}^\sharp$ be a short exact transformation of homological bases. For every N-category B we have an associated functor between categories of theories (Section 6.1.1)

$$F^*\colon \mathrm{Th}(\mathsf{E}^\sharp, \mathsf{B}) \to \mathrm{Th}(\mathsf{X}^\sharp, \mathsf{B}), \qquad F^*(H) = (H_n F, \partial_n^*). \qquad (6.10)$$

The procedure $F \mapsto F^*$ is consistent with composition (and identities):

$$(FG)^* = G^* F^*, \qquad\qquad (6.11)$$

where $G\colon \mathsf{Y}^\sharp \to \mathsf{X}^\sharp$ and $F\colon \mathsf{X}^\sharp \to \mathsf{E}^\sharp$ are short exact transformations.

Short exact and exact transformations produce the same notion of *isomorphism* of homological bases, namely an isomorphism $F\colon \mathsf{X} \to \mathsf{E}$ of semiexact categories that establishes a bijective correspondence between theories on $\mathsf{E}^\sharp$ and $\mathsf{X}^\sharp$. Two bases (E, R, Σ) and $(\mathsf{E}, R', \Sigma')$ on the same category E are said to be *equivalent* if they have the same homology theories, or - in other words - if the identity of E induces an isomorphism of bases.

6.1.6 Left transformations

More generally, a *left transformation* $F\colon \mathsf{X}^\sharp \to \mathsf{E}^\sharp$ will be given by a functor $F\colon \mathsf{X} \to \mathsf{E}$ such that

(a) if (x', x'') is a short exact sequence of X, then $Fx' \sim \ker Fx''$,

(b) if $H = (H_n, \partial_n)$ is a homology theory over $\mathsf{E}^\sharp$, then for every short exact sequence $x = (x', x'') = (X' \rightarrowtail X \twoheadrightarrow X'')$ of X, the canonical map $u\colon \mathrm{Cok}\,Fx' \to FX''$

$$
\begin{array}{ccccc}
FX' & \xrightarrow{\ Fx'\ } & FX & \xrightarrow{\ Fx''\ } & FX'' \\[4pt]
\Big\| & & \Big\| & & \Big\uparrow{\scriptstyle u} \\[4pt]
FX' & \longrightarrow & FX & \xrightarrow[\mathrm{cok}\,Fx']{} & \mathrm{Cok}\,Fx'
\end{array}
\qquad (6.12)
$$

induces isomorphisms in homology (i.e. $H_n u$ is an isomorphism of B, for every n) and produces a homology theory over X

$$F^*(H) = (H_n F, \partial_n^*),$$

$$\partial_n^* = \partial_n (H_n u)^{-1} \colon H_n(FX'') \to H_n(\operatorname{Cok} Fx') \to H_{n-1}(FX'). \tag{6.13}$$

Note that, if F is short exact, the map u in (6.12) is an isomorphism and the two procedures give the same result.

More particularly, a *left exact transformation* is a left exact functor satisfying (b). This notion is closed under composition (while the former is not), produces the same isomorphisms of homological bases considered above, in 6.1.5, and yields a formula analogous to (6.11)

$$(FG)^* = G^* F^*, \tag{6.14}$$

where $G \colon \mathsf{Y}^\sharp \to \mathsf{X}^\sharp$ and $F \colon \mathsf{X}^\sharp \to \mathsf{E}^\sharp$ are left exact transformations.

Dually, one can define *right* and *right exact* transformations, getting the new differential via the canonical map $FX' \to \operatorname{Ker} Fx''$.

An *equivalence* of homological bases will be a pair of transformations, in *some* of the above senses (not necessarily the same), providing an equivalence between the categories of theories

$$F \colon \mathsf{X}^\sharp \rightleftarrows \mathsf{E}^\sharp \colon G, \quad G^* F^* \cong 1, \quad F^* G^* \cong 1. \tag{6.15}$$

6.1.7 The single-space transformation

Coming back to our concrete situation 6.1.4, the exact functor P is a strict exact transformation of homological bases $\mathsf{RHTop}_2 \to \mathsf{Top}_\bullet$ (cf. 6.1.5).

First, P preserves the homotopy relation. If $h \colon f \to g \colon (X, A) \to (Y, B)$ is a homotopy of maps of pairs, defined by a map $h \colon (X \times I, A \times I) \to (Y, B)$, then $Ph \colon (X \times I)/(A \times I) \to Y/B$ defines a homotopy $(X/A) \times I \to Y/B$, from Pf to Pg.

Second, P preserves the pre-isomorphisms by definition, as P transforms a relative homeomorphism into a pointed homeomorphism.

Third, P also preserves the standard point, up to (a unique) isomorphism

$$P(\{*\}, \emptyset) = \{*\}/\emptyset = \{*\}^+ \cong S^0. \tag{6.16}$$

On the other hand, $J \colon \mathsf{Top}_\bullet \to \mathsf{RHTop}_2$ is a left exact transformation of homological bases (as defined in 6.1.6).

First, J is a left exact functor. Second, it satisfies the cokernel condition 6.1.6(b), since every short exact sequence of $\mathsf{Top}_\bullet$ yields a map u (see (6.12)) that is easily seen to belong to Σ_P (either by explicitly writing down the diagram or as a formal consequence of (5.35)); hence, u is made invertible by any theory over RHTop_2.

Finally, the functor J trivially preserves the homotopy relation and pre-isomorphisms; the obvious map

$$i\colon (\{*\}, \emptyset) \to J(S^0),$$

induces an isomorphism in homology, by excision.

Now, $PJ \cong 1$ and the unit $r\colon 1 \to JP$ is carried to a functorial isomorphism by any RH-theory. The functors P and J thus produce an equivalence of homological bases

$$J^*\colon \mathrm{Th}(\mathsf{RHTop}_2, \mathsf{B}) \rightleftarrows \mathrm{Th}(\mathsf{Top}_\bullet, \mathsf{B}) : P^*,$$
$$P^*J^* \cong 1, \qquad J^*P^* \cong 1. \tag{6.17}$$

This allows us to identify:

- weakly exact or exact, ordinary or generalised homology theories over $\mathsf{Top}_\bullet$,

- the analogous theories over RHTop_2,

- the analogous theories over Top_2 that satisfy the relative homeomorphism axiom.

6.1.8 Compact and locally compact spaces

These arguments can be restricted to compact Hausdorff spaces. We get an N-adjunction $P \dashv J$, with P exact and $PJ \cong 1$

$$J\colon \mathsf{Cph}_\bullet \to \mathsf{Cph}_2, \qquad X \mapsto (X, \{0_X\}) \qquad \text{(left exact)},$$
$$P\colon \mathsf{Cph}_2 \to \mathsf{Cph}_\bullet, \qquad (X, A) \mapsto X/A \qquad \text{(exact)}, \tag{6.18}$$

and an equivalence of homological bases

$$J^*\colon \mathrm{Th}(\mathsf{RHCph}_2, \mathsf{B}) \rightleftarrows \mathrm{Th}(\mathsf{Cph}_\bullet, \mathsf{B}) : P^*,$$
$$P^*J^* \cong 1, \qquad J^*P^* \cong 1. \tag{6.19}$$

Moreover, the p-homological category $\mathsf{Cph}_\bullet$ of pointed compact Hausdorff spaces can be replaced, up to categorical equivalence, with the p-homological subcategory $\mathcal{T}_{LC}$ of $\mathcal{T}$, that consists of locally compact Hausdorff spaces and partial proper continuous mappings, defined on open subsets (cf. 1.7.4(d)).

The (obvious) equivalence consists of the Alexandroff one-point compactification $S \mapsto S^+$ and the functor depriving a pointed compact space X of its base point

$$(-)^+\colon \mathcal{T}_{LC} \to \mathsf{Cph}_\bullet, \qquad S^+ = (S \cup \{\infty_S\}, \infty_S),$$
$$(-)^-\colon \mathsf{Cph}_\bullet \to \mathcal{T}_{LC}, \qquad X^- = X \setminus \{0_X\}. \tag{6.20}$$

The previous adjunction can be rewritten as $P^- \dashv J^+$

$$
\begin{aligned}
J^+ &: \mathcal{T}_{LC} \to \mathsf{Cph}_2, & S &\mapsto (S^+, \{\infty_S\}) & \text{(left exact)}, \\
P^- &: \mathsf{Cph}_2 \to \mathcal{T}_{LC}, & (X, A) &\mapsto X \setminus A & \text{(exact)}.
\end{aligned}
\tag{6.21}
$$

providing an equivalence of homological bases between RHCph_2 and $\mathcal{T}_{LC}$.

Again, everything restricts to the full subcategories of the metric case: Cpm_2 (pairs of compact metric spaces), $\mathsf{Cpm}_\bullet$ (pointed compact metric spaces) and $\mathcal{T}_{LCM}$ (locally compact metric spaces).

Summarising, the following notions are equivalent, as stated in [ES], X, 7.1 - 7.4

- a single space homology theory for compact Hausdorff (resp. metric) spaces,

- a single space homology theory for locally compact Hausdorff (resp. metric) spaces,

- a homology theory over Cph_2 (resp. Cpm_2), invariant up to relative homeomorphisms.

We have also shown that $\mathcal{T}_{LC} \simeq \mathsf{Cph}_\bullet \simeq \Sigma_R^{-1}\mathsf{Cph}_2$. Notice also that the equivalence of homological bases (6.21) gives the usual description of relative homeomorphisms of *compact* Hausdorff pairs as those mappings $f\colon (X, A) \to (Y, B)$ which restrict to a homeomorphism from $X \setminus A \to Y \setminus B$; such a characterisation does not hold in the general case.

6.1.9 Examples of single space theories

Classical examples have been given in the compact case, via Čech homology and cohomology. Strong homology seems to be the first example of a (nontrivial) *covariant* and *exact* theory satisfying the relative homeomorphism axiom for all topological pairs, as proved in [MaM].

(a) Čech cohomology gives a homology theory $\mathrm{RHCph}_2 \to \mathsf{Ab}^{\mathrm{op}}$, or equivalently a cohomology theory over $\mathsf{Cph}_\bullet$, or also over $\mathcal{T}_{LC}$. Čech homology gives a *weak* homology theory over RHCph_2, with values in Ab ([ES], X.5.4).

(b) The same holds for general pairs, i.e. over RHTop_2, provided that one utilises Čech cohomology and homology groups in the sense of Morita [Mta], based on normal open covers ([Wa], Theorems 27, 28).

(c) Steenrod homology [St] is an *exact* homology theory $\mathrm{RHCpm}_2 \to \mathsf{Ab}$ [Mi2].

(d) Strong homology, an extension of the former to arbitrary pairs (introduced by Lisitsa and Mardešić [LiM]), is an exact homology theory $\mathrm{RHTop}_2 \to \mathsf{Ab}$, or equivalently over $\mathsf{Top}_\bullet$ [MaM].

6.2 Singular homology for fibrations

We show now that a *singular homology theory* for fibrations $p\colon X \to X'$ can be defined by means of the following chain complex in $\mathrm{Ch}_+\mathsf{Ab}$

$$\mathrm{Ch}_+(p) = \mathrm{Ker}\, p_\sharp\colon \mathrm{Ch}_+(X) \twoheadrightarrow \mathrm{Ch}_+(X').$$

This theory forms an exact sequence of functors over an N-category with 'distinguished' short exact sequences, a notion that will not be developed here.

The full subcategory of covering maps is homological, and the previous singular homology forms there an exact sequence of functors.

6.2.1 Singular homology for fibrations

Let us start by recalling that singular homology for a pair (X, A) is produced by the monomorphism of singular chain complexes in $\mathrm{Ch}_+\mathsf{Ab}$

$$m_\sharp\colon \mathrm{Ch}_+(A) \rightarrowtail \mathrm{Ch}_+(X), \tag{6.22}$$

associated to the inclusion $m\colon A \to X$, by means of its cokernel, namely: $\mathrm{Ch}_+(X, A) = \mathrm{Cok}\, m_\sharp$.

Dually, in a loose sense, a surjective Serre fibration $p\colon X \to X'$ produces an *epimorphism* of singular chain complexes

$$p_\sharp\colon \mathrm{Ch}_+(X) \twoheadrightarrow \mathrm{Ch}_+(X'), \tag{6.23}$$

since every singular simplex of X' (being homotopic to a constant map) can be lifted to X; more generally, let us say that p is a *weak fibration* if the associated $p_\sharp$ is epi.

It seems natural to define the complex of *singular chains* (in $\mathrm{Ch}_+\mathsf{Ab}$) and the *singular homology groups* of the weak fibration p as

$$\mathrm{Ch}_+(p) = \mathrm{Ker}\,(p_\sharp), \qquad H_n(p) = H_n(\mathrm{Ch}_+(p)). \tag{6.24}$$

Now, the short exact sequence of complexes

$$\mathrm{Ch}_+(p) \rightarrowtail \mathrm{Ch}_+(X) \twoheadrightarrow \mathrm{Ch}_+(X'), \tag{6.25}$$

yields an *exact homology sequence* for p, corresponding to the usual homology sequence of a pair

$$\begin{aligned} \ldots\, H_n(p) &\to H_n(X) \to H_n(X') \to\, \ldots \\ &\to H_0(p) \to H_0(X) \to H_0(X') \to 0. \end{aligned} \tag{6.26}$$

The differential on $[z] \in H_n(X')$ is $[\partial\hat{z}] \in H_{n-1}(p)$, where $\hat{z}$ is any chain of X lifting z; then $p_\sharp(\hat{z}) = z$ and $\partial\hat{z}$ is a chain of the fibration, as $p_\sharp(\partial\hat{z}) = \partial p_\sharp(\hat{z}) = 0$.

Note that $\mathrm{Ch}_n(p)$ is free abelian, as a subgroup of $\mathrm{Ch}_n(X)$; this is essential in order to treat the chains of p with coefficients in some abelian group G, in the usual way

$$\mathrm{Ch}_+(p; G) = \mathrm{Ch}_+(p) \otimes G. \tag{6.27}$$

6.2.2 Exactness

In analogy with the homology sequence for a *triple* of spaces, i.e. for a composition of two inclusions, the composition of two weak fibrations $r = qp\colon X \to X' \to X''$ is a weak fibration and gives a diagram of complexes

$$
\begin{array}{ccccc}
\mathrm{Ch}_+(p) & =\!=\!= & \mathrm{Ch}_+(p) & \longrightarrow & 0 \\
\downarrow & & \downarrow & & \downarrow \\
\mathrm{Ch}_+(r) & \rightarrowtail & \mathrm{Ch}_+(X'') & \longrightarrow\!\!\!\!\to & \mathrm{Ch}_+(X) \\
\downarrow & & \downarrow & & \| \\
\mathrm{Ch}_+(q) & \rightarrowtail & \mathrm{Ch}_+(X') & \longrightarrow\!\!\!\!\to & \mathrm{Ch}_+(X)
\end{array}
\qquad . \tag{6.28}
$$

The first column is short exact, by the 3×3-Lemma (in $\mathrm{Ch}_+\mathsf{Ab}$); we thus have an exact homology sequence for the 'triple' $r = qp\colon X \to X' \to X''$

$$
\begin{aligned}
&\ldots H_n(p) \to H_n(r) \to H_n(q) \to \ldots \\
&\to H_0(p) \to H_0(r) \to H_0(q) \to 0.
\end{aligned}
\tag{6.29}
$$

6.2.3 The N-category of weak fibrations

An object of the category Wfb is a weak fibration $p\colon X \to X'$, in the previous sense (of 6.2.1). A morphism $f = (f, f')\colon p \to q$ is a fibre-preserving map $f\colon X \to Y$, i.e. a continuous mapping that induces a commutative square in Top

$$
\begin{array}{ccc}
X & \xrightarrow{\ f\ } & Y \\
p\downarrow & \nearrow{\scriptstyle g} & \downarrow q \\
X' & \xrightarrow[\ f'\ ]{} & Y'
\end{array}
\qquad . \tag{6.30}
$$

This morphism is defined to be *null* if there is a (unique) map g that fills-in commutatively (as in 2.5.9). The null objects are the trivial coverings, i.e. the homeomorphisms $X \to X'$.

Our ideal is closed, since g exists if and only if (f, f') factorises through

$1\colon Y \to Y$ as in the left diagram below (or equivalently through $1\colon X' \to X'$, as in the right diagram)

$$
\begin{array}{ccccc}
X & \xrightarrow{\ f\ } & Y & = & Y \\
p\downarrow & & \| & & \downarrow q \\
X' & \xrightarrow[g]{} & Y & \xrightarrow[q]{} & Y'
\end{array}
\qquad
\begin{array}{ccccc}
X & \xrightarrow{\ p\ } & X' & \xrightarrow{\ g\ } & Y \\
p\downarrow & & \| & & \downarrow q \\
X' & = & X' & \xrightarrow[f']{} & Y'
\end{array}
\qquad . \qquad (6.31)
$$

This category, likely, does not have (all) kernels and cokernels. But, in correspondence with any triple $r = qp = (X \to X' \to X'')$ of weak fibrations, we do have a short exact sequence, i.e. a pair of maps (m, p) with $m \sim \ker p$ and $p \sim \operatorname{cok} m$ (with respect to the given ideal of null maps)

$$
\begin{array}{ccccc}
X & = & X & \xrightarrow{\ p\ } & X' \\
p\downarrow & & \downarrow r & & \downarrow q \\
X' & \xrightarrow[q]{} & X'' & = & X''
\end{array}
\qquad . \qquad (6.32)
$$

Therefore, the functors H_n that we have introduced may be thought to form a homology theory over a category equipped with 'distinguished' short exact sequences. We shall not study this notion here.

6.2.4 The category of covering maps

Restricting our attention to *covering maps* (i.e. fibre bundles with discrete fibre), we get the full N-subcategory Cvm $\subset$ Wfb of *coverings*, which we now show to be homological (with respect to the restricted ideal). For the sake of simplicity, *we assume that all spaces are path connected and locally path connected.*

Recall that a covering map is necessarily open and surjective, whence a topological quotient. It follows that the morphism (6.30) is null if and only if $R_p \subset R_f$ (where R_p denotes the equivalence relation of X associated to the mapping p), if and only if f is constant on each fibre.

Given a morphism $f\colon p \to q$, it is easy to see that the equivalence relation $R = R_f \cap R_p$ of X satisfies the following conditions:

(i) $R \subset R_p$,

(ii) if σ and σ' are paths of X with $p\sigma = p\sigma'$, then $\sigma(0) \; R \; \sigma'(0)$ implies $\sigma(1) \; R \; \sigma'(1)$.

Let us call *p-congruence* such a relation on X; one can easily verify that the factorisation of p through X/R is composed of two *covering maps*

$X \to X/R \to X'$. Therefore, the kernel and cokernel of a morphism $f\colon p \to q$ are expressed as follows

$$
\begin{array}{ccccccc}
X & \!\!=\!\!=\!\! & X & \xrightarrow{\;f\;} & Y & \longrightarrow & Y/S \\
\downarrow & & {\scriptstyle p}\downarrow & & {\scriptstyle q}\downarrow & & \downarrow \\
X/R & \longrightarrow & X' & \xrightarrow[f']{} & Y' & \!\!=\!\!=\!\! & Y'
\end{array}
\qquad (6.33)
$$

where $R = R_f \cap R_p$ (a p-congruence), while S is the q-congruence of Y generated by the relation S_0: if $px = px'$ then $f'x\, S_0\, f'x'$. The normal factorisation of $f\colon p \to q$ is:

$$
\begin{array}{ccccccc}
X & \longrightarrow & X/R & \longrightarrow & Y & \!\!=\!\!=\!\! & Y \\
{\scriptstyle p}\downarrow & & \downarrow & & \downarrow & & \downarrow{\scriptstyle q} \\
X' & \!\!=\!\!=\!\! & X' & \longrightarrow & Y/S & \longrightarrow & Y'
\end{array}
\qquad (6.34)
$$

Therefore a *general* short exact sequence in Cvm is a diagram (6.32), determined by a 'triple' $r = qp = (X \to X' \to X'')$ of coverings, or also by the central object $r\colon X \to X''$ and any r-congruence R of the latter, by setting $X' = X/R$. This characterises normal monos and normal epis and shows that they are stable under composition.

Cvm is homological. Indeed if, in the following commutative diagram

$$
\begin{array}{ccccc}
Y & \!\!=\!\!=\!\! & Y & \xrightarrow{\;q'\;} & Y/S \\
{\scriptstyle q}\downarrow & & \downarrow{\scriptstyle r} & & \downarrow{\scriptstyle p'} \\
Y/R & \xrightarrow[p]{} & X & \!\!=\!\!=\!\! & X
\end{array}
\qquad (6.35)
$$

the normal mono $(1, p)\colon q \rightarrowtail r$ is greater than $(1, p') = \ker(r \twoheadrightarrow p')$, then q factorises through q', i.e. $S \subset R$ and the covering $Y/S \to Y/R$ is the required subquotient.

The exact sequence of a triple $r = qp$ (in (6.29)) proves that the singular homology of weak fibrations is indeed an exact sequence of functors over Cvm.

6.3 Homotopy spectral sequences for path-connected spaces

We introduce here the p-homological category Ngp of *normalised groups*, in which Gp embeds so that all the homomorphisms of groups become exact morphisms of Ngp.

The homotopy spectral sequence of a tower of fibrations *between path connected spaces* can then be obtained as the spectral sequence of an exact couple in the p-homological category Ngp, following the theory developed in Section 3.5.

The extension to arbitrary topological spaces will be given in Section 6.5.

6.3.1 A sketch of the construction

We shall construct the p-homological category $\mathsf{Ngp} = \mathsf{Q}/\mathcal{R}$, of *normalised groups*, as a quotient of a homological category $\mathsf{Q} = \mathsf{Gp}_2'$ containing Gp_2. The construction is inspired by the category of *homogeneous spaces* introduced by Lavendhomme [La].

We shall also see (in 6.3.6) that Ngp is a homological category of fractions (Section 5.2) of Gp_2

$$\mathsf{Ngp} = \Sigma^{-1}\mathsf{Gp}_2, \qquad P\colon \mathsf{Gp}_2 \to \Sigma^{-1}\mathsf{Gp}_2, \tag{6.36}$$

obtained by 'excision' of the invariant subgroups - a procedure which eliminates a manifest redundancy in the category of pairs of groups. In other words, Σ is the set of maps

$$p\colon (S, S_0) \to (T, T_0), \qquad p(S) = T, \quad S_0 = p^{-1}(T_0), \tag{6.37}$$

and these maps coincide up to isomorphism with the canonical projections $(S, S_0) \to (S/N, S_0/N)$, where N is an invariant subgroup of S contained in S_0 (take $N = \mathrm{Ker}\, p$). Notice that Σ does not satisfy the two-out-of-three property (see 5.5.1), so that the isokernel of P is certainly larger than Σ itself.

The canonical functor obtained from the embedding $I\colon \mathsf{Gp} \to \mathsf{Gp}_2$ (see 5.3.5)

$$J = PI\colon \mathsf{Gp} \to \mathsf{Ngp}, \qquad J(G) = (G, 0), \tag{6.38}$$

is a left exact and short exact embedding, whose properties are studied in Theorem 6.3.7. In particular, we prove there a crucial fact for our applications to homotopy spectral sequences: all the morphisms $f\colon G \to H$ of Gp become *exact morphisms* in Ngp.

As a more complicated alternative, one could rewrite this section using the calculus of fractions of Gabriel and Zisman ([GaZ], I.2), not directly in Gp_2, where the hypotheses are not satisfied, but in an auxiliary quotient (like, for instance, in the construction of the derived category of an abelian one).

(One considers the congruence $f\mathcal{R}g$ of Gp_2 given, for $f, g\colon (S, S_0) \to (T, T_0)$, by $fs - gs \in T_0$ for all $s \in S$. This congruence is *implicit* in Σ, in the sense that every functor defined over Gp_2, that makes all Σ-maps invertible, identifies all pairs of $\mathcal{R}$-equivalent maps - this can be proved as below, in (6.3.6)-(6.3.6). Then the category of general fractions of Gp_2 with respect to Σ trivially coincides with the category of general fractions of $\mathsf{Gp}_2/\mathcal{R}$, with respect to the image $\overline{\Sigma}$ of Σ; and it is easy to see that the latter category of fractions has a *right* calculus. This can be used to prove modified versions of Theorems 6.3.6 and 6.3.7, below.)

6.3.2 The homotopy exact couple of a tower of fibrations

The motivation for our construction is clear. Let us start from a tower of fibrations of pathwise connected pointed spaces ([BouK], p. 258)

$$\ldots\ X_s \xrightarrow{\ f_s\ } X_{s-1} \longrightarrow \ldots\ X_0 \xrightarrow{\ f_0\ } X_{-1} = \{*\} \tag{6.39}$$

and write $i_s \colon F_s \to X_s$ the fibre of the fibration $f_s \colon X_s \to X_{s-1}$. (The fibre is not assumed to be path connected, of course.)

Consider the exact homotopy sequence of $f_{-p} \colon X_{-p} \to X_{-p-1}$ $(p \leqslant 0)$, in Gp_2, writing $H_p = (f_{-p})_*(\pi_1 X_{-p})$

$$
\begin{array}{ccccccc}
\ldots\ \pi_{n+1}X_{-p} & \to & \pi_{n+1}X_{-p-1} & \to & \pi_n F_{-p} & \to & \pi_n X_{-p} \ \ldots \\
\| & & \| & & \| & & \| \\
\ldots\ D_{n,p-1} & \xrightarrow{u_{np}} & D_{np} & \xrightarrow{v_{np}} & E_{np} & \xrightarrow{\partial_{np}} & D_{n-1,p-1}\ \ldots
\end{array}
$$

$$
\begin{array}{ccccccc}
\ldots\ \pi_1 X_{-p} & \to & \pi_1 X_{-p-1} & \to & (\pi_1 X_{-p-1}, H_p) & \to & 0 \\
\| & & \| & & \| & & \| \\
\ldots\ D_{0,p-1} & \xrightarrow{u_{0p}} & D_{0p} & \xrightarrow{v_{0p}} & E_{0p} & \longrightarrow & 0
\end{array}
\tag{6.40}
$$

All these sequences produce a bigraded exact couple of type 1 in the homological category Ngp (cf. 3.5.5), with indices $n \geqslant 0 \geqslant p$

$$
\begin{aligned}
D_{np} &= D_{np}^1 = \pi_{n+1}X_{-p-1}, \\
E_{np} &= E_{np}^1 = \pi_n F_{-p} \quad (n > 0), \\
E_{0p} &= (\pi_1 X_{-p-1}, (f_{-p})_*(\pi_1 X_{-p})),
\end{aligned}
\tag{6.41}
$$

$$
\begin{aligned}
u_{np} &= \pi_{n+1}(f_{-p}) \colon \pi_{n+1}X_{-p} \to \pi_{n+1}X_{-p-1}, \\
v_{np} &\colon \pi_{n+1}X_{-p-1} \to \pi_n F_{-p} \quad (n > 0), \\
v_{0p} &\colon \pi_1 X_{-p-1} \to (\pi_1 X_{-p-1}, (f_{-p})_*(\pi_1 X_{-p})), \\
\partial_{np} &= \pi_n(i_{-p}) \colon \pi_n F_{-p} \to \pi_n X_{-p} \quad (n > 0).
\end{aligned}
\tag{6.42}
$$

The undefined objects are the zero object of Ngp.

Indeed, all these morphisms are group homomorphisms (embedded in Ngp) or normal epimorphisms (all v_{0p}); therefore *all the morphisms are exact in* Ngp (Section 6.3.1), and the same is true of all the compositions of morphisms u_{np}, consistently with our definition of an exact couple in a homological category (in 3.5.1).

Of course it is important to know that $J\colon$ Gp $\to$ Ngp is an embedding, so that we are not losing essential information about homotopy groups.

6.3.3 Additive ordered combinations

We now prepare the ground for a concrete construction of Ngp. Some notation concerning general groups will be useful.

Finite $\mathbb{Z}$-linear combinations $\sum_i \lambda_i s_i$ for an abelian group S also make sense for an arbitrary group in additive notation; of course we are no longer allowed to reorder terms, and repetitions of the elements s_i must be permitted.

It is thus simpler to consider *additive (ordered) combinations* $\sum_i \varepsilon_i s_i$ where ε_i stays for ± 1 and the index i varies in a finite *totally ordered* set, say $\{1, ...n\}$. Thus

$$- (\textstyle\sum_i \varepsilon_i s_i) = \sum_i (-\varepsilon_{n-i}).s_{n-i},$$

and the empty additive combination gives the identity 0 of the group.

(a) If S is a group, the subgroup $\langle X \rangle$ spanned by a subset X is the set of additive combinations $\sum_i \varepsilon_i x_i$, with $x_i \in X$.

(b) A mapping $f\colon S \to T$ between two groups is a homomorphism if and only if it preserves all the additive combinations $(f(\sum \varepsilon_i s_i) = \sum \varepsilon_i.f s_i)$, if and only if

$$(\textstyle\sum \varepsilon_i s_i = 0) \;\Rightarrow\; (\sum \varepsilon_i.f s_i = 0), \quad \text{for } s_i \in S, \; \varepsilon_i = \pm 1. \tag{6.43}$$

(c) The free group generated by a set X can be described as the set FX of formal additive combinations $\sum \varepsilon_i \hat{x}_i$ of the elements of X, provided that two such formulas are identified when they have the same *reduced* combination, obtained by suppressing all occurrences of type $+\hat{x} - \hat{x}$ or $-\hat{x} + \hat{x}$; the sum in FX is obvious.

(d) If S is a group, we write $ES = F|S|$ the free group generated by the underlying set $|S|$ and

$$e = e_S\colon ES \to S, \qquad e(\textstyle\sum \varepsilon_i \hat{s}_i) = \sum \varepsilon_i s_i, \tag{6.44}$$

the canonical homomorphism, given by the *evaluation* of a formal additive combination as an actual additive combination in S. (As well-known, these homomorphisms $e_S\colon F|S| \to S$ form the counit of the adjunction between the free-group functor F and the forgetful functor $|-|\colon \mathsf{Gp} \to \mathsf{Set}$; the unit is given by the embeddings $i_X\colon X \to |FX|$, $x \mapsto \hat{x}$.)

(e) Every pair of groups (S, S_0) is linked to a *free* pair $(ES, \overline{S}_0)$ by a Σ-map

$$e\colon (ES, \overline{S}_0) \to (S, S_0),$$
$$\overline{S}_0 = e^{-1}(S_0) = \{\textstyle\sum \varepsilon_i \hat{x}_i \in ES \mid \sum \varepsilon_i x_i \in S_0\}. \tag{6.45}$$

6.3.4 Quasi homomorphisms

Consider the category $\mathsf{Q} = \mathsf{Gp}_2'$ of *pairs of groups* (the objects of Gp_2) and *quasi homomorphisms* $f\colon (S, S_0) \to (T, T_0)$.

By definition, the latter are *mappings* $f\colon |S| \to |T|$ between the underlying sets, such that the following equivalent conditions hold (same notation as in 6.3.3, with $s, s', s_i \in S$ and $\varepsilon_i = \pm 1$):

(a) $(\sum \varepsilon_i s_i \in S_0) \Rightarrow (\sum \varepsilon_i . f s_i \in T_0)$,

(b) $Ef(\overline{S}_0) \subset \overline{T}_0$,

(c) $f S_0 \subset T_0$, $\quad f(\sum \varepsilon_i s_i) - (\sum \varepsilon_i . f s_i) \in T_0$,

(c') $f S_0 \subset T_0$, $\quad -f(\sum \varepsilon_i s_i) + \sum \varepsilon_i . f s_i \in T_0$,

(d) $f S_0 \subset T_0$, $\quad f(s + \varepsilon s') - \varepsilon f s' - f s \in T_0$,

(d') $f S_0 \subset T_0$, $\quad -f(\varepsilon s + s') + \varepsilon f s + f s' \in T_0$.

(In fact, (a) $\Leftrightarrow$ (b) is obvious. For (a) $\Rightarrow$ (d), take $\sum \varepsilon_i s_i = (s + \varepsilon s') - \varepsilon s' - s = 0 \in S_0$. Then (d) $\Rightarrow$ (c) can be proved by induction on the length of additive combinations: if (c) holds for $\sum \varepsilon_i s_i$, there exist $t, t' \in T_0$ such that $f(\sum \varepsilon_i s_i + \varepsilon s) = t + f(\sum \varepsilon_i s_i) + \varepsilon f s = t + t' + \sum \varepsilon_i f(s_i) + \varepsilon f s$. Finally, for (c) $\Rightarrow$ (a), let $s = \sum \varepsilon_i s_i \in S_0$. Then $f s \in T_0$ and $\sum \varepsilon_i . f s_i = (\sum \varepsilon_i . f s_i - f(\sum \varepsilon_i s_i)) + f s \in T_0$. In the same way one proves that (a) $\Rightarrow$ (d') $\Rightarrow$ (c') $\Rightarrow$ (a).)

Furthermore, it is easy to see that $f^{-1} T_0$ is a subgroup of S: if all $s_i \in f^{-1} T_0$, then $f(\sum \varepsilon_i s_i) \in T_0 + \sum \varepsilon_i . f s_i = T_0$.

Gp_2 is a subcategory of Q, and the embedding of Gp in Q is *full* (every quasi homomorphism with values in a pair $(T, 0)$ is a homomorphism).

The map $f\colon (S, S_0) \to (T, T_0)$ of Q is assumed to be null if the *set* $f S$ is contained in T_0. Q is semiexact, with the following normal factorisation

$$
\begin{array}{ccccc}
(f^{-1} T_0, S_0) \rightarrowtail (S, S_0) & \xrightarrow{\ f\ } & (T, T_0) & \twoheadrightarrow & (T, T_1) \\[4pt]
\quad\ \ p\big\downarrow & & \big\uparrow m & & \\[4pt]
(S, f^{-1} T_0) & \underset{g}{\dashrightarrow} & (T_1, T_0) & &
\end{array}
\tag{6.46}
$$

where $T_1 = \langle T_0 \cup fS \rangle$ denotes the subgroup of T spanned by the subset $T_0 \cup fS$.

The normal subobjects and normal quotients of (S, S_0) in $\mathbf{Q}$ are the same as in Gp_2, determined by subgroups M with $S_0 \subset M \subset S$; every short exact sequence of $\mathbf{Q}$ is of the following type, up to isomorphism:

$$(M, S_0) \rightarrowtail (S, S_0) \twoheadrightarrow (S, M) \qquad (S_0 \subset M \subset S). \qquad (6.47)$$

Therefore $\mathbf{Q}$ is a homological category, with the same description of subquotients as in Gp_2; the latter is a conservative homological subcategory of $\mathbf{Q}$ (it contains all the isomorphisms of the latter).

The properties (d) and (d′) show that $\mathbf{Q}$ is contained in the intersection $\mathcal{P}_g \cap \mathcal{P}_d$ of two (homological) categories considered by Lavendhomme [La]; the 'left-extended category of group pairs' $\mathcal{P}_g$ is defined as above by the properties

$$fS_0 \subset T_0, \qquad f(s + s') - fs' - fs \in T_0,$$

while the 'right-extended category of group pairs' $\mathcal{P}_d$ is defined by the properties

$$fS_0 \subset T_0, \qquad -f(s + s') + fs + fs' \in T_0.$$

It can be noted that $\mathcal{P}_g$ and $\mathcal{P}_d$ are isomorphic, by associating to each group the opposite one.

6.3.5 Normalised groups

The category $\mathsf{Ngp} = \mathbf{Q}/\mathcal{R}$ of *normalised groups* is the quotient of $\mathbf{Q}$ up to the congruence of categories $f\mathcal{R}g$ defined by the following equivalent conditions (for $f, g \colon (S, S_0) \to (T, T_0)$)

(a) for every $s \in S$, $fs - gs \in T_0$,

(b) for every $s \in S$, $-fs + gs \in T_0$,

(c) for all $s_i \in S$ and $\varepsilon_i = \pm 1$, $\sum \varepsilon_i . fs_i - \sum \varepsilon_i . gs_i \in T_0$.

It suffices to show that (a) $\Rightarrow$ (c); using the quasi homomorphism property of f and g, there are some $t, t' \in T_0$ such that

$$\sum \varepsilon_i . fs_i - \sum \varepsilon_i . gs_i = t + f(\sum \varepsilon s_i) - (t' + g(\sum \varepsilon_i s_i)$$
$$= t + (f(\sum \varepsilon_i s_i) - g(\sum \varepsilon_i s_i)) - t' \in T_0.$$

A map $[f]$ in Ngp is assumed to be null if and only if f is null in $\mathbf{Q}$, independently of the choice of a representative. The null objects are the pairs (S, S), as in $\mathbf{Q}$ and Gp_2. But Ngp is pointed, with zero object $0 =$

$(0,0) \cong (S,S)$, since the map $0\colon (S,S) \to (S,S)$ is $\mathcal{R}$-equivalent to the identity.

Ngp has kernels and cokernels, that have the same description as in Q (independently of the representative we choose for [f]).

Therefore Ngp is p-homological and the canonical functor

$$P\colon \mathsf{Gp_2} \to \mathsf{Ngp}, \qquad (S,S_0) \mapsto (S,S_0), \quad f \mapsto [f], \qquad (6.48)$$

given by the composition $\mathsf{Gp_2} \to \mathsf{Q} \to \mathsf{Ngp}$ is exact, nsb-faithful and nsb-full.

(Again, Ngp is a subcategory of the Lavendhomme category $\mathcal{H}_g = \mathcal{P}_g/\mathcal{R}_g$ of *left homogeneous spaces* obtained by the congruence $\mathcal{R}_g$ described in (a); it is also a subcategory of the category $\mathcal{H}_d = \mathcal{P}_d/\mathcal{R}_d$ of *right homogeneous spaces*, obtained by the congruence $\mathcal{R}_d$ described in (b). Indeed, both congruences restrict to our relation $\mathcal{R}$ over Q. See [La], Section 3.)

6.3.6 Theorem

The functor $P\colon \mathsf{Gp_2} \to \mathsf{Ngp}$ 'is' the category of fractions $\Sigma^{-1}\mathsf{Gp_2}$, i.e. it solves the universal problem of making the morphisms of Σ (defined in (6.37)) invertible, within (arbitrary) categories and functors.

Furthermore, P is exact and also solves this universal problem within semiexact categories and exact functors (or left exact, or right exact, or short exact functors).

Proof (a) First we prove that P carries each map of Σ to an isomorphism of Ngp.

Given $p\colon (S,S_0) \to (S/N, S_0/N)$ in Σ, choose a *mapping* $j\colon S/N \to S$ such that $p.j = 1$. Then $j\colon (S/N, S_0/N) \to (S,S_0)$ is a morphism of Q, since it satisfies 6.3.4(a): if $\sum \varepsilon_i.p(s_i) \in S_0/N$, then

$$p(\sum \varepsilon_i.jp(s_i)) \in S_0/N, \qquad \sum \varepsilon_i.jp(s_i) \in S_0.$$

Further, $jp\,\mathcal{R}\,1$ (and $[j][p] = 1$) because, for every $s \in S$, we have $jp(s) - s \in N \subset S_0$.

(b) Every functor $G\colon \mathsf{Gp_2} \to \mathsf{C}$ which makes each map of Σ invertible in C can be uniquely extended to Ngp.

Indeed, given a Q-map $f\colon (S,S_0) \to (T,T_0)$, let us write, as in 6.3.3, $ES = F|S|$, $e\colon ES \to S$ the canonical evaluation epimorphism and

$$\overline{S}_0 = e^{-1}(S_0) = \{\sum \varepsilon_i \hat{x}_i \mid \sum \varepsilon_i x_i \in S_0\}.$$

Then the group homomorphism $f'\colon ES \to T$ defined by the mapping

$|f|$ gives a map of $\mathsf{Gp_2}$ and the following commutative diagram in Q, with $e \in \Sigma$

$$
\begin{array}{ccc}
(S, S_0) & \xrightarrow{\;f\;} & (T, T_0) \\
{\scriptstyle e}\big\uparrow & \nearrow{\scriptstyle f'} & \\
(ES, \overline{S}_0) & &
\end{array}
\tag{6.49}
$$

$$
f' = e_T.Ef, \quad f'(\overline{S}_0) = e_T.Ef(\overline{S}_0) \subset e_T(\overline{T}_0) = e_T e_T^{-1}(T_0) = T_0,
$$

$$
[f] = [f'].[e]^{-1}, \quad \text{in } \mathsf{Ngp}.
$$

Therefore any functor G' that extends G on Ngp is uniquely determined, as follows:

$$
G' : \mathsf{Ngp} \to \mathsf{C},
$$
$$
G'(S, S_0) = (S, S_0), \qquad G'[f] = (Gf').(Ge)^{-1}.
\tag{6.50}
$$

We now prove that G' is well defined by these formulas, and indeed a functor. Firstly, we verify that $f\mathcal{R}g$ in Q implies $Gf' = Gg'$. Let

$$
U = \langle fS \cup gS \rangle, \quad U_0 = T_0 \cap U,
$$
$$
N = \langle \textstyle\sum \varepsilon_i.fs_i - \sum \varepsilon_i.gs_i \mid s_i \in S,\ \varepsilon_i = \pm 1 \rangle.
$$

N is a subgroup of U_0, invariant in U, as it follows from the following computation (together with the similar one concerning the inner automorphism produced by $\varepsilon.gs$ instead of $\varepsilon.fs$)

$$
\varepsilon.fs + \left(\textstyle\sum \varepsilon_i.fs_i - \sum \varepsilon_i.gs_i \right) - \varepsilon.fs
$$
$$
= \varepsilon.fs + \textstyle\sum \varepsilon_i fs_i - \sum \varepsilon_i gs_i - \varepsilon.gs + \varepsilon.gs - \varepsilon.fs
$$
$$
= \left(\varepsilon.fs + \textstyle\sum \varepsilon_i.fs_i \right) - \left(\varepsilon.gs + \sum \varepsilon_i.gs_i \right) + \varepsilon.gs - \varepsilon.fs \in N.
$$

Considering the following diagram in $\mathsf{Gp_2}$, where f'' and g'' are the restrictions of f' and g', the arrow Gs is an isomorphism and $sf'' = sg''$

$$
\begin{array}{ccc}
(ES, \overline{S}_0) & = = & (ES, \overline{S}_0) \\
{\scriptstyle f'}\big\downarrow\big\downarrow{\scriptstyle g'} & & {\scriptstyle f''}\big\downarrow\big\downarrow{\scriptstyle g''} \\
(T, T_0) & \xleftarrow{\;i\;} (U, U_0) & \xrightarrow{\;s\;} (U/N, U_0/N)
\end{array}
$$

we get that $Gf' = Gg'$.

Secondly, the following computations show that G' is a functor

$$
\begin{array}{ccccc}
(ES,\overline{S}_0) & \xrightarrow{\;Ef\;} & (ET,\overline{T}_0) & \xrightarrow{\;Eg\;} & (EU,\overline{U}_0) \\
\left\downarrow{\scriptstyle e}\right. & \searrow{\scriptstyle f'} & \left\downarrow{\scriptstyle e}\right. & \searrow{\scriptstyle g'} & \left\downarrow{\scriptstyle e}\right. \\
(S,S_0) & \xdashrightarrow[{[f]}]{} & (T,T_0) & \xdashrightarrow[{[g]}]{} & (U,U_0)
\end{array}
$$

$$(gf)' = e_U.E(gf) = e_U.Eg.Ef,$$
$$G'[gf] = (G(gf')).(Ge_S)^{-1} = Ge_U.GEg.GEf.(Ge_S)^{-1}$$
$$= (Ge_U.GEg.(Ge_T)^{-1}).((Ge_T).GEf.(Ge_S)^{-1}) = G'[g].G'[f].$$

(c) We already know (by 6.3.5) that P is exact, nsb-faithful and nsb-full. The last assertion now follows easily from the fact that every map of **Ngp** factorises as $\varphi = Pf.(Pe)^{-1}$ for some f in **Gp$_2$** and some $e \in \Sigma$ (cf. (6.49)). Indeed, if **C** is semiexact and the functor $G\colon \mathsf{Gp_2} \to \mathsf{C}$ is left exact, take $m = \ker f$ in **Gp$_2$**; then

$$Pm = \ker Pf, \quad Ps.Pm = \ker Pf.(Ps)^{-1} = \ker \varphi,$$
$$G'(\ker \varphi) = Gs.Gm = \ker Gf.(Gs)^{-1} = \ker G'(Pf.(Ps)^{-1}) = \ker G'(\varphi).$$

The right exact case is proved similarly, as well as the short exact case (taking into account the fact that P is nsb-full). $\qquad\square$

6.3.7 Theorem (Normalising groups)

The canonical functor

$$J = PI\colon \mathsf{Gp} \to \mathsf{Ngp}, \qquad S \mapsto (S,0), \quad f \mapsto [f], \tag{6.51}$$

is an embedding. It satisfies the following properties:

*(a) J is a functor from **Gp** to a semiexact category,*

(b) J is short exact,

(c) J takes every monomorphism to a normal mono.

*Moreover, J is universal for such properties, i.e. every functor $G\colon \mathsf{Gp} \to$ **E** that also satisfies (a) - (c) factorises uniquely as $G = G'J$, where $G'\colon \mathsf{Ngp} \to \mathsf{E}$ is a short exact functor.*

Furthermore, in the presence of (a) and (b), the property (c) is equivalent to:

(c$'$) J is left exact and carries every morphism to an exact morphism.

Proof J is trivially an embedding, because of the definition of the congruence $\mathcal{R}$ in 6.3.5.

Property (b) is an easy consequence of Theorem 6.3.6. Indeed, a short exact sequence of Gp, say $N \rightarrowtail S \twoheadrightarrow S/N$, is transformed by I into the following (solid) left exact sequence of $\mathsf{Gp_2}$

$$(N, 0) \rightarrowtail (S, 0) \longrightarrow (S/N, 0)$$

$$\begin{array}{c} \nwarrow \quad \uparrow u \\ (S, N) \end{array} \qquad (6.52)$$

The latter becomes short exact in Ngp, because the projection u belongs to Σ.

As to (c), all the monomorphisms of Gp become normal monos in $\mathsf{Gp_2}$, and are preserved as such by the exact functor $P\colon \mathsf{Gp_2} \to \mathsf{Ngp}$.

Now, let the functor $G\colon \mathsf{Gp} \to \mathsf{E}$ satisfy (a) - (c). It is easy to see that G extends uniquely to a short exact functor $G^+\colon \mathsf{Gp_2} \to \mathsf{E}$ defined as

$$G^+(S, S_0) = \mathrm{Cok}_{\mathsf{E}}(G(S_0 \to S)).$$

Furthermore, an arbitrary short exact sequence of groups

$$(m, p) = (N \rightarrowtail S \twoheadrightarrow S/N)$$

produces in $\mathsf{Gp_2}$ a diagram

$$\begin{array}{ccc} (N, 0) & \xrightarrow{m'} (S, 0) & \xrightarrow{p'} (S/N, 0) \\ \| & \| & \uparrow u \\ (N, 0) & \rightarrowtail (S, 0) & \longrightarrow (S, N) \end{array} \qquad (6.53)$$

This is transformed by G^+ into a commutative diagram whose *upper* row is exact (since $G = G^+I$ is short exact) as well as the *lower* row (because G^+ itself is short exact). Therefore G^+ carries all the maps of type u in (6.52) to isomorphisms; these maps u are particular maps of Σ (those with codomain of type $I(T) = (T, 0)$).

Given now an arbitrary Σ-map $p\colon (S, S_0) \to (S/N, S_0/N)$, the commutative diagram of $\mathsf{Gp_2}$, with short exact rows

$$\begin{array}{ccc} (S_0, N) & \rightarrowtail (S, N) & \longrightarrow (S, S_0) \\ v\downarrow & u\downarrow & p\downarrow \\ (S_0/N, 0) & \rightarrowtail (S/N, 0) & \twoheadrightarrow (S/N, S_0/N) \end{array} \qquad (6.54)$$

is transformed by G^+ into a commutative diagram of E, with short exact rows; since G^+u and G^+v are isomorphisms, so is G^+p (that is the result of applying the cokernel functor $\mathrm{Cok}\colon \mathsf{E^2} \to \mathsf{E}$ to the map (v, u)).

Therefore, there is precisely one short exact functor $G'\colon \Sigma^{-1}\mathsf{Gp}_2 \to \mathsf{E}$ such that $G'P = G^+$. Finally $G'J = G'PI = G^+I = G$.

On the other hand, if G' is short exact and $G'J = G$, then $G'P$ is short exact and extends G to Gp_2, so that $G'P$ coincides with G^+ and G' is uniquely determined.

Finally, we verify the last assertion. If (c) holds and $k = \ker f$, $p = \operatorname{cok} k$ in Gp, then $f = ip$ with i mono; therefore, in Ngp, the map Ji is a normal mono and (Jk, Jp) is a short exact sequence. Then $Jf = Ji.Jp$ is an exact morphism and $\ker Jf = \ker Jp$ is equivalent to Jk.

The converse follows from the fact that every exact morphism with a null kernel is a normal mono. □

6.3.8 Normalised rings

In a similar way one can construct the p-homological category Nrn of normalised rings, solving a similar problem for the semiexact category Rng of (possibly non-unital) rings.

The additive combinations of 6.3.3 are replaced with *non-commutative polynomials*, with coefficients in $\mathbb{Z}$ *and degree* $\geqslant 1$. More precisely, every set W generates a free ring FW, whose elements are non-commutative polynomials over W, i.e. $\mathbb{Z}$-linear combinations of elements of the free semigroup generated by S. For example

$$p(X, Y, Z) = 2.XYXZ - ZX^2Y - XYZX \tag{6.55}$$

is a (non-null!) non-commutative polynomial over $W = \{X, Y, Z\}$.

If x, y, z belong to a ring R, $p(x, y, z)$ denotes of course the element $2.xyxz - zx^2y - xyzx$ of R. The evaluation map

$$e\colon E(R) = F|R| \to R$$

carries each formal non-commutative polynomial with indeterminates in the *set* $|R|$ to its 'value' in R.

We now define, as in 6.3.4, $\mathsf{Nrn} = \mathbf{R}/\mathcal{R}$. Here, $\mathbf{R}$ is the category of *pairs of rings* and *quasi homomorphisms* $f\colon (R, R_0) \to (S, S_0)$, i.e. mappings $f\colon |R| \to |S|$ between the underlying sets such that the following equivalent conditions hold (p denotes an arbitrary non-commutative polynomial and $x_i \in R$):

(a) $p(x_1, ..., x_n) \in R_0 \Rightarrow p(fx_1, ..., fx_n) \in S_0$,

(b) $Ef(\overline{R}_0) \subset \overline{S}_0$,

(c) $f(R_0) \subset S_0$, $\quad fp(x_1, ..., x_n) - p(fx_1, ..., fx_n) \in S_0$.

The relation $f\mathcal{R}g$ means again that $fx - gx \in S_0$, for all $x \in R$.

6.4 Actions and homotopy theory

This section is concerned with the category Act of actions of groups on pointed sets [G13], or actions for short, in which Set$_\bullet$, Gp and Gp$_2$ can be naturally embedded.

We prove that Act is homological and show that the homotopy sequences of a pair of spaces or of a fibration can be interpreted as exact sequences in this category. Homotopy theory can thus be viewed as a homology theory over the category Top$_{\bullet 2}$ of pairs of pointed spaces, with values in Act.

6.4.1 The category of actions

An *action* is a pair (X, S) where S is a group (always in additive notation) and X is a pointed set (whose base-point is written 0 or 0_X) equipped with a right action of S on X, written as a sum $x + s$ (for $x \in X$, $s \in S$) and satisfying the usual axioms

$$x + 0_S = x, \qquad (x + s) + s' = x + (s + s') \qquad (x \in X; \; s, s' \in S). \quad (6.56)$$

If $x + s = x'$, we say that the operator s *links* x and x'. Notice that the base point is *not* assumed to be fixed under S; we often write

$$S_0 = \mathrm{Fix}_S(0_X) = \{s \in S \mid 0_X + s = 0_X\},$$

the subgroup of operators that leave the base point fixed.

A morphism of actions $f = (f', f''): (X, S) \to (Y, T)$ consists of a morphism $f': X \to Y$ of pointed sets and a group-homomorphism $f'': S \to T$ consistent with the former:

$$f'0_X = 0_Y, \quad f''(s + s') = f''s + f''s', \quad f'(x + s) = f'x + f''s. \quad (6.57)$$

We shall often write fx and fs instead of $f'x$ and $f''s$. The composition is plain.

This defines the N-category Act, where a morphism $f = (f', f'')$ is assumed to be *null* whenever f' is a zero-morphism in the category of pointed sets, i.e. $fx = 0_Y$ for all $x \in X$.

The null objects are thus the pairs $(0, S)$, where 0 is the pointed singleton; the ideal of null morphisms is closed, because f is null if and only if it factorises through the null object $(\{0_Y\}, T_0)$, where $T_0 = \mathrm{Fix}_T(0_Y)$. There is a zero object, the pair $(0, 0)$ formed by the null group acting on the pointed singleton; it is null, but the null morphisms do not reduce to the zero morphisms.

The kernel of the morphism $f = (f', f''): (X, S) \to (Y, T)$ is the following embedding:

$$(X_1, S_1) \rightarrowtail (X, S),$$
$$X_1 = f^{-1}\{0_Y\} = \operatorname{Ker} f',$$
$$S_1 = f^{-1}T_0 = \{s \in S \mid 0_Y + fs = 0_Y\} \tag{6.58}$$
$$= \{s \in S \mid X_1 + s \subset X_1\} = \{s \in S \mid X_1 + s = X_1\}$$
$$= \{s \in S \mid s \text{ links two points of } X_1\}.$$

(Note that S_1 is determined by X_1.) The cokernel of f is the natural projection:

$$(Y, T) \twoheadrightarrow (Y/R, T), \tag{6.59}$$

where R is the T-congruence of Y generated by identifying all the elements of fX.

This proves that Act is semiexact; we need further work to show that it is homological.

6.4.2 Normal subobjects and quotients

A normal subobject of the object (X, S) in Act can be characterised as the embedding of a pair (X_1, S_1) where:

(a) S_1 is a subgroup of S and X_1 is a pointed subset of X stable under S_1,

(b) if $s \in S$ links two points of X_1 then $s \in S_1$ (*normality* condition);

or equivalently:

(a′) X_1 is a pointed subset of X, and if $s \in S$ links two points of X_1 then $X_1 + s \subset X_1$,

(b′) $S_1 = \{s \in S \mid X_1 + s \subset X_1\} = \{s \in S \mid \exists x, x' \in X_1 : x + s = x'\}$,

or also:

(a″) X_1 is a pointed subset of X and S_1 is a subgroup of S,

(b″) if $x \in X_1$ and $s \in S$, then: $x + s \in X_1 \Leftrightarrow s \in S_1$.

Indeed, it is easy to verify that these conditions are equivalent and that every kernel is of this type.

Conversely, given an action (X_1, S_1) satisfying these conditions, consider the following relation in the set X:

$$xRx' \Leftrightarrow (x = x' \text{ or } x = x_1 + s, \ x' = x_1' + s') \tag{6.60}$$
$$(\text{where } x_1, x_1' \in X_1 \text{ and } s - s' \in S_1).$$

It is an equivalence relation: if we also have $x' = x_2' + t$ and $x'' = x_2' + t'$, with $x_2, x_2' \in X_1$ and $t - t' \in S_1$, then $s' - t \in S_1$ (because it links two points of X_1, namely x_1' and $x_2' = x' - t = x_1' + s' - t$), whence $s - t' = (s - s') + (s' - t) + (t - t') \in S_1$ and $x = x_1 + s \, R \, x'' = x_2' + t'$.

Therefore, R is the S-congruence of X generated by identifying all the points in X_1; further, $x \, R \, 0_X$ if and only if $x \in X_1$. It follows immediately that the natural projection:

$$p \colon (X, S) \to (X/R, S), \qquad p(x) = [x], \quad p(s) = s, \qquad (6.61)$$

is a morphism of Act, with kernel (X_1, S_1).

It also follows that the cokernel of $m \colon (X_1, S_1) \to (X, S)$ is

$$p \colon (X, S) \to (X/R, S),$$

where R is described in (6.60): indeed, if $f \colon (X, S) \to (Y, T)$ annihilates on m, then $f'(X_1) = \{0_Y\}$. Thus, we have also characterised the normal quotients of (X, S).

By our characterisation, the ordered set $\mathrm{Nsb}(X, S)$ of normal subobjects of (X, S) can be identified with the set of parts $X_1 \subset X$ satisfying the condition (a$'$) above. Since this set is plainly closed under arbitrary intersections, $\mathrm{Nsb}(X, S)$ is a complete lattice.

6.4.3 Normal factorisation

We can now describe the normal factorisation of f in Act

$$
\begin{array}{ccccc}
(X_1, S_1) & \rightarrowtail & (X, S) & \xrightarrow{\ f\ } & (Y, T) \longrightarrow (Y/R', T) \\
 & & \downarrow{\scriptstyle p} & & \uparrow{\scriptstyle m} \\
 & & (X/R, S) & \xrightarrow[\ g\ ]{} & (Y_1, T_1)
\end{array}
\qquad (6.62)
$$

Let us recall that the kernel (X_1, S_1) has been computed in (6.58):

$$X_1 = f^{-1}\{0_Y\} = \mathrm{Ker}\, f',$$
$$
\begin{aligned}
S_1 = f^{-1}T_0 &= \{s \in S \mid 0_Y + fs = 0_Y\} \\
&= \{s \in S \mid X_1 + s \subset X_1\} = \{s \in S \mid X_1 + s = X_1\} \\
&= \{s \in S \mid s \text{ links two points of } X_1\}.
\end{aligned}
\qquad (6.63)
$$

and the normal coimage $(X/R, S) = \mathrm{Cok}\,(\ker f)$ in (6.60):

$$
\begin{aligned}
x R x' \quad &\Leftrightarrow \quad (x = x' \ \text{or} \ x = x_1 + s, \ x' = x_1' + s'), \\
&\text{where } x_1, x_1' \in X_1, \ s - s' \in S_1.
\end{aligned}
\qquad (6.64)
$$

The normal image $\mathrm{Nim}\,(f) = (Y_1, T_1)$ is the least normal subobject of (Y, T) through which f factorises and can be characterised as follows:

(i) T_1 is the subgroup of T spanned by the elements $t \in T$ which link the elements of $f'(X)$, i.e. $fx + t = fx'$, for some $x, x' \in X$,

(ii) $Y_1 = f'X + T_1$.

Indeed this pair (Y_1, T_1) satisfies the conditions 6.4.2(a), (b) for a normal subobject of (Y, T): Y_1 is stable under T_1 and, if $y = fx + t_1 \in Y_1$, $t \in T$ and $y + t \in Y_1$, then $y + t = fx' + t_1'$, whence

$$fx + t_1 + t = fx' + t_1', \qquad t_1 + t - t_1' \in T_1. \tag{6.65}$$

It follows that $t \in T_1$. It is now easy to check that (Y_1, T_1) is the least normal subobject through which f factorises.

Last, we know from 6.4.1 that $\operatorname{Cok} f = (Y/R', T)$ is determined by the least T-congruence R' of Y that identifies the elements of $f'X$; but $\operatorname{Cok} f = \operatorname{Cok} m$, so that R' can be more concretely presented as the congruence produced by (Y_1, T_1)

$$
\begin{aligned}
yR'y' \quad &\Leftrightarrow \quad (y = y' \text{ or } y = y_1 + t, \ y' = y_1' + t'), \\
&\qquad \text{where } y_1, y_1' \in Y_1, \ t, t' \in T, \ t - t' \in T_1, \\
&\Leftrightarrow \quad (y = y' \text{ or } y = fx + t, \ y' = fx' + t'), \\
&\qquad \text{where } x, x' \in X, \ t, t' \in T, \ t - t' \in T_1.
\end{aligned}
\tag{6.66}
$$

On the other hand, since $m = \ker \operatorname{cok} f$

$$T_1 = \operatorname{Fix}_T(0_{Y/R'}). \tag{6.67}$$

6.4.4 Theorem

Act *is a homological category.*

Proof First, the normal subobjects are closed under composition; given

$$(X_2, S_2) \rightarrowtail (X_1, S_1) \rightarrowtail (X, S), \tag{6.68}$$

if $s \in S$ links the elements $x, x' \in X_2 \subset X_1$ (i.e. $x + s = x'$), it follows that $s \in S_1$, and then $s \in S_2$.

Second, let the normal quotients p and q be given, with $Y = X/R$ and $Z = Y/R'$

$$
\begin{array}{ccc}
(X, S) & \xrightarrow{\ p\ } & (Y, S) \\
{\scriptstyle h}\downarrow & & \downarrow{\scriptstyle q} \\
(X/R'', S) & \underset{g}{\ -\!\!\!-\!\!\!\rightarrow\ } & (Z, S)
\end{array}
\tag{6.69}
$$

and let $h = \operatorname{ncm} qp = \operatorname{cok} \ker qp$. There is a unique morphism g which

makes the diagram commutative, and we have to show that it is an isomorphism. Let

$$(X_1, S_1) = \operatorname{Ker} qp = (p^{-1}q^{-1}\{0_Z\}, p^{-1}q^{-1}(\operatorname{Fix}_S(0_Z)))$$
$$= (p^{-1}q^{-1}\{0_Z\}, \operatorname{Fix}_S(0_Z)).$$

The invertibility of g amounts to the injectivity of its component g' (since
g' is trivially surjective and $g'' = 1_S$), or also to the following condition:
if $x, x' \in X$ and $qp(x) = qp(x')$ then $x\, R''\, x'$. Actually, $px\, R'\, px'$, whence
$px = y_1 + s$, $px' = y_1' + s'$ with

$$y_1, y_1' \in Y_1 = q^{-1}(\{0_Z\}) = p(X_1),$$
$$s - s' \in \{s \in S \mid 0_Z + s = 0_Z\} = S_1.$$

Therefore $y_1 = px_1$, $y_1' = px_1'$, and $x\, R\, (x_1 + s)\, R''\, (x_1' + s')\, R\, x'$, so that
$x\, R''\, x'$.

As to the subquotient axiom, let us start from the central row of the
following diagram

$$\begin{array}{ccccc}
(X_2, S_2) & \xrightarrowtail{n} & (X, S) & \xrightarrow{p} & (X/R_1, S) \\
{\scriptstyle u}\big\downarrow & & \big\| & & \big\uparrow{\scriptstyle v} \\
(X_1, S_1) & \xrightarrowtail{m} & (X, S) & \xrightarrow{q} & (X/R_2, S) \\
{\scriptstyle h}\big\downarrow & & & & \big\uparrow{\scriptstyle k} \\
(X_1/\overline{R}_2, S_1) & \dashrightarrow[f]{} & & & (Y, S')
\end{array} \qquad (6.70)$$

with $m \geqslant n = \ker q$ and $q \geqslant p = \operatorname{cok} m$. Let $n = mu$ and $p = vq$,
so that $u = \ker qm$ and $v = \operatorname{cok} qm$ (by the general theory, or by direct
verification); let $h = \operatorname{cok} u = \operatorname{ncm} qm$, $k = \ker v = \operatorname{nim} qm$ and write f the
induced morphism.

We have to prove that $f = (f', f'')$ is an isomorphism. The mapping f'
is surjective, as

$$Y = v^{-1}0_{X/R}1 = q(q^{-1}v^{-1}0_{X/R_1}) = qX_1 = \operatorname{Im} q'm'.$$

It is injective, because $qm(x_1) = qm(x_1')$ if and only if $x_1 R_2 x_1'$, if and
only if they are equal or $x_1 = x_2 + s$, $x_1' = x_2' + s'$ with $s - s' \in S_2$, if and
only if $x_1 \overline{R}_2 x_1'$.

Finally, the inclusion $f'' \colon S_1 \to S'$ is the identity, since

$$S' = v^{-1}\operatorname{Fix}_S(0_{X/R_1}) = \operatorname{Fix}_S(0_{X/R_1}) = S_1.$$

$\square$

6.4.5 Pointed sets as actions of null groups

The category $\mathsf{Set}_{\bullet}$ of pointed sets is equivalent to the full homological subcategory Act_N of Act consisting of the actions of null groups (on pointed sets).

Indeed, let us start from the adjoint functors $V \dashv U$

$$
\begin{aligned}
U \colon \mathsf{Set}_{\bullet} &\to \mathsf{Act}, & U(Z) &= (Z, 0), \\
V \colon \mathsf{Act} &\to \mathsf{Set}_{\bullet}, & V(X, S) &= X/S.
\end{aligned}
\tag{6.71}
$$

U associates to a pointed set Z the action on Z of the null group, V associates to an action (X, S) its orbit-set X/S, pointed in the orbit $0_X + S$ of the base point, and

$$
VU \cong 1, \qquad \mathsf{Set}_{\bullet}(X/S, Z) \cong \mathsf{Act}((X, S), (Z, 0)).
\tag{6.72}
$$

Plainly, U restricts to an equivalence $\mathsf{Set}_{\bullet} \to \mathsf{Act}_N$ that preserves the null morphisms; therefore, U is exact.

V too is exact: given a morphism $g \colon (X, S) \to (Y, T)$, V preserves its cokernel (by adjointness) and also its kernel

$$
\begin{aligned}
\operatorname{Ker} g &= (g^{-1}0_Y, g^{-1}T_0), \\
V(\operatorname{Ker} g) &= (g^{-1}0_Y)/(g^{-1}T_0) = \operatorname{Ker}(Vg \colon X/S \to Y/T).
\end{aligned}
\tag{6.73}
$$

6.4.6 Pairs of groups as transitive actions

The full subcategory Act_T of Act consisting of the *transitive* actions is again a homological subcategory, which we now prove to be equivalent to Gp_2. It follows that the semiexact category Gp has a left exact embedding in Act.

In fact, there is an adjoint retraction $F \dashv G$, where F is exact while G is left exact and short exact

$$
\begin{aligned}
F \colon \mathsf{Gp}_2 &\to \mathsf{Act}, & F(S, S_0) &= (S/S_0, S), \\
G \colon \mathsf{Act} &\to \mathsf{Gp}_2, & G(X, S) &= (S, S_0) & (S_0 = \mathrm{Fix}_S 0_X).
\end{aligned}
\tag{6.74}
$$

F associates to a pair (S, S_0) the canonical (right) action of the group S over the pointed set S/S_0 of the right cosets of S_0 in S: $(S_0 + s) + s' = S_0 + (s + s')$, so that

$$
\begin{aligned}
GF(S, S_0) &= (S, S_0), \\
\mathsf{Act}((S/S_0, S), (Y, T)) &\cong \mathsf{Gp}_2((S, S_0), (T, T_0)).
\end{aligned}
\tag{6.75}
$$

Further, Gp_2 is equivalent to the full homological subcategory Act_T of Act, by means of the restriction $F' \colon \mathsf{Gp}_2 \to \mathsf{Act}_T$ of F. Indeed $F(S, S_0) = (S/S_0, S)$ is a transitive action, F' is fully faithful and for every transitive action (X, S), the counit ε is an isomorphism. This also proves that F is exact, sparing us a longer direct proof.

The functor G is not exact, but is left exact (as a right adjoint) and also short exact (i.e. it preserves short exact sequences): given a short exact sequence and its G-image

$$
\begin{aligned}
(X_1, S_1) &\rightarrowtail (X, S) \twoheadrightarrow (X/R, S), \\
(S_1, S_0) &\rightarrowtail (S, S_0) \twoheadrightarrow (S, \mathrm{Fix}0_{X/R}),
\end{aligned}
\tag{6.76}
$$

by (6.67) we have: $\mathrm{Fix}(0_{X/R}) = S_1$, whence $(S, \mathrm{Fix}(0_{X/R})) = (S, S_1) = \mathrm{Cok}\,((S_1, S_0) \rightarrowtail (S, S_0))$.

Last, the category Gp is a retract of Act, as it follows by composing two retractions already considered, in 5.3.5 and above

$$
\begin{aligned}
FI\colon \mathsf{Gp} &\to \mathsf{Act}, & S &\mapsto (|S|, S), \\
KG\colon \mathsf{Act} &\to \mathsf{Gp}, & KG(X, S) &= S/\overline{S}_0.
\end{aligned}
\tag{6.77}
$$

The left exact functor FI associates to a group S the canonical (right) action $x + s$ of S over the underlying pointed set $|S|$. The functor KG associates to an action (X, S) the quotient of S modulo the invariant closure of $S_0 = \mathrm{Fix}_S(0_X)$ in S; KG preserves normal quotients.

Note that also here (as in 1.6.8), a sequence of groups, embedded via FI, is exact in Act if and only if it is exact in the usual sense (im $f = \ker g$).

6.4.7 Exact homotopy sequences in Act or Gp$_2$

Homotopy sequences can be seen as exact sequences in the homological category Act; or also in Gp_2, for the pathwise connected case.

For instance, let (X, A) be a pair of pointed spaces. The classical homotopy sequence of this pair can be read as a sequence in Act

$$
\begin{aligned}
&\ldots\ \pi_n A \to \pi_n X \to \pi_n(X, A) \to \pi_{n-1} A \to\ \ldots \\
&\ldots \to \pi_1 X \to (\pi_1(X, A), \pi_1 X) \to \pi_0 A \to \pi_0 X \to \pi_0(X, A) \to 0,
\end{aligned}
\tag{6.78}
$$

and this sequence is *exact* in Act, as follows from Theorem 6.4.9.

All the terms down to $\pi_1 X$ are groups (embedded in Act, by (6.77)).

$\pi_1(X, A)$ is the pointed set of paths $\sigma\colon I \to X$ with $\sigma(0) \in A$, $\sigma(1) = 0_X = 0_A$, modulo homotopy with first end in A and second end in 0_X; the group $\pi_1 X$ acts on the right on $\pi_1(X, A)$ in the natural, standard way.

The three last terms are pointed sets (embedded in Act, (6.71)); in particular, $\pi_0(X, A)$ is the cokernel of the pointed mapping $\pi_0 A \to \pi_0 X$, i.e. the set of path-components of X modulo the relation identifying all the components that intersect A (pointed in this class).

More generally, given a triple of pointed spaces $B \subset A \subset X$, we get an exact sequence in Act

$$... \to \pi_2(A, B) \to \pi_2(X, B) \to \pi_2(X, A) \to$$

$$(\pi_1(A, B), \pi_1 A) \to (\pi_1(X, B), \pi_1 X) \to (\pi_1(X, A), \pi_1 X) \to \qquad (6.79)$$

$$\pi_0(A, B) \to \pi_0(X, B) \to \pi_0(X, A) \to 0,$$

whose terms are groups in degree $\geqslant 2$, 'general actions' in degree 1 and pointed sets in degree 0.

Thus, the usual relative homotopy theory, or more generally every homotopy theory satisfying the axioms of Milnor [Mi1], can be viewed as a homology theory over $\mathsf{Top}_{\bullet}2$, with values in the homological category Act: homotopy and points are standard, while the *excision* maps $p \colon (X, p^{-1}B_0) \to (B, B_0)$ are induced by fibrations.

It can be noted that, if all the spaces are path-connected, the last three terms in (6.79) annihilate and all actions become transitive; by 6.4.6, the sequence (6.79) can be realised as an exact sequence in Gp_2.

6.4.8 The homotopy sequence of a fibration

Analogously, if $p \colon X \to B$ is a Serre fibration of pointed spaces, with fibre $F = p^{-1}\{0_B\}$, its homotopy sequence can be viewed as a sequence in Act

$$... \pi_1 F \to \pi_1 X \to \pi_1 B \to (\pi_0 F, \pi_1 B) \to \pi_0 X \to \pi_0 B, \qquad (6.80)$$

and this sequence is *exact* in Act, again by Theorem 6.4.9. Here

- $(\pi_0 F, \pi_1 B)$ is the usual action of the group $\pi_1 B$ over the pointed set $\pi_0 F$,
- all the terms at its left are groups, embedded in Act (cf. (6.77)),
- the last two terms are pointed sets, embedded in Act (cf. (6.71)).

If X and B are pathwise connected, $\pi_0 X = \pi_0 B = 0$ and the action of $\pi_1 B$ on $\pi_0 F$ is transitive. By the equivalence recalled in 6.4.6, we can replace (6.80) with an exact sequence of Gp_2, replacing the transitive action $(\pi_0 F, \pi_1 B)$ with the pair of groups $(\pi_1 B, H)$ where $H = p_*(\pi_1 X)$ is the subgroup of $\pi_1 B$ formed by the operators that leave $[0_F]$ fixed

$$...\pi_1 F \to \pi_1 X \to \pi_1 B \to (\pi_1 B, p_*(\pi_1 X)) \to 0. \qquad (6.81)$$

Now, a tower of fibrations produces a set of such exact sequences. Their morphisms, generally, are not exact in Gp_2 (except the last one, a normal epi) and we would not get an exact couple according to definition 3.5.1. (Recall that a morphism $f \colon (X, A) \to (Y, B)$ of Gp_2 is exact if and only if f is injective and $fX \supset B$, cf. 1.6.7). But these maps become exact in the category Ngp of normalised groups, defined in Section 6.3.

In the general case, without assuming path-connectedness, a similar result will be obtained with the category of 'normalised actions', in Section 6.5.

6.4.9 Theorem (Exactness from groups to pointed sets)

Consider the following sequence in Act

$$... H \xrightarrow{u} G \xrightarrow{v} S \xrightarrow{f} (X, S) \xrightarrow{g} Y \xrightarrow{h} Z \qquad (6.82)$$

where

- *H, G, S, u, v are in* Gp, *viewed in* Act *as $(|H|, H)$ etc., by the left exact embedding FI of (6.74),*
- *(X, S) is an action and $f = (f', \text{id} S)$ with $f'(s) = 0_X + s$ (for $s \in |S|$),*
- *Y, Z, h are in* Set$_\bullet$, *viewed in* Act *as $(Y, 0)$ etc., via the exact embedding U of (6.71),*
- *$g : X \to Y$ is a map of pointed sets such that $g(x + s) = g(x)$ for all $x \in X$, $s \in S$.*

   Then:

(a) the sequence is exact in G if and only if $\operatorname{Im} u = \operatorname{Ker} v$,

(b) the sequence is exact in S if and only if $\operatorname{Im} v = \operatorname{Ker} f' = \operatorname{Fix}_S\{0_X\}$,

(c) the sequence is exact in (X, S) if and only if $0_X + S = g^{-1}\{0_Y\}$,

(d) the sequence is exact in Y if and only if $g'(X) = h^{-1}\{0_Z\}$,

(e) the morphism f is necessarily exact,

(f) the morphism g is necessarily right modular (cf. 2.3.1).

Note. The classical properties of exact homotopy sequences coincide with the present ones, except in (X, S), where the classical property is stronger: for $x, x' \in X$, $g(x) = g(x')$ if and only if $x = x' + s$ for some $s \in S$.

Proof Point (a) follows from 1.6.8. Points (b) and (e) follow from the normal factorisation of f (see 6.4.3), expressed as follows, with $S_0 = \operatorname{Fix}_S\{0_X\}$ and $X_0 = 0_X + S$

$$(|S_0|, S_0) \rightarrowtail (|S|, S) \xrightarrow{f} (X, S) \twoheadrightarrow (X/X_0, S) \qquad (6.83)$$
$$\downarrow \qquad\qquad \uparrow$$
$$(|S|/S_0, S) \to (X_0, S)$$

   Here, X/X_0 is the pointed set obtained by collapsing the orbit $X_0 = 0_X + S$ to a (base) point. Moreover, $|S|/S_0$ denotes the set of right cosets

$S_0 + x$. Indeed, the relation R in $|S|$ described in (6.60) for the normal coimage of f becomes now

$$xRx' \quad \Leftrightarrow \quad (x = x_1 + s, \; x' = x_1' + s'),$$
$$\text{where } x_1, x_1' \in S_0, \; s - s' \in S_0. \tag{6.84}$$

that is easily seen to be equivalent to $x - x' \in S_0$. (If this is the case, just let $x = 0 + x$ and $x' = 0 + x'$; conversely, given (6.84), we have $x - x' = x_1 + s - s' - x_1' \in S_0$.)

Points (c) and (d) follow from the normal factorisation of g, where $X_1 = g^{-1}\{0_Y\} = \operatorname{Ker} g'$

$$
\begin{array}{ccccccc}
(X_1, S) & \rightarrowtail & (X, S) & \xrightarrow{\;g\;} & (Y, 0) & \longrightarrow & (Y/g'(X), 0) \\
& & \downarrow & & \uparrow & & \uparrow \\
& & (X/X_1, S) & \rightarrow & (g'(X), 0) & &
\end{array}
\tag{6.85}
$$

Finally, for (f), let Y' be a pointed subset of Y (i.e. a normal subobject in $\mathsf{Set_{\bullet}}$ and Act). Then $g^{-1}(Y')$ is stable under the action of S, and we have

$$g_* g^*(Y') = g_*(g^{-1}(Y'), S) = g(g^{-1}(Y')) = Y' \cap g(X). \tag{6.86}$$

$\square$

6.5 Homotopy spectral sequences and normalised actions

We introduce a homological category of fractions $\mathsf{Nac} = \Sigma^{-1}\mathsf{Act}$, which will be called *the category of normalised action*, and proved to be adequate for homotopy sequences of (possibly) *not path-connected spaces*. This matter has been developed, quite recently, in [G19].

6.5.1 Quasi-exact couples

Extending 3.5.5, a *bigraded quasi exact couple* $C = (D, E, u, v, \partial)$ *of type 1* in the homological category E will be a system of objects and morphisms

$$
\begin{aligned}
D_{np} \; (n \geqslant 0, \, p \leqslant 0), \qquad & E_{np} \; (n \geqslant 1, \, p \leqslant 0), \\
u = u_{np} \colon D_{n,p-1} \to D_{np} \qquad & (n \geqslant 0, \, p \leqslant 0), \\
v = v_{np} \colon D_{np} \to E_{np} \qquad & (n \geqslant 1, \, p \leqslant 0), \\
\partial = \partial_{np} \colon E_{np} \to D_{n-1,p-1} \qquad & (n \geqslant 1, \, p \leqslant 0),
\end{aligned}
\tag{6.87}
$$

such that:

(a) the following sequences are exact

$$\ldots E_{n+1,p} \xrightarrow{\partial} D_{n,p-1} \xrightarrow{u} D_{np} \xrightarrow{v} E_{np} \xrightarrow{\partial} D_{n-1,p-1} \ldots \quad (6.88)$$

(b) all the morphisms $u_{n,p}^r = u_{np}\ldots u_{n,p-r+1} \colon D_{n,p-r} \to D_{np}$ are exact, for $r \geqslant 1$ and $n > 0$,

(c) v_{np} is left modular on $\mathrm{Ker}\,(u_{n,p+r}^r \colon D_{np} \to D_{n,p+r})$, for $r \geqslant 1$,

(d) ∂_{np} is right modular on $\mathrm{Nim}\,(u_{np}^r \colon D_{n,p-r} \to D_{np})$, for $r \geqslant 1$.

Apart from the restriction on the indices n, p, the real interest of the extension is that *we are not requiring the exactness of* the last morphism in sequence (6.88), namely $u_{0p} \colon D_{0,p-1} \to D_{0p}$, because this is not satisfied in our application below (in 6.5.2).

However, the present hypotheses are sufficient to obtain the associated derived couples and spectral sequence, as in 3.5.6; indeed, in the construction of the derived couple, the morphism

$$v_{np}^{(r)} = (D_{n,p+r-1}^r \cong D_r^{np} \to E_{np}^r), \quad (6.89)$$

is only needed for $n \geqslant 1$. Therefore, we only need the exactness of the morphism $u^{r-1} \colon D_{np} \to D_{n,p+r-1}$ (and the induced isomorphism $i \colon D_r^{np} \to D_{n,p+r-1}^r$) for $n \geqslant 1$.

6.5.2 A category of fractions

Similarly to the construction of the category of normalised groups, we are now interested in the category of fractions $\mathsf{Nac} = \Sigma^{-1}\mathsf{Act}$ obtained by 'excision of the invariant subgroups of operators which act trivially'. In other words, the set Σ is formed - up to isomorphism - by all the natural projections $p \colon (X, S) \to (X, S/N)$, where N is an invariant subgroup of S which acts trivially on X.

The motivation comes from a tower of fibrations of *arbitrary* pointed spaces ([BouK], p. 258)

$$\ldots X_s \xrightarrow{f_s} X_{s-1} \to \ldots X_0 \xrightarrow{f_0} X_{-1} = \{*\}\,. \quad (6.90)$$

Again, we write $i_s \colon F_s \to X_s$ the fibre of the fibration $f_s \colon X_s \to X_{s-1}$. Consider the exact homotopy sequence of the fibration

$$f_{-p} \colon X_{-p} \to X_{-p-1} \quad (p \leqslant 0),$$

in Act (see 6.4.8):

$$
\ldots \; \pi_2 X_{-p-1} \;\rightarrow\; \pi_1 F_{-p} \;\rightarrow\; \pi_1 X_{-p} \;\rightarrow\; \pi_1 X_{-p-1} \;-\!\!\!-
$$

$$
\ldots \; D_{2p} \;\xrightarrow{v}\; E_{2p} \;\xrightarrow{\partial}\; D_{1,p-1} \;\xrightarrow{u}\; D_{1p} \;-\!\!\!-\!\!\!-
$$

$$
\rightarrow (\pi_0 F_{-p}, \pi_1 X_{-p-1}) \;\rightarrow\; \pi_0 X_{-p} \;\rightarrow\; \pi_0 X_{-p-1}
$$

$$
\xrightarrow{v}\; E_{1p} \;\xrightarrow{\partial}\; D_{0,p-1} \;\xrightarrow{u}\; D_{0p} \tag{6.91}
$$

Here $(\pi_0 F_{-p}, \pi_1 X_{-p-1})$ is the well-known canonical action, the preceding objects are groups (embedded in Act, by (6.77)), and the last two terms are pointed sets (embedded in Act, by (6.71)).

All these sequences, for $p \leqslant 0$, produce a semiexact couple in Act:

$$
\begin{aligned}
D_{np} &= D^1_{np} = \pi_n X_{-p-1}, & (n \geqslant 0), \\
E_{np} &= E^1_{np} = \pi_{n-1} F_{-p}, \quad E_{1p} = (\pi_0 F_{-p}, \pi_1 B) & (n > 1),
\end{aligned} \tag{6.92}
$$

$$
\begin{aligned}
u_{np} &= \pi_n(f_{-p}) \colon \pi_n X_{-p} \to \pi_n X_{-p-1} & (n \geqslant 0), \\
v_{np} &\colon \pi_n X_{-p-1} \to \pi_{n-1} F_{-p} & (n > 1), \\
v_{1p} &\colon \pi_1 X_{-p-1} \to (\pi_0 F_{-p}, \pi_1 X_{-p-1}), \\
\partial_{np} &= \pi_{n-1}(i_{-p}) \colon \pi_{n-1} F_{-p} \to \pi_{n-1} X_{-p} & (n > 1), \\
\partial_{1p} &= \pi_0(i_{-p}) \colon (\pi_0 F_{-p}, \pi_1 X_{-p-1}) \to \pi_0 X_{-p}.
\end{aligned} \tag{6.93}
$$

It becomes a *quasi exact couple* in Nac. Indeed, all these morphisms fall in the following situations:

- group homomorphisms (embedded in Act and) projected to exact morphisms in Nac,

- morphisms $v_{1p} \colon \pi_1 X_{-p-1} \to (\pi_0 F_{-p}, \pi_1 X_{-p-1})$ which are exact in (Act and) Nac, by 6.4.9(e), whence also left modular,

- morphisms $\partial_{1p} = \pi_0(i_{-p}) \colon (\pi_0 F_{-p}, \pi_1 X_{-p-1}) \to \pi_0 X_{-p}$, which are right modular by 6.4.9(f).

We *only* have a *quasi* exact couple because the morphisms of pointed sets

$$
u_{0p} = \pi_0(f_{-p}) \colon \pi_0 X_{-p} \to \pi_0 X_{-p-1}
$$

need not be exact.

The sequel is devoted to construct Nac and verify the desired properties.

6.5.3 A larger category of actions

We shall construct $\Sigma^{-1}\mathsf{Act}$ as a quotient of a larger homological category Act' of actions, following the same line as for the extension $\mathsf{Q} = \mathsf{Gp}_2'$, in Section 6.3.

An object of Act' is a triple (X, S, S_0), where (X, S) belongs to Act, (S, S_0) is a pair of groups, and S_0 acts trivially on X.

A morphism

$$f = (f', f''): (X, S, S_0) \to (Y, T, T_0), \tag{6.94}$$

consists of a map of pointed sets $f': X \to Y$ and a Q-morphism $f'': (S, S_0) \to (T, T_0)$ that are consistent: $f'(x + s) = f'x + f''s$, for $x \in X$ and $s \in S$. Composition is obvious.

The map $f = (f', f'')$ of Act' (in (6.94)) is assumed to be null if $f': X \to Y$ is a zero-map of pointed sets, i.e. $f'(X) = \{0_Y\}$.

Act' is semiexact, with the following normal factorisation of f (obtained from the normal factorisation (6.62) of a morphism in Act)

$$
\begin{array}{ccccc}
(X_1, S_1, S_0) \rightarrowtail (X, S, S_0) & \xrightarrow{f} & (Y, T, T_0) \twoheadrightarrow (Y/R', T, T_0) \\
 \downarrow{\scriptstyle p} & & \uparrow{\scriptstyle m} \\
(X/R, S, S_0) & \underset{g}{\to} & (Y_1, T_1, T_0)
\end{array} \tag{6.95}
$$

The definition of $X_1, S_1, R, R', T_1, Y_1$ is the same as in 6.4.3 (even if f'' is just a quasi homomorphism); notice that $S_1 = \{s \in S \mid X_1 + s \subset X_1\}$ contains S_0 and T_1 (the subgroup of T spanned by the elements which link the elements of $f'(X)$) contains T_0.

One proves that Act' is homological, as in Theorem 6.4.4. There is an exact embedding

$$\mathsf{Act} \to \mathsf{Act}', \qquad (X, S) \mapsto (X, S, S_0), \tag{6.96}$$

where $S_0 = \mathrm{Fix}_S(0_X)$.

(Notice that replacing S_0 with $\mathrm{Fix}_S(X)$ would not give a functor: from $s \in \mathrm{Fix}_S(X)$ we can only deduce that $f(s) \in \mathrm{Fix}_T(fX)$.)

6.5.4 Normalised actions

The category $\mathsf{Nac} = \mathsf{Act}'/\mathcal{R}$ of *normalised actions* is the quotient modulo the congruence of categories $f\mathcal{R}g$ defined by $f' = g'$ and $f''\mathcal{R}g''$ in Q (see 6.3.5). The latter amounts to the equivalent conditions

(a) for every $s \in S$, $f''s - g''s \in T_0$,

(b) for every $s \in S$, $-f''s + g''s \in T_0$,

(c) for all $s_i \in S$ and $\varepsilon_i = \pm 1$, $\sum \varepsilon_i.f''s_i - \sum \varepsilon_i.g''s_i \in T_0$.

A map $[f]$ in Nac is assumed to be *null* if and only if f' is null in Act. The null objects are the triples $(0, S, S_0)$.

Nac has kernels and cokernels, with the same description as in Act$'$ (independently of the representative we choose for $[f]$).

Therefore Nac is homological and the canonical functor

$$P \colon \mathsf{Act} \to \mathsf{Nac},$$

$$(X, S) \mapsto (X, S, S_0), \quad f \mapsto [f] \quad (S_0 = \mathrm{Fix}_S(0_X)), \tag{6.97}$$

given by the composition Act $\to$ Act$'$ $\to$ Nac is exact, nsb-faithful and nsb-full.

6.5.5 Theorem

The functor $P \colon \mathsf{Act} \to \mathsf{Nac}$ 'is' the category of fractions $\Sigma^{-1}\mathsf{Act}$, i.e. it solves the universal problem of making the morphisms of Σ (defined in 6.5.2) invertible (within arbitrary categories and functors).

Furthermore, P is exact and also solves this universal problem within semiexact categories and exact functors (or left exact, or right exact, or short exact functors).

Proof The proof follows the same line as for Theorem 6.3.6, with suitable modifications.

(a) P carries each map of Σ to an isomorphism of Nac.

Given $p \colon (X, S) \to (X, S/N)$ in Σ (with N a normal subobject of S which acts trivially on X), we want to prove that the associated map of Act$'$

$$\hat{p} \colon (X, S, S_0) \to (X, S/N, S_0/N) \quad (S_0 = \mathrm{Fix}_S(0_X)),$$

becomes an isomorphism in Nac.

Choose a *mapping* $j \colon S/N \to S$ such that $p.j = 1$. Then

$$\hat{j} = (\mathrm{id}X, j) \colon (X, S/N, S_0/N) \to (X, S, S_0),$$

is a morphism of Act$'$, since:

- if $\sum \varepsilon_i.p(s_i) \in S_0/N$, then $p(\sum \varepsilon_i.jp(s_i)) \in S_0/N$ and $\sum \varepsilon_i.jp(s_i) \in S_0$,
- if $x \in X$ and $s \in S$, $x + jp(s) = x + s = x + p(s)$, because $-s + pj(s) \in N \subset \mathrm{Fix}_S(X)$.

By the same reason, $\hat{j}\hat{p} \, \mathcal{R} \, 1$.

(b) Every functor $G \colon \mathsf{Act} \to \mathsf{C}$ which makes each Σ-map invertible in C can be uniquely extended to Nac.

In fact, given an Act$'$-morphism $f = (f', f'')\colon (X, S, S_0) \to (Y, T, T_0)$, we write, as in 6.3.3, $ES = F|S|$ and $e\colon ES \to S$ the canonical evaluation epimorphism; ES acts on X in the obvious way, by evaluating ES in S, and $\overline{S}_0 = e^{-1}(S_0) = \{\Sigma\varepsilon_i\hat{s}_i \mid \Sigma\varepsilon_i s_i \in S_0\}$ acts trivially on X.

Then the *group homomorphism* $f_1'' = e_T.Ef\colon ES \to T$ defined by the *mapping* $|f''|\colon |S| \to |T|$ gives a map $f_1 = (f', f_1'')\colon (X, ES) \to (Y, T)$ of Act, that can be viewed in Act$'$

$$
\begin{array}{ccc}
(X, S, S_0) & \xrightarrow{\ f\ } & (Y, T, T_0) \\[2pt]
{\scriptstyle e}\big\uparrow & \nearrow & \\[-4pt]
& {\scriptstyle f_1} & \\[2pt]
(X, ES, \overline{S}_0) & &
\end{array}
\tag{6.98}
$$

$$
f_1 = (f', f_1'') = (f', e_T.Ef), \qquad [f] = [f_1].[e]^{-1} \text{ in Nac},
$$
$$
f_1''(\overline{S}_0) = e_T.Ef(\overline{S}_0) \subset e_T(\overline{T}_0) = e_T e_T^{-1}(T_0) = T_0.
$$

Therefore any functor G' which extends G on Nac is uniquely determined, as follows:

$$
G'\colon \mathsf{Nac} \to \mathsf{C},
$$
$$
G'(X, S, S_0) = (X, S, S_0), \qquad G'[f] = (Gf_1).(Ge)^{-1}.
\tag{6.99}
$$

We now prove that G' is well defined by theses formulas, and indeed a functor. Firstly, we verify that $f\,\mathcal{R}\,g$ in Act$'$ implies $Gf_1 = Gg_1$. Let

$$
T' = \langle fS \cup gS \rangle, Y' = (fX \cup gX) + T',
$$
$$
T'' = T_0 \cap T', \quad N = \langle \Sigma\,\varepsilon_i.fs_i - \Sigma\,\varepsilon_i.gs_i \mid s_i \in S,\ \varepsilon_i = \pm 1 \rangle.
$$

Now, N is a subgroup of T'', invariant in T' (with the same proof as in 6.3.6). Therefore the following diagram in Act$'$, where f_2 and g_2 are the restrictions of f_1 and g_1

$$
\begin{array}{ccc}
(X, ES, \overline{S}_0) & = & (X, ES, \overline{S}_0) \\[2pt]
{\scriptstyle f_1}\big\downarrow\big\downarrow{\scriptstyle g_1} & & {\scriptstyle f_2}\big\downarrow\big\downarrow{\scriptstyle g_2} \\[2pt]
(Y, T, T_0) & \xleftarrow[i]{} \ (Y', T', T'') \ \xrightarrow[p]{} & (Y'/N, T'/N, T''/N)
\end{array}
$$

shows that $Gf_1 = Gg_1$ (because $G(p)$ is an isomorphism and $pf_2 = pg_2$).

Finally, G' is a functor, with the same computations as in 6.3.6, based

on a slightly different diagram

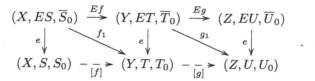

(c) One applies the same argument as in point (c) of the proof of Theorem 6.3.6.   □

7

Homological theories and biuniversal models

After studying, in Section 7.1, some homological subcategories of Set_2 where we shall construct our biuniversal models, we define in the next section homological and g-exact theories, similar to the *proper* EX-theories of Part I, and their biuniversal models.

Then, in the next three sections, we study three homological theories:

- the *modular bifiltration*,

- the *modular sequence of morphisms*,

- the *modular exact couple*,

and construct their biuniversal model. This also produces the biuniversal model of the corresponding p-exact theory, which agrees with the corresponding result of Part I.

As in Part I, a crucial point consists of a well-known Birkhoff theorem on free modular lattices, recalled in 7.3.1. In 7.3.5 we give a generalisation of this theorem, based on modular elements in a lattice, that can be applied - for instance - to groups equipped with two filtrations, one of them formed of invariant subgroups.

7.1 Categories for biuniversal models

We here study the categories

$$\mathcal{I}^\sharp = \mathsf{Inj}_2, \qquad \mathcal{I}_0^\sharp = \mathsf{Inc}_2, \qquad \mathcal{J}^\sharp, \qquad \mathcal{J}_0^\sharp,$$

where we shall construct some biuniversal models of distributive homological theories; these categories will play a role analogous to that of $\mathcal{I}$, $\mathcal{I}_0$, $\mathcal{J}$, $\mathcal{J}_0$ for p-exact theories, respectively (Part I). The latter categories are also described here, in 1.4.3 and 5.1.9.

7.1.1 Injections and inclusions

We shall write $\mathcal{I}^\sharp = \mathsf{Inj}_2$ for the subcategory of Set_2 formed by all the pairs of sets (X, A) and the *injective* mappings (of pairs), already considered in 1.7.4(b) and 2.2.9.

It is a homological subcategory of Set_2, nsb-faithful and nsb-full. It is *distributive* by 2.3.2.

We have also considered in 1.7.4(b) the subcategory $\mathcal{I}_0^\sharp = \mathsf{Inc}_2$ of pairs of sets and *inclusions* (of pairs):

$$f\colon (X, A) \to (Y, B) \qquad (A \subset X \subset Y, \ A \subset B \subset Y; \ f\colon X \subset Y). \quad (7.1)$$

It is an nsb-faithful and nsb-full homological subcategory of both, again distributive. The normal factorisation of f, and the direct and inverse images of normal subobjects along f, can now be written as follows

$$
\begin{array}{ccccccc}
(X{\cap}B, A) & \rightarrowtail & (X, A) & \xrightarrow{\ f\ } & (Y, B) & \twoheadrightarrow & (Y, X{\cup}B) \\
& & {\scriptstyle p}\big\downarrow & & \big\uparrow{\scriptstyle m} & & \\
& & (X, X{\cap}B) & \xrightarrow[g]{} & (X{\cup}B, B) & &
\end{array}
\quad (7.2)
$$

$$f_*(X', A) = (X' \cup B, B), \quad f^*(Y', B) = (Y' \cap X, A),$$
$$(A \subset X' \subset X, \ B \subset Y' \subset Y). \quad (7.3)$$

We prove in the next proposition that the exact quotient (see 5.3.1) of $\mathcal{I}^\sharp = \mathsf{Inj}_2$ is equivalent to the (distributive, p-exact) category $\mathcal{I}$ of sets and partial bijections. In the same way one proves that the exact quotient of $\mathcal{I}_0^\sharp = \mathsf{Inc}_2$ is equivalent to the category $\mathcal{I}_0$ of sets and partial identities.

7.1.2 Proposition

The exact quotient (Section 5.3.1) of the category Inj_2 can be realised as the p-exact category $\mathcal{I}$ of sets and partial bijections, by means of the exact functor

$$P\colon \mathsf{Inj}_2 \to \mathcal{I}, \qquad (X, A) \mapsto X \setminus A, \quad f \mapsto Pf, \quad (7.4)$$

where Pf is the partial bijection obtained by restricting the morphism $f\colon (X, A) \to (Y, B)$ of Inj_2 to $X \setminus f^{-1}(B) \subset X \setminus A$.

Proof The functor P is exact, since the kernel and cokernel of f are taken by P to:

$$P(\operatorname{Ker} f) = P(f^{-1}(B), A) = f^{-1}(B) \setminus A = \operatorname{Ker} Pf,$$

$$P(\operatorname{Cok} f) = P(Y, fX \cup B) = Y \setminus (fX \cup B) = \operatorname{Cok} Pf.$$

Let $F\colon \mathsf{Inj}_2 \to \mathsf{E}$ be an exact functor with values in a g-exact category; we show that F factorises uniquely as $F = GP$, via an exact functor $G\colon \mathcal{I} \to \mathsf{E}$.

The uniqueness is easy to show: given $f\colon X \to Y$ in $\mathcal{I}$, consider the following diagrams, in Inj_2 and in $\mathcal{I}$, respectively

$$
\begin{array}{ccc}
(X, \emptyset) & & (Y, \emptyset) \\
& \searrow^{f'} & \downarrow^{i} \\
& & (Y + H, H)
\end{array}
\qquad
\begin{array}{ccc}
X & \xrightarrow{\ f\ } & Y \\
& \searrow^{Pf'} & \downarrow^{Pi} \\
& & Y + H \setminus H
\end{array}
\tag{7.5}
$$

where $H = \operatorname{Ker} f = X \setminus \operatorname{Def} f$, $Y + H$ is the disjoint union of the sets Y and H, and

$$f'(x) = f(x) \in Y \text{ for } x \notin H, \qquad f'(x) = x \text{ for } x \in H,$$
$$i(y) = y.$$
$$\tag{7.6}$$

Note that $i^*(0) = 0$, $i_*(1) = 1$, so that Fi is invertible in E (and in particular Pi is invertible in $\mathcal{I}$). Then

$$G(X) = GP(X, \emptyset) = F(X, \emptyset),$$
$$G(f) = (Fi)^{-1}.Fi.Gf = (Fi)^{-1}.GPi.Gf = (Fi)^{-1}.G(Pi.f)$$
$$= (Fi)^{-1}.G(Pf') = (Fi)^{-1}.Ff'.$$

As to the existence, let G be defined by the previous formulas, where i and f' are as in (7.5) and (7.6)

$$G(X) = F(X, \emptyset), \qquad G(f) = (Fi)^{-1}.Ff'. \tag{7.7}$$

Then G is a functor: given a second morphism $g\colon Y \to Z$ in $\mathcal{I}$ and their composite $h = gf$, we form the following commutative diagram in Inj_2, where the left-hand part concerns the calculus of $Gg.Gf$, the right-hand part the calculus of Gh and all the vertical arrows are inclusions, taken by F to isomorphisms of E

$$
\begin{array}{ccccccc}
(X, \emptyset) & & (Y, \emptyset) & & (Z, \emptyset) & = & (Z, \emptyset) & & (X, \emptyset) \\
& \searrow^{f'} & \downarrow^{i} & \searrow^{g'} & \downarrow^{j} & & \downarrow^{a} \quad \searrow^{h'} & \\
& & (Y + H, H) & & (Z + K, K) & & (Z + L, L) & \\
& & & \searrow^{g''} & \downarrow^{r} & b \downarrow & & \\
& & & & (Z + K + L, K + L) & & &
\end{array}
$$

$$H = X \setminus \operatorname{Def} f, \qquad K = Y \setminus \operatorname{Def} g, \qquad L = Z \setminus \operatorname{Def} h,$$

$$g'(y) = g(y) \in Z \text{ for } y \notin K, \qquad g'(y) = y \text{ for } y \in K,$$

$$g''(y) = g'(y) \text{ for } y \in Y, \quad g''(h) = h \in L \text{ for } h \in H,$$

$$F(ba).(Gg.Gf) = F(ba).(Fj)^{-1}.Fg'.(Fi)^{-1}.Ff'$$

$$= F(ba).(Fj)^{-1}.(Fr)^{-1}.Fg''.Ff' = F(g''f'),$$

$$F(ba).(Gh) = F(ba).(Fa)^{-1}.Fh' = Fb.Fh' = F(g''f').$$

Finally, to prove that G is exact, we start from the kernel and cokernel of $f: X \to Y$ in $\mathcal{I}$

$$m = \ker f: X \setminus \operatorname{Def} f \rightarrowtail X, \quad p = \operatorname{cok} f: Y \twoheadrightarrow Y \setminus fX.$$

Then, in the following commutative diagram of Inj_2, we note that $m = m' = \ker f'$ and $p'' = \operatorname{cok} f'$

$$
\begin{array}{cccc}
(H,\emptyset) \xrightarrow{\;m\;} (X,\emptyset) & (Y,\emptyset) & (Y \setminus fX, \emptyset) \\
\searrow^{m'} \; \big\| & \searrow^{f'} \; \downarrow^{i} & \searrow^{p'} \quad \downarrow^{j} \\
(X,\emptyset) & (Y+H,H) & (Y+fX,fX) \\
& \searrow^{p''} & \downarrow^{r} \\
& & (Y+H, fX+H)
\end{array}
$$

Therefore:

$$Gm = Fm = F(\ker f') \sim \ker{}_{\mathsf{E}} Ff' = \ker{}_{\mathsf{E}} Fi.Gf = \ker{}_{\mathsf{E}} Gf,$$

$$Gp = (Fj)^{-1}.Fp' \sim Fp' \sim Fp''.Fi \sim F(\operatorname{cok} f').Fi$$

$$\sim (\operatorname{cok}{}_{\mathsf{E}} Ff').Fi \sim \operatorname{cok}{}_{\mathsf{E}}((Fi)^{-1}.Ff') = \operatorname{cok}{}_{\mathsf{E}} Gf.$$

$\square$

7.1.3 Pairs of semitopological spaces

The homological categories Set_2, of 'pairs' of sets, and Top_2, of 'pairs' of topological spaces (of common use in algebraic topology), have been analysed in 1.4.2 and 1.6.7.

We need here an (obvious) extension of Top_2: the homological category Stp_2 of *pairs of semitopological spaces* (cf. 5.1.9).

Its objects are the 'pairs' of (small) semitopological spaces (X, A), where A is a subspace of X. A morphism $f: (X, A) \to (Y, B)$ is a continuous mapping from X into Y, that takes A into B (where 'continuous mapping' obviously means that the inverse image of any closed subset is closed). It is again a homological category, with a structure analogous to that of Set_2 and Top_2.

In particular, the kernel, cokernel and normal factorisation of f are computed as in Set_2 (see (1.44)), replacing subsets with semitopological subspaces. As in Top_2, the morphism f is exact if and only if it is a homeomorphism onto $f(X) \supset B$. Every normal subobject and every normal quotient of (X, A) is determined by a semitopological subspace $M \subset X$ that contains A

$$(M, A) \overset{m}{\rightarrowtail} (X, A) \overset{p}{\twoheadrightarrow} (X, M) \tag{7.8}$$

and each short exact sequence in Stp_2 is of this type, up to isomorphism - determined by a *triple* of semitopological spaces $X \supset M \supset A$.

7.1.4 Closed embeddings

Let $\mathcal{J}^\sharp$ be the subcategory of Stp_2 formed by the pairs of semitopological spaces (X, A), where A is a *closed* subspace of X, whereas a morphism $f \colon (X, A) \to (Y, B)$ is a *closed embedding* of X into Y (as a closed subspace), that takes A into B. Again, we get a *distributive* nsb-full homological subcategory of Stp_2.

The subcategory $\mathcal{J}_0^\sharp$ of *closed inclusions* is an nsb-full homological subcategory of $\mathcal{J}^\sharp$.

These categories are also distributive and one can prove (as in 7.1.1, 7.1.2) that *their exact quotients are equivalent to $\mathcal{J}$ and $\mathcal{J}_0$, respectively.*

7.1.5 Collections of pairs of sets or spaces

Let Σ be a small set of pairs $S_i = (X_i, A_i)$ of (small) sets. We write $\mathcal{I}^\sharp\langle\Sigma\rangle$ for the full subcategory of $\mathcal{I}^\sharp$ whose objects are the pairs (X, A) where

$$A_i \subset A \subset X \subset X_i, \tag{7.9}$$

for some pair (X_i, A_i) belonging to Σ. Analogously we set

$$\mathcal{I}_0^\sharp\langle\Sigma\rangle = \mathcal{I}_0^\sharp \cap \mathcal{I}^\sharp\langle\Sigma\rangle.$$

These categories are nsb-full homological subcategories of $\mathcal{I}^\sharp$ and $\mathcal{I}_0^\sharp$, respectively.

Let now Σ be a small set of pairs $S_i = (X_i, A_i)$ of semitopological spaces, where A_i is closed in X_i. We write $\mathcal{J}^\sharp\langle\Sigma\rangle$ for the full subcategory of $\mathcal{J}^\sharp$ whose objects are the pairs (X, A) for which there exists a pair (X_i, A_i) belonging to Σ so that (7.9) is a sequence of closed subspaces. We let

$\mathcal{J}_0^\sharp\langle\Sigma\rangle = \mathcal{J}_0^\sharp \cap \mathcal{I}^\sharp\langle\Sigma\rangle$. They are nsb-full homological subcategories of $\mathcal{J}^\sharp$ and $\mathcal{J}_0^\sharp$.

The exact quotients of these categories are respectively equivalent to the following distributive exact categories, produced by the set Σ' of sets (or semitopological spaces) $X_i \setminus A_i$:

- $\mathcal{I}\langle\Sigma'\rangle$, the full subcategory of $\mathcal{I}$ whose objects are the subsets of some set belonging to Σ',

- $\mathcal{I}_0\langle\Sigma'\rangle$, the full subcategory of $\mathcal{I}_0$ whose objects are the subsets of some set belonging to Σ',

- $\mathcal{J}\langle\Sigma'\rangle$, the full subcategory of $\mathcal{J}$ whose objects are the locally closed subspaces of some space belonging to Σ',

- $\mathcal{J}_0\langle\Sigma'\rangle$, the full subcategory of $\mathcal{J}_0$ whose objects are the locally closed subspaces of some space belonging to Σ' (already considered in I.5.7.1).

7.1.6 Order spaces

The semitopological spaces that we shall actually use are obtained from ordered sets (as in I.5.8.1).

If I is a totally ordered (small) set, its *order semitopology* has closed subsets of the form

$$\downarrow i = \{j \in I \mid j \leqslant i\} \qquad (i \in I), \tag{7.10}$$

together with $\emptyset$ and I. (It is less fine than the *order topology* on I, whose closed subsets are the downward closed ones; it coincides with the latter when I is finite.)

The interval

$$]i_1, i_2] = \{j \in I \mid i_1 < j \leqslant i_2\} \qquad (i_1 < i_2 \text{ in } I), \tag{7.11}$$

is locally closed in I.

If I and J are both totally ordered, the *crossword space* $I \times J$ will be their cartesian product endowed with the *product semitopology*. The closed subsets are thus the finite unions of products of closed subsets of I and J

$$H = \bigcup_k I_k \times J_k \qquad (I_k \text{ closed in } I, \ J_k \text{ closed in } J). \tag{7.12}$$

The subset H has a unique non-redundant expression of this form.

Again, the *order topology* on $I \times J$ is finer than our semitopology, but induces the same topology on every finite subset of $I \times J$.

7.2 Homological and g-exact theories

We define here theories with values in homological or g-exact categories, and their biuniversal models, following a pattern similar to the EX-theories of I.5.6, but *without using the categories of relations* - not available for homological categories.

The present theories can thus be compared with the *proper* EX-theories of Part I. Δ is always a small graph.

7.2.1 Definition

A *homological theory*, or *ex3-theory*, T on a small graph Δ associates to every homological category E a set $T(\mathsf{E})$ of graph-morphisms $t\colon \Delta \to \mathsf{E}$, called the *models* of T in E, so that:

(ex3t.1) (*Naturality*) if $F\colon \mathsf{E} \to \mathsf{E}'$ is an exact functor between homological categories and $t \in T(\mathsf{E})$, then $Ft \in T(\mathsf{E}')$;

(ex3t.2) (*Reflection*) if $F\colon \mathsf{E} \to \mathsf{E}'$ is a faithful conservative exact functor between homological categories and $t\colon \Delta \to \mathsf{E}$ is a morphism of graphs, then $Ft \in T(\mathsf{E}')$ implies $t \in T(\mathsf{E})$;

(ex3t.3) (*Products*) if $t_i \in T(\mathsf{E}_i)$ for every index i in a small set I, then the morphism $(t_i) \in T(\Pi \mathsf{E}_i)$.

Let us recall that a conservative exact functor reflects the null maps and is nsb-faithful (cf. 1.7.2); it will be useful to know that it reflects exact and modular morphisms.

We shall also speak of a *theory*, leaving 'homological' understood, when the context is clear and no confusion (e.g. with EX-theories) may arise.

The set $T(\mathsf{E})$ will also be viewed as a category, whose morphisms are the natural transformations $\tau\colon t_1 \to t_2\colon \Delta \to \mathsf{E}$ of models.

Analogously we define a *g-exact theory* (or *ex4-theory*), and a *p-exact theory* (or *ex6-theory*). In the last case, the Reflection axiom (ex6t.2) can be simplified, because a faithful ex6-functor is always conservative (by 2.2.8).

For the sake of simplicity we will not treat the ex5-case, very similar to the p-exact one.

A p-exact theory is the same as a *proper EX-theory*, in the sense of I.5.6.

7.2.2 Biuniversal models

A *biuniversal model* for the homological theory T is a model $t_0\colon \Delta \to \mathsf{E}_0$ such that:

(i) if $t\colon \Delta \to \mathsf{E}$ is a model of T, there is some exact functor $F\colon \mathsf{E}_0 \to \mathsf{E}$ such that $t \cong Ft_0$,

(ii) if $F\colon \mathsf{E}_0 \to \mathsf{E}$ and $G\colon \mathsf{E}_0 \to \mathsf{E}$ are ex3-functors and $\tau\colon Ft_0 \to Gt_0\colon \Delta \to \mathsf{E}$ is a natural transformation, there is precisely one natural transformation $\varphi\colon F \to G\colon \mathsf{E}_0 \to \mathsf{E}$ such that $\tau = \varphi t_0$.

The functor F of the condition (ii) is determined up to isomorphism and the biuniversal model itself is determined up to equivalence of homological categories; we prove below (Theorem 7.2.3) that every theory has such a model. The homological category E_0 (determined up to equivalence) will be called the *classifying* category of the theory T.

We show in 7.2.4 that one can always modify $t_0\colon \Delta \to \mathsf{E}_0$ (up to equivalence) so that property (i) can be realised in a strict way: $t = Ft_0$, and *we will always use it in this form*, for the sake of simplicity. (Notice, however, that we are not requiring the uniqueness of t_0, nor we could; even in this stricter form, t_0 is still a *biuniversal* model, determined up to equivalence of categories.)

In the same way one defines the biuniversal model and the classifying category of a g-exact or p-exact theory.

We say that a homological theory T is *g-exact* (resp. *p-exact*) if its classifying homological category is of this type. In that case, the biuniversal model of T as a homological theory is also the biuniversal model of its restriction to g-exact (resp. p-exact) categories.

Similarly, we say that a homological theory T is *modular* (resp. *distributive*) if its classifying homological category is of this type. These properties are characterised below by equivalent formulations.

Two homological theories (possibly defined on different graphs) are said to be *equivalent* if their classifying categories are.

7.2.3 Theorem (Existence of the biuniversal model)

Every homological (or g-exact, or p-exact) theory on the graph Δ has a biuniversal model, with a small classifying category E_0 satisfying

$$\mathrm{card}(\mathrm{Ob}\mathsf{E}_0) \leqslant \mathrm{card}(\mathrm{Mor}\mathsf{E}_0) \leqslant \omega(\Delta)) = \max(\mathrm{card}\Delta, \aleph_0). \qquad (7.13)$$

Proof It is a consequence of a general result on the existence of a biuniversal model ([BeG], Theorem 6.6). In fact, it is easy to verify that the 2-category EX_3 of homological categories, exact functors and natural transformations is 'well-adapted for theories', in the sense of that paper, with 'bounding function' $\omega(\Delta) = \max(\mathrm{card}\Delta, \aleph_0)$.

One proceeds in the same way for EX_4 and EX_6; actually the p-exact case is explicitly treated in [BeG], Section 8.2. $\qquad \square$

7.2.4 Lemma (Biuniversal models)

Let $t_0 \colon \Delta \to E_0$ be the biuniversal model of a homological (or g-exact, or p-exact) theory T. One can always modify it, up to equivalence of categories, so that property (i) of the definition (in 7.2.2) can be realised in a strict way: $t = F t_0$ (for every model t).

Proof We construct an equivalent category E_1 by adding new objects to those of E_0, namely a new copy $t_1(x)$ for each object x of Δ, linked by a specified (new) isomorphism $i_x \colon t_0(x) \to t_1(x)$ to the corresponding old object of E_0; of course, we also add the new morphisms generated by the old ones with the new isos, so that the embedding $U \colon E_0 \subset E_1$ is an equivalence of (homological) categories.

We define $t_1 \colon \Delta \to E_1$ on the arrows of Δ so that the family $i = (i_x)$ becomes an isomorphism $U t_0 \to t_1$ of graph-morphisms; in other words, for $u \colon x \to y$ in Δ, we let $t_1(u) = i_y . t_0(u) . i_x^{-1}$

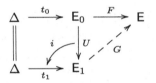

Now, given a model $t \colon \Delta \to E$ and an isomorphism $\varphi \colon F t_0 \to t$, we extend the exact functor $F \colon E_0 \to E$ to $G \colon E_1 \to E$, defining it on the new objects $t(x)$ and generators i_x as

$$G(t_1(x)) = t(x), \qquad G(i_x) = \varphi(x) \colon F t_0(x) \to t(x).$$

Therefore, for $u \colon x \to y$ in Δ, the commutative square $t(u) . \varphi(x) = \varphi(y) . F t_0(u)$ shows that

$$G(t(u)) = \varphi(y) . F t_0(u) . \varphi(x)^{-1} = t(u),$$

and finally $G t_1 = t$. □

7.2.5 Theorem (Properties of theories)

Consider a homological theory T, with biuniversal model $t_0 \colon \Delta \to E_0$.

(a) T is modular (resp. distributive) if and only if every model $t \colon \Delta \to E$ factorises through the modular expansion $\mathrm{Mdl}\, E \to E$ (resp. the distributive expansion $\mathrm{Dst}\, E \to E$; cf. 5.1.7), if and only if every model factorises through an exact functor defined on a modular (resp. distributive) homological category.

(b) T is g-exact (resp. distributive g-exact) if and only if every model $t: \Delta \to \mathsf{E}$ factorises through the exact expansion $\mathrm{Ex}\,\mathsf{E} \to \mathsf{E}$ (resp. the distributive exact expansion $\mathrm{Dx}\,\mathsf{E} \to \mathsf{E}$; cf. 5.1.8), if and only if every model factorises through an exact functor defined on a g-exact (resp. distributive g-exact) category.

Proof We verify the modular case in (a); the other three cases are similarly proved.

If T is modular, every model factorises through the modular homological category E_0.

If every model $t: \Delta \to \mathsf{E}$ factorises through an exact functor $F: \mathsf{E}' \to \mathsf{E}$ defined on a suitable modular homological category, then it also factorises through the modular expansion $\mathrm{Mdl}\,\mathsf{E} \to \mathsf{E}$ (by the universal property of the latter, cf. 5.1.7).

Finally, the last property implies that the biuniversal model $t_0: \Delta \to E_0$ factorises as $F t_1: \Delta \to \mathrm{Mdl}\,\mathsf{E}_0 \to \mathsf{E}_0$; but then $t_1 = G t_0: \Delta \to \mathsf{E}_0 \to \mathrm{Mdl}\,\mathsf{E}_0$ and $t_0 = FG t_0$, whence $FG = 1$. Therefore, E_0 has an exact embedding into $\mathrm{Mdl}\,\mathsf{E}_0$, and is also modular. □

7.2.6 Restricted theories

Every homological theory T defines a *restricted p-exact theory* RT on the same graph Δ: the models of the latter are just the models of T with values in some p-exact category.

Furthermore, if $t_0: \Delta \to \mathsf{E}_0$ is the biuniversal model of T, the biuniversal model of the restricted p-exact theory RT is the composite

$$P t_0: \Delta \to \mathsf{E}_0 \to \mathsf{E}_1, \qquad \mathsf{E}_1 = \mathrm{Exq}\,\mathsf{E}_0, \tag{7.14}$$

where $\mathrm{Exq}\,\mathsf{E}_0$ is the exact quotient of the homological category E_0 and $P: \mathsf{E}_0 \to \mathsf{E}_1$ is the universal arrow (see 5.3.1).

Plainly, the homological theory T is p-exact, as defined above (in 7.2.2), if and only if this universal arrow $\mathsf{E}_0 \to \mathsf{E}_1$ is an isomorphism of categories.

7.2.7 Theories and regular induction

Let Δ be a small graph, T a homological theory on Δ (Section 7.2.1) and let $S: \Delta \to \mathsf{E}_0$ be its biuniversal model. (We want to extend to homological theories the global representation functor dealt with in Part I, see I.5.6.6.)

For each model $t: \Delta \to \mathsf{E}$, let us choose precisely one *representative* exact functor $F_t: \mathsf{E}_0 \to \mathsf{E}$, so that $F_t.S = t$. If $\tau: t \to t': \Delta \to \mathsf{E}$ is a natural transformation of models, let us write $F_\tau\colon F_t \to F_{t'}\colon \mathsf{E}_0 \to \mathsf{E}$ for the unique natural transformation such that $F_\tau.S = u$.

Thus every object E of E_0 produces a functor:

$$E\colon T(\mathsf{E}) \to \mathsf{E}, \qquad E(t) = F_t(E),$$
$$E(\tau\colon t \to t') = F_\tau(E)\colon F_t(E) \to F_t(E). \tag{7.15}$$

Analogously, every map $u\colon E \to E'$ in E_0 yields a natural transformation $u\colon E \to E'\colon T(\mathsf{E}) \to \mathsf{E}$, with components

$$u(t) = F_t(u)\colon E(t) \to E'(t), \tag{7.16}$$

$$
\begin{array}{ccc}
E(t) \xrightarrow{\ E_\tau\ } E(t') & \qquad & F_t(E) \xrightarrow{\ F_\tau(E)\ } F_{t'}(E) \\
\ \downarrow{\scriptstyle ut} \qquad \downarrow{\scriptstyle ut'} & & \ \downarrow{\scriptstyle F_t(u)} \qquad \downarrow{\scriptstyle F_{t'}(u)} \\
E'(t) \xrightarrow[\ E'_\tau\ ]{} E'(t') & & F_t(E') \xrightarrow[\ F_\tau(E')\ ]{} F_{t'}(E')
\end{array}
$$

because the left square above coincides with the right one, by definition, and the latter commutes, by the naturality of $F_\tau\colon F_t \to F_t$.

Now, if $s = mq'^\sharp = q^\sharp m'\colon E' \to E$ ($E' = M/N$) is a subquotient of E (see 3.1.1), the morphism $E(\tau)\colon E(t) \to E(t')$ takes the subobject $M(t)$ into the subobject $M(t')$, because of the commutative square below, and similarly $N(t)$ into $N(t')$.

$$
\begin{array}{ccc}
M(t) \xrightarrow{\ m\ \rightarrowtail} E(t) \xrightarrow{\ q\ \twoheadrightarrow} E(t)/N(t) \\
\ \downarrow{\scriptstyle M\tau} \qquad \downarrow{\scriptstyle E\tau} \qquad\qquad \downarrow{\scriptstyle (E\tau)''} \quad . \\
M(t') \underset{m}{\rightarrowtail} E(t') \underset{q}{\twoheadrightarrow} E(t')/N(t')
\end{array}
$$

Therefore, by 3.1.3, it *regularly induces* the morphism $E'(\tau)\colon E'(t) \to E'(t')$, as displayed in the following diagram

$$
\begin{array}{ccc}
M(t) \xrightarrow{\ q'\ \twoheadrightarrow} E'(t) \xrightarrow{\ m'\ \rightarrowtail} E(t)/N(t) \\
\ \downarrow{\scriptstyle M\tau} \qquad \downarrow{\scriptstyle E'\tau} \qquad\qquad \downarrow{\scriptstyle (E\tau)''} \quad . \\
M(t') \underset{q'}{\twoheadrightarrow} E'(t') \underset{m'}{\rightarrowtail} E(t')/N(t')
\end{array}
$$

7.3 The modular bifiltered object

We begin our concrete examples by studying the ex3-theory of the *modular bifiltration*, i.e. an object equipped with a bifiltration that generates a modular sublattice, and construct its biuniversal model in Inj_2.

A crucial point consists of a well-known Birkhoff theorem on free modular lattices, recalled in 7.3.1 and generalised below (Theorem 7.3.5), so that our result can be applied to groups equipped with two filtrations, one of them made of invariant subgroups (Corollary 7.3.6).

7.3.1 The Birkhoff theorem on free modular lattices

Let us recall the statement of a crucial theorem by Garrett Birkhoff ([Bi], III.7, Theorem 9), that has been proved in I.1.7 and will also be proved below, in a generalised form.

The theorem states that the free modular lattice $M(m, n)$ generated by two chains (with minimum and maximum)

$$0 = x_0 < x_1 < \ldots < x_m = 1, \qquad 0 = y_0 < y_1 < \ldots < y_n = 1, \qquad (7.17)$$

is distributive and finite. It can be realised as the lattice of parts of $[0, m] \times [0, n]$ (in $\mathbb{R}^2$ or $\mathbb{Z}^2$) generated by the subsets

$$x_i =]0, i] \times]0, n], \qquad y_j =]0, m] \times]0, j]. \qquad (7.18)$$

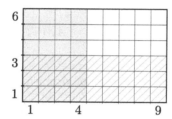

Every element of $M(m, n)$ can be uniquely written in the following form, with $i_1 > \ldots > i_p$ and $j_1 < \ldots < j_p$

$$(x_{i_1} \wedge y_{j_1}) \vee \ldots \vee (x_{i_p} \wedge y_{j_p})$$
$$= x_{i_1} \wedge (y_{j_1} \vee x_{i_2}) \wedge \ldots \wedge (y_{j_{p-1}} \vee x_{i_p}) \wedge y_{j_p}. \qquad (7.19)$$

Dually, every element can be uniquely described in the following form, again with $i_1 > \ldots > i_p$ and $j_1 < \ldots < j_p$

$$(y_{j_1} \vee x_{i_1}) \wedge \ldots \wedge (y_{j_p} \vee x_{i_p})$$
$$= y_{j_1} \vee (x_{i_1} \wedge y_{j_2}) \vee \ldots \vee (x_{i_{p-1}} \wedge y_{j_p}) \vee x_{i_p}. \qquad (7.20)$$

The theorem can be immediately extended to arbitrary chains, possibly infinite. The free modular lattice $M(I, J)$ generated by two totally ordered sets I, J is distributive and isomorphic to a sublattice L of $\mathcal{P}(I' \times J')$, where the ordinal sums $I' = I + \top$, $J' = J + \top$ consist of the ordered sets I and J with the addition of a *common* maximum.

The chains I, J are embedded in L as

$$
\begin{aligned}
i \mapsto x_i &= {\downarrow} i \times J' \quad (i \in I), \\
j \mapsto y_j &= I' \times {\downarrow} j \quad (j \in J),
\end{aligned}
\tag{7.21}
$$

and L is generated by these downward closed subsets of $I' \times J'$.

7.3.2 The homological theory of modular bifiltrations

Let I and J be two small totally ordered sets. In order to consider an object provided with a bifiltration of normal subobjects, indexed over $I \times J$, let Δ be the graph having the following (distinct) vertices and arrows:

$$
(i, \top) \to \top, \qquad (\top, j) \to \top \qquad (i \in I, \ j \in J),
\tag{7.22}
$$

where $\top$ does not belong to I nor to J.

Consider now the ex3-theory T whose models $A_* \colon \Delta \to \mathsf{E}$, in a homological category E, are the graph-morphisms $A_* = (A, (h_i), (k_j))$

$$
A = A_*(\top), \qquad h_i = A_*((i, \top) \to \top), \qquad k_j = A_*((\top, j) \to \top),
\tag{7.23}
$$

such that:

(a) the morphisms h_i and k_j are normal monomorphisms, and

$$
h_i \prec h_{i'} \text{ for } i \leqslant i' \text{ in } I, \qquad k_j \prec k_{j'} \text{ for } j \leqslant j' \text{ in } J,
$$

(b) the sublattice of $\mathrm{Nsb}A$ generated by the two chains $(\mathrm{nim}\, h_i)$ and $(\mathrm{nim}\, k_j)$ is modular (hence distributive, by the Birkhoff theorem recalled above, in 7.3.1).

Notice that $\mathrm{nim}\, h_i$ and $\mathrm{nim}\, k_j$ are just the normal *subobjects* corresponding to the normal *monomorphisms* h_i and k_j (and might be identified with the latter).

7.3.3 The biuniversal model

We now construct the biuniversal homological model of this theory.

Let $I' = I + \{\top\}$, $J' = J + \{\top\}$ be the ordinal sums of I and J with a singleton: $\top$ is the maximum in both sums; we provide both these totally ordered sets with the order semitopology (see 7.1.6), whose non-empty closed subsets are the downward sections ${\downarrow} x$.

The cartesian product $S = I' \times J'$ is equipped with the product semi-topology, whose closed subsets are the finite unions of elementary closed sets $\downarrow i \times \downarrow j$ ($i \in I'$, $j \in J'$).

Let $\mathsf{S} = \mathcal{J}_0^\sharp \langle S \rangle$ be the distributive, homological category produced by the pair $(S, \emptyset)$ (cf. 7.1.5), i.e. the full subcategory of $\mathcal{J}_0^\sharp$ whose objects are the pairs (X, A) of closed subspaces of S; let $S_* = (S, (h_i), (k_j))$ be the obvious model of T in S, with $i \in I$, $j \in J$ (as represented above, in (7.18)):

$$h_i \colon X_i = \downarrow i \times J' \subset S, \qquad k_j \colon Y_j = I' \times \downarrow j \subset S. \tag{7.24}$$

We want to prove that S_* is the biuniversal model of the ex3-theory T, which is therefore distributive.

We know that the lattice $\mathrm{Cls}(S)$ of the closed subsets of S is the free modular lattice generated by the chains I and J, embedded in $\mathrm{Cls}(S)$ by $i \mapsto X_i$ and $j \mapsto Y_j$ (see 7.3.1).

Given a model $A_* = (A, (h_i), (k_j)) \colon \Delta \to \mathsf{E}$, let us assume, for the sake of simplicity, that all the normal monomorphisms h_i and k_j are normal *subobjects* and let $R \colon \mathrm{Cls}S \to \mathrm{Nsb}A$ be the (unique) homomorphism of lattices taking every X_i to h_i and every Y_j to k_j.

Then, we define:

$$R \colon S \to \mathsf{E}, \qquad R(X, A) = RX/RA,$$
$$R((X, A) \subset (Y, B)) = (RX/RA \to RY/RB), \tag{7.25}$$

where RX/RA is a subquotient of A in E, while the right-hand morphism above is the canonical morphism between subquotients of A in E, *regularly induced* by id$A$ (3.1.3-3.1.6).

R is a functor, by the composition property of regularly induced morphisms, and $RS_* = A_*$. From the characterisation of kernels and cokernels of induced morphisms (in 3.2.2) it follows easily that R is exact. It is also easy to show that R satisfies the second universal property 7.2.2(ii).

Furthermore, the universal model of the reduced g-exact (or p-exact) theory is given by the composition of S_* with the universal arrow $\mathsf{S} \to \mathrm{Exq}\,\mathsf{S}$ (cf. 7.2.6). This yields again the p-exact biuniversal model of the bifiltration, as constructed in I.6.1.

7.3.4 Modular elements in a lattice

We now want to extend Birkhoff's theorem, also in order to apply the previous representation to bifiltered groups.

Let us recall that an element x of a lattice X is said to be *modular* (cf. Fujiwara-Murata [FuM]) if, for all $y, t \in X$:

(i) $t \geqslant x \ \Rightarrow \ (x \vee y) \wedge t = x \vee (y \wedge t)$,

(ii) $t \geqslant y \ \Rightarrow \ (x \vee y) \wedge t = (x \wedge t) \vee y$.

In other words, every join $x \vee y$ is preserved by the meet with any t bigger than x or y.

Plainly, in the lattice $L(G)$ of all subgroups of a group G, *every invariant subgroup is a modular element*. We also recall that $L(G)$ is the lattice of normal subobjects of $G = (G, 0)$ in Gp_2 (Section 1.6.8). Similarly, in the lattice of all subrings of a ring, *every bilateral ideal is a modular element*.

A reader interested in the structure of the lattices of subgroups can see Schmidt's book [Sc].

7.3.5 Theorem (An extension of a Birkhoff Theorem)

Let L be a lattice with two chains:

$$0 = x_0 < x_1 < ... < x_m = 1, \qquad 0 = y_0 < y_1 < ... < y_n = 1, \qquad (7.26)$$

and suppose that the first chain consists of modular elements *(in the sense of 7.3.4).*

Then the sublattice X of L generated by the chains is finite and distributive: a quotient of the free modular lattice $M(m, n)$ generated by the chains - which is finite, distributive and described by the Birkhoff theorem 7.3.1.

Note. A similar result holds for arbitrary chains (except the finiteness of $M(m, n)$ and X, of course).

Proof We follow the proof of the Birkhoff theorem cited above ([Bi], Chapter III, Theorem 9 and Lemmas 1-3), modifying some points in order to use the present 'local' hypothesis of modularity.

First, let us define (for $i = 0, ..., m$ and $j = 0, ..., n$):

$$u(i, j) = x_i \wedge y_j, \qquad v(i, j) = x_i \vee y_j. \qquad (7.27)$$

By Birkhoff's Lemma 1, every join of elements $u(i, j)$ in the lattice L can be (easily) written in the following form, using the absorbing properties of joins and meets

$$u(i_1, j_1) \vee ... \vee u(i_p, j_p), \qquad i_1 > ... > i_p; \quad j_1 < ... < j_p. \qquad (7.28)$$

As in Birkhoff's Lemma 2, we now prove that, if $a_1 \geqslant ... \geqslant a_p$ and $b_1 \leqslant ... \leqslant b_p$ are chains in L and *the first consists of modular elements*, then

$$(a_1 \wedge b_1) \vee ... \vee (a_p \wedge b_p) = a_1 \wedge (b_1 \vee a_2) \wedge ... \wedge (b_{p-1} \vee a_p) \wedge b_p, \qquad (7.29)$$

$$(b_1 \vee a_1) \wedge \ldots \wedge (b_p \vee a_p) = b_1 \vee (a_1 \wedge b_2) \vee \ldots \vee (a_{p-1} \wedge b_p) \vee a_p. \qquad (7.30)$$

The proof is slightly longer here, also because our hypotheses are not selfdual. Let us assume that (7.29) holds for $p - 1$ and let

$$y = a_1 \wedge (b_1 \vee a_2) \wedge \ldots \wedge (b_{p-2} \vee a_{p-1}),$$

$$z = (a_1 \wedge b_1) \vee \ldots \vee (a_{p-1} \wedge b_{p-1}) = y \wedge b_{p-1},$$

using the inductive hypothesis in the last equality. Then, applying the modular property of a_p

$$
\begin{aligned}
(a_1 \wedge b_1) \vee \ldots \vee (a_p \wedge b_p) &= z \vee (a_p \wedge b_p) \\
&= (z \vee a_p) \wedge b_p && \text{(by } b_p \geqslant z) \\
&= ((y \wedge b_{p-1}) \vee a_p) \wedge b_p \\
&= y \wedge (b_{p-1} \vee a_p) \wedge b_p && \text{(by } y \geqslant a_p) \\
&= a_1 \wedge (b_1 \vee a_2) \wedge \ldots \wedge (b_{p-1} \vee a_p) \wedge b_p.
\end{aligned}
$$

Similarly, we now assume that (7.30) holds for $p - 1$ and let (using the inductive hypothesis)

$$y = b_1 \vee (a_1 \wedge b_2) \vee \ldots \vee (a_{p-2} \wedge b_{p-1}),$$

$$z = (b_1 \vee a_1) \wedge \ldots \wedge (b_{p-1} \vee a_{p-1}) = y \vee a_{p-1}.$$

Then, applying the modular property of a_p and a_{p-1}

$$
\begin{aligned}
(b_1 \vee a_1) \wedge \ldots \wedge (b_p \vee a_p) &= z \wedge (b_p \vee a_p) \\
&= (z \wedge b_p) \vee a_p && \text{(by } z \geqslant a_p), \\
&= ((y \vee a_{p-1}) \wedge b_p) \vee a_p \\
&= y \vee (a_{p-1} \wedge b_p) \vee a_p && \text{(by } b_p \geqslant y), \\
&= b_1 \vee (a_1 \wedge b_2) \vee \ldots \vee (a_{p-1} \wedge b_p) \vee a_p.
\end{aligned}
$$

(We could now prove, as in Birkhoff's Lemma 3, that the joins of all $u(i,j)$ form a sublattice of L. In fact, by Lemmas 1, 2, any join of such elements can be rewritten as a meet of $v(i,j)$, using (7.29); therefore, any meet of such joins is a meet of meets of $v(i,j)$, hence a meet of $v(i,j)$, that can be rewritten as a join of $u(i,j)$, using (7.30). But this fact will also follow from the argument below.)

By the original Birkhoff theorem, recalled in 7.3.1, the free modular lattice $M(m,n)$ generated by the chains (7.26) is distributive and can be identified with the sublattice of parts of $[0,m] \times [0,n]$ generated by the subsets

$$X_i = \,]0,i] \times \,]0,n], \qquad Y_j = \,]0,m] \times \,]0,j].$$

The mapping sending X_i to x_i and Y_j to y_j has a unique extension to a homomorphism of lattices $f\colon M(m,n) \to L$, defined as follows (for $i_1 > ... > i_p$ and $j_1 < ... < j_p$):

$$f((X_{i_1} \wedge Y_{j_1}) \vee ... \vee (X_{i_p} \wedge Y_{j_p})) = (x_{i_1} \wedge y_{j_1}) \vee ... \vee (x_{i_p} \wedge y_{j_p}),$$
$$f((Y_{j_1} \vee X_{i_1}) \wedge ... \wedge (Y_{j_p} \vee X_{i_p})) = (y_{j_1} \vee x_{i_1}) \wedge ... \wedge (y_{j_p} \vee x_{i_p}). \tag{7.31}$$

Indeed, these definitions are consistent, because of formulas (7.29), (7.30). Moreover, the first shows that f preserves joins, while the second shows that f preserves meets.

It follows that the sublattice of L spanned by the chains (7.26) is a homomorphic image of $M(m,n)$. Therefore, it is distributive and finite, and all its elements can be represented in each of the two forms at the right hand of (7.31). (But notice that each of these representations need not be unique.) $\square$

7.3.6 Corollary

(a) In an object A of the homological category E, let two chains of normal subobjects be given, and suppose that one of them consists of modular elements of the lattice $\mathrm{Nsb}(A)$.

Then the sublattice X of $\mathrm{Nsb}(A)$ generated by the chains is distributive, and finite if both chains are.

(b) Let G be a group equipped with two chains of subgroups, and suppose that one of them consists of invariant subgroups of G.

Then the sublattice X of $L(G)$ generated by the chains is distributive, and finite if both chains are.

Proof Both points are an immediate consequence of the previous theorem. One can notice that point (b) is a particular instance of (a), viewing $G = (G,0)$ in the homological category $\mathsf{E} = \mathsf{Gp}_2$. $\square$

7.3.7 Theorem (A Jordan-Hölder theorem for lattices)

Let X be a lattice generated by two finite chains, as above, and suppose that one of them consists of modular elements. If both chains are strictly increasing and maximal, then they have the same length: $m = n$.

Proof The proof, by the usual argument based on 'Schreier refinements', can be drawn on the free modular lattice $M(m,n)$ generated by the chains, of which X is a quotient. For an index $1 \leqslant i \leqslant m$, the second chain yields a (weakly increasing) chain between x_{i-1} and x_i

$$x_{i-1} = x_{i0} \leqslant x_{i1} \leqslant \dots \leqslant x_{in} = x_i,$$
$$x_{ij} = x_{i-1} \vee (y_j \wedge x_i) = (x_{i-1} \vee y_j) \wedge x_i. \tag{7.32}$$

By the maximality of the chain (x_i), there is one index $j = \varphi(i)$ such that

$$x_{i-1} = x_{i,j-1} < x_{ij} = x_i. \tag{7.33}$$

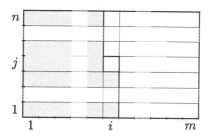

Symmetrically, for an index $1 \leqslant j \leqslant n$, there is one index $i = \psi(j)$ such that

$$y_{j-1} = y_{j,i-1} < y_{ji} = y_j,$$
$$y_{ji} = y_{j-1} \vee (x_i \wedge y_j) = (y_{j-1} \vee x_i) \wedge y_j. \tag{7.34}$$

It suffices now to prove that φ, ψ are inverse mappings between $[1, m]$ and $[1, n]$.

Indeed, the condition $x_{i,j-1} = x_{ij}$ implies $y_{j,i-1} = y_{ji}$ (and viceversa), because in the distributive lattice X we have:

$$y_{j,i-1} \vee (x_{ij} \wedge y_j) = y_{j-1} \vee (x_{i-1} \wedge y_j) \vee (x_i \wedge y_j) = y_{ji},$$
$$y_{j,i-1} \vee (x_{i,j-1} \wedge y_j) = y_{j-1} \vee (x_{i-1} \wedge y_j) \vee (x_{i-1} \wedge y_j) \vee (x_i \wedge y_{j-1})$$
$$= y_{j,i-1}.$$

Therefore $x_{i,j-1} < x_{ij}$ is equivalent to $y_{j,i-1} < y_{ji}$, i.e. $j = \varphi(i)$ is equivalent to $i = \psi(j)$. $\qquad\square$

7.3.8 Corollary

(a) In an object A of the homological category E, let two finite chains of normal subobjects be given, and suppose that one of them consists of modular elements of the lattice $\mathrm{Nsb}(A)$.

If both chains are strictly increasing and maximal, then they have the same length.

(b) Let G be a group equipped with two finite chains of subgroups, and suppose that one of them consists of invariant subgroups of G.

Then the sublattice X of L(G) generated by the chains is finite and distributive. If both chains are strictly increasing and maximal, the same conclusion follows.

Note. In both cases, it is sufficient to assume that the chains are maximal *in the sublattice X generated by the chains themselves.* Therefore point (b) also applies to two chains of invariant subgroups that are maximal among invariant chains, since then X must consist of invariant subgroups. The present result extends thus a part of the classical Jordan-Hölder theorem for chief series, in group theory (see [Su]).

Proof It is an immediate consequence of the previous theorem.   □

7.4 The modular sequence of morphisms

We study now the homological theory of a *modular sequence of morphisms.*

7.4.1 The theory

Let I be a small, totally ordered set and Δ the associated category. Consider the ex3-theory T whose models $A_* : \Delta \to \mathsf{E}$, in the homological category E, are the systems $A_* = ((A_i), (u_{ij}))$ such that

(a) $u_{ij} : A_i \to A_j$, $\quad u_{ii} = \mathrm{id} A_i$, $\quad u_{jk}.u_{ij} = u_{ik}$ $\quad (i \leqslant j \leqslant k$ in $I)$,

(b) for every index i, the sublattice X_i of $\mathrm{Nsb} A_i$ generated by the bifiltration:

$$'F_\alpha A_i = \mathrm{Ker}\, u_{i\alpha} \ (\alpha \geqslant i), \qquad ''F_\beta A_i = \mathrm{Nim}\, u_{\beta i} \ (\beta \leqslant i), \qquad (7.35)$$

is *modular* (and therefore distributive),

(c) for $i \leqslant j$, the morphism u_{ij} is left-modular on the elements of X_i and right-modular on those of X_j (cf. 2.3.1).

7.4.2 The biuniversal model

We now construct the biuniversal ex3-model of this theory. Let $I' = I + \top$ be the ordered set obtained by adding a (new) maximum to I.

We equip $S = I' \times I$ with the product semitopology (Section 7.1.6) and consider the set Σ of the following pairs S_i, for $i \in I$

$$S_i = (X_i, Y_i),$$
$$X_i = I' \times {\downarrow}i \subset S, \qquad Y_i = {\downarrow}i \times {\downarrow}i \subset X_i. \qquad (7.36)$$

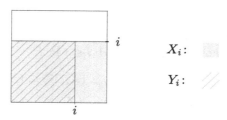

(Notice that, without the addition of $\top$, if I has a maximum k the corresponding object S_k would be a null pair.)

Let $\mathsf{S} = \mathcal{J}_0^\sharp \langle \Sigma \rangle$ be the *distributive*, homological category produced by the set Σ (Section 7.1.5): by definition, it is the full subcategory of $\mathcal{J}_0^\sharp$ whose objects are the pairs (X, Y) of closed subspaces of S such that there exist some pair $S_i = (X_i, Y_i)$ in Σ satisfying the conditions

$$Y_i \subset Y \subset X \subset X_i. \qquad (7.37)$$

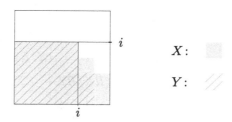

It will be important to notice that, assuming that $X \neq \emptyset$ in (7.37):
(i) the pair (X, Y) determines the index $i \in I$ (and S_i) as

$$i = \max\{i' \in I \mid (i', i') \in X\},$$

(ii) if $(X, Y) \to (X', Y')$ is a morphism of S and

$$Y_i \subset Y \subset X \subset X_i, \qquad Y_j \subset Y' \subset X' \subset X_j,$$

then $i \leqslant j$.

Indeed, for the second fact, the existence of that $\mathcal{J}_0^\sharp$-map proves that $X \subset X'$, whence $(i, i) \in X'$ and $i \leqslant j = \max\{i' \in I \mid (i', i') \in X'\}$.

There is an obvious model $S_* = ((S_i), (s_{ij}))$ of T in S, already drawn above

$$s_{ij} \colon (X_i, Y_i) \subset (X_j, Y_j) \qquad (i \leqslant j \text{ in } I); \tag{7.38}$$

we now prove that it is the universal model of the ex3-theory T, which is therefore distributive.

The lattice $\mathrm{Nsb}S_i$ consists of the closed subsets of X_i which contain Y_i, and is the free modular lattice generated by the following two chains of normal subobjects of S_i, for $\alpha \geqslant i$ and $\beta \leqslant i$ in I

$$\begin{aligned}
{}'F_\alpha S_i &= (X_i \cap Y_\alpha, Y_i) = \mathrm{Ker}\,(s_{i\alpha} \colon (X_i, Y_i) \subset (X_\alpha, Y_\alpha)), \\
{}''F_\beta S_i &= (X_\beta \cup Y_i, Y_i) = \mathrm{Nim}\,(s_{\beta i} \colon (X_\beta, Y_\beta) \subset (X_i, Y_i)).
\end{aligned} \tag{7.39}$$

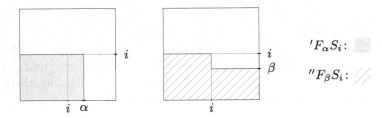

Now, given a model $A_* = (A_i, (a_{ij})) \colon \Delta \to \mathsf{E}$, let $R \colon \mathrm{Nsb}S_i \to \mathrm{Nsb}A_i$ be the (unique) homomorphism of lattices taking ${}'F_\alpha S_i$ to ${}'F_\alpha A_i$ and ${}''F_\beta S_i$ to ${}''F_\beta A_i$, for all α and β (see (7.35)).

We define:

$$\begin{aligned}
R \colon \mathsf{S} \to \mathsf{E}, \qquad R(X, Y) &= RX/RY, \\
R((X, Y) \subset (X', Y')) &= (RX/RY \to RX'/RY').
\end{aligned} \tag{7.40}$$

For (X, Y) in S, we know that there is precisely one index $i \in I$ such that $Y_i \subset Y \subset X \subset X_i$ (closed subspaces); then RX/RY is a subquotient of A_i in E.

For a morphism $(X, Y) \subset (X', Y')$ of S, we know that $Y_j \subset Y' \subset X' \subset X_j$ for a unique $j \geqslant i$, and we define the right-hand morphism above as regularly induced by $u_{ij} \colon A_i \to A_j$ in E. R is a functor, by the composition property of regularly induced morphisms (Section 3.1.4), and $RS_* = A_*$.

From the characterisation of kernels and cokernels of induced morphisms (in 3.2.2) it follows easily that R is exact. Again, it is also easy to show that R satisfies the second universal property.

Finally, the universal model of the reduced g-exact (or p-exact) theory is given by the composition of S_* with the universal arrow $S \to \mathrm{Exq}\,S$. This will be drawn below, in the finite case.

7.4.3 The finite modular sequence of morphisms

If I is finite, so is the biuniversal ex3-model. For instance let us take, in the theory described in Section 7.4.1

$$I = [0, 5] \subset \mathbb{Z}, \qquad I' = [0, 6].$$

Then the biuniversal model can be drawn as follows in the discrete plane (always representing a point by a unit square of the real plane)

$$S_i = (X_i, Y_i) = ([0, 6] \times [0, i], [0, i]^2) \subset \mathbb{Z} \times \mathbb{Z} \quad (i = 0, ..., 5), \qquad (7.41)$$

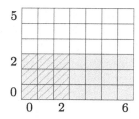

$$X_2 = [0, 6] \times [0, 2]:$$

$$Y_2 = [0, 2]^2:$$

The biuniversal ex4-model S'_* is the composite of S_* with the universal arrow $\mathsf{S} \to \mathsf{Exq\,S}$, and consists of the following subspaces S'_i with the induced morphisms $u_{ij} \colon S'_i \to S'_j$, for $i \leqslant j$

$$S_i = X_i \setminus Y_i = [i + 1, 6] \times [0, i] \subset \mathbb{Z} \times \mathbb{Z}, \qquad (7.42)$$

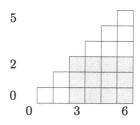 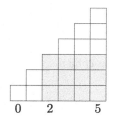

(The gray rectangle is $S'_2 = X_2 \setminus Y_2$.) Then we reindex the model in a more convenient form, as at the right hand, by a shift on the first coordinate. It agrees thus with the universal model of the corresponding RE-theory, in I.6.2.5.

7.4.4 Sequences of group homomorphisms

In particular, let $A_* = ((A_i), (u_{ij}))$ be a system of groups and homomorphisms indexed on the totally ordered set I, and viewed in the homological category Gp_2.

Assume that the functorial condition 7.4.1(a) holds: $u_{ii} = \mathrm{id}\,A_i$ and $u_{jk}.u_{ij} = u_{ik}$, for $i \leqslant j \leqslant k$ in I.

The remaining conditions (b) and (c) of 7.4.1 hold *automatically*, by Sections 7.3.6 and 1.6.8. Therefore our system is a modular sequence in Gp_2, and can be represented by the previous biuniversal model.

7.5 The modular exact couple

Exact couples in a homological category E have been studied in Section 3.5. We recall here, from Part I, the biuniversal model of their p-exact theory; then we construct the biuniversal model of the corresponding homological theory.

7.5.1 Review

As defined in 3.5.5, a *bigraded exact couple* $C = (D, E, u, v, \partial)$ *of type 1* in the *homological* category E is a system of morphisms in E, indexed on $n, p \in \mathbb{Z}$

$$u = u_{np}\colon D_{n,p-1} \to D_{np}, \qquad v = v_{np}\colon D_{np} \to E_{np},$$
$$\partial = \partial_{np}\colon E_{np} \to D_{n-1,p-1}, \tag{7.43}$$

such that:

(a) the following sequences are exact

$$\ldots E_{n+1,p} \xrightarrow{\partial} D_{n,p-1} \xrightarrow{u} D_{np} \xrightarrow{v} E_{np} \xrightarrow{\partial} D_{n-1,p-1} \ldots \tag{7.44}$$

(b) all the morphisms $u^r_{n,p} = u_{np}\ldots u_{n,p-r+1}\colon D_{n,p-r} \to D_{np}$ are exact, for $r \geqslant 1$,

(c) v_{np} is left modular on $\mathrm{Ker}\,(u^r_{n,p+r}\colon D_{np} \to D_{n,p+r})$, for $r \geqslant 1$,

(d) ∂_{np} is right modular on $\mathrm{Nim}\,(u^r_{np}\colon D_{n,p-r} \to D_{np})$, for $r \geqslant 1$.

The r-th derived couple $C^r = (D^r, E^r)$ is defined as follows, for $r \geqslant 1$ (cf. 3.5.6)

$$D^r_{n,p} = \mathrm{Nim}\,(u^{r-1}\colon D_{n,p-r+1} \to D_{np}) \in \mathrm{Nsb}(D_{np}),$$
$$E^r_{n,p} = \partial^*(D^r_{n-1,p-1})/v_*(\mathrm{Ker}\,(u^{r-1}\colon D_{np} \to D_{n,p+r-1})), \tag{7.45}$$

We shall say that C is a *modular exact couple* if it also satisfies the following additional condition:

(e) the sublattice of $\mathrm{Nsb}(D_{np})$ generated by the chains

$$
\begin{aligned}
'F_{p'} D_{np} &= \mathrm{Ker}\,(D_{np} \to D_{np'}) && (p' \geqslant p), \\
''F_{p''} D_{np} &= \mathrm{Nim}\,(D_{np''} \to D_{np}) && (p'' \leqslant p),
\end{aligned}
\tag{7.46}
$$

is modular (whence distributive, by 7.3.1).

Here we will only consider this stronger theory, which is distributive and therefore admits a biuniversal model in $\mathcal{J}_0^\sharp$. Notice that all the conditions (b)-(e) are automatically satisfied if E is g-exact.

Notice also that the additional condition (e) is certainly satisfied when the normal subobjects of the first filtration ($'F_{p'} D_{np}$) are modular elements of $\mathrm{Nsb}(D_{np})$ (by 7.3.5); this certainly holds when the morphisms $D_{np} \to D_{np'}$ are group-homomorphisms, viewed in the homological category $\mathsf{Gp_2}$ or in Ngp.

7.5.2 The theory

Consider the graph Δ, with objects

$$
(n, p)', \quad (n, p)'', \qquad \text{(for } (n, p) \in \mathbb{Z} \times \mathbb{Z}\text{)},
$$

and arrows

$$
u_{np} \colon (n, p-1)' \to (n, p)', \qquad v_{np} \colon (n, p)' \to (n, p)'',
$$

$$
\partial_{np} \colon (n, p)'' \to (n-1, p-1)'.
$$

The homological theory T of the *modular exact couple*, on the graph Δ, has for models in a homological category E all the graph morphisms

$$
C = ((D_{np}), (E_{np}), (u), (v), (\partial)) \colon \Delta \to E,
$$

that satisfy the conditions (a)-(e) of 7.5.1. (Such conditions are reflected by conservative exact functors, by 7.2.1.)

7.5.3 A crossword space

We recall now, from I.6.7.5, the construction of the semitopological space which has been used as the ground of the biuniversal model of the corresponding p-exact theory, and will be used below for the present homological theory.

Let $\Gamma = \{-\infty\} + \mathbb{Z}$, an ordinal sum. Let $\mathbf{\Gamma}$ be the ordinal sum of countably many copies of Γ, indexed on the ordered set of integers; in other words, Γ is the set $\mathbb{Z} \times \Gamma$ equipped with the (total) lexicographic order:

$$(n,p) \leqslant (n',p') \quad \text{if} \quad (n < n' \text{ or } (n = n' \text{ and } p \leqslant p')). \tag{7.47}$$

The set Γ will be equipped with the order semitopology, with (non-trivial) closed subsets $\downarrow (n,p)$, and $\Gamma \times \Gamma$ with the product semitopology, with (non-trivial) closed subsets given by the finite unions of subsets

$$\downarrow (n,p) \times \downarrow (n',p').$$

The group $\mathbb{Z}$ acts on Γ by translations: $k + (n,p) = (k+n,p)$; the group $\mathbb{Z} \times \mathbb{Z}$ acts similarly on $\Gamma \times \Gamma$.

We fix an embedding of ordered sets $\lambda \colon \Gamma \to [0,1[$, with $\lambda(-\infty) = 0$. Then we embed Γ in the ordered real line, by the mapping:

$$\lambda \colon \Gamma \to \mathbb{R}, \qquad \lambda(n,p) = n + \lambda(p) \in [n, n+1[, \tag{7.48}$$

that sends the point $(n,-\infty)$ to n. The point (n,p) will often be written as p in the interval $[n, n+1[$ of the real line.

But actually these points are isolated (for $p \in \mathbb{Z}$), and it is more convenient to represent the point (n,p) by an *interval*, as large as suitable for placing a label.

Similarly we embed $\Gamma \times \Gamma$ in the ordered plane $\mathbb{R} \times \mathbb{R}$; the transposition of coordinates is written as $\sigma \colon \Gamma \times \Gamma \to \Gamma \times \Gamma$.

7.5.4 The biuniversal p-exact model

We recall now, from I.6.8.5, the description of the biuniversal p-exact couple $\hat{C}$.

Writing $[-]$ for the integral part of a real number, we consider the following points of the discrete plane

$$\alpha_n = (-[n/2], -[(n+1)/2]) \in \mathbb{Z} \times \mathbb{Z}, \tag{7.49}$$

and the following subspaces of the semitopological space $\Gamma \times \Gamma$ (for $(p,q) \in \mathbb{Z} \times \mathbb{Z}$)

$$T'(p) =]0,(0,p)] \times [(-1,p+1),0] \subset]0,1[\times]-1,0],$$
$$T''(p) = \{(0,p)\} \times [(-1,p+1),(0,p-1)] \subset]0,1[\times]-1,1[, \qquad (7.50)$$
$$\hat{D}_{np} = \alpha_n + \sigma^n(T'(p)), \qquad \hat{E}_{np} = \alpha_n + \sigma^n(T''(p)).$$

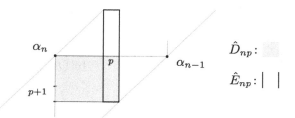

Notice that the point α_n belongs to the diagonal $t_1 = t_2$ for n even, and to the parallel line $t_1 = t_2 + 1$ for n odd; therefore, *the diagram above is correct for n even*. However, the theory is invariant up to translations of the degree n and we shall feel free of placing α_n, $\hat{D}_{np}$ and $\hat{E}_{np}$ in a position which, strictly speaking, would require an *even* (or an *odd*) index n.

These data define the model $\hat{C}$, in the p-exact category $\mathcal{J}_0 < \Sigma >$ determined by the set Σ of all the subspaces $\hat{D}_{np}$ and $\hat{E}_{np}$ of $\Gamma \times \Gamma$; the proper morphisms u, v, ∂ are the canonical ones, induced by the identity of $\Gamma \times \Gamma$.

7.5.5 Comments

(a) The exact sequence (7.44) can be read on the biuniversal model

$$\ldots D_{n,p-1} \xrightarrow{u} D_{np} \xrightarrow{v} E_{np} \xrightarrow{\partial} D_{n-1,p-1} \longrightarrow \ldots \qquad (7.51)$$

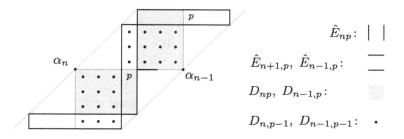

(b) The object $\hat{D}_{np}$ is bifiltered by the following chains

$$'F_{p'}\hat{D}_{np} = \mathrm{Ker}\,(\hat{D}_{np} \to \hat{D}_{np'}) \qquad (p' \geqslant p),$$
$$''F_{p''}\hat{D}_{np} = \mathrm{Im}\,(\hat{D}_{np''} \to \hat{D}_{np}) \qquad (p'' \leqslant p).$$
$$(7.52)$$

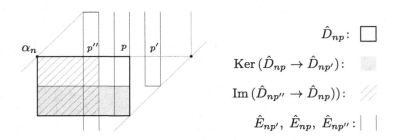

$$\hat{D}_{np}: \quad \square$$
$$\mathrm{Ker}\,(\hat{D}_{np} \to \hat{D}_{np'}):$$
$$\mathrm{Im}\,(\hat{D}_{np''} \to \hat{D}_{np})):$$
$$\hat{E}_{np'},\ \hat{E}_{np},\ \hat{E}_{np''}: \mid \ \mid$$

On the other hand, the object $\hat{E}_{np}$ is filtered by the following chain

$$0 = \,'F_p\hat{E}_{np} \subset \,'F_{p+1}\hat{E}_{np} \subset \ldots \subset \,''F_{p+1}\hat{E}_{np} \subset \,''F_p\hat{E}_{np} = \hat{E}_{np},$$

$$'F_{p'}\hat{E}_{np} = v_*(\mathrm{Ker}\,(D_{np} \to D_{np'})) \qquad (p' \geqslant p),$$
$$''F_{p''}\hat{E}_{np} = \partial^*\mathrm{Im}\,(\hat{D}_{n-1,p''} \to \hat{D}_{n-1,p}) \quad (p'' \leqslant p).$$
$$(7.53)$$

(c) For the derived couple $\hat{C}^r$, the locally closed subspaces $\hat{D}^r_{np}$ and $\hat{E}^r_{np}$ are obtained from $\hat{D}_{np}$ and $\hat{E}_{np}$ by cutting out, respectively, one or two strips of height r.

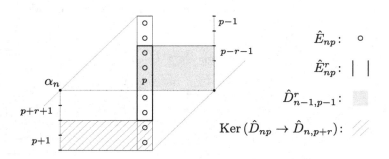

$$\hat{E}_{np}: \quad \circ$$
$$\hat{E}^r_{np}: \mid \ \mid$$
$$\hat{D}^r_{n-1,p-1}:$$
$$\mathrm{Ker}\,(\hat{D}_{np} \to \hat{D}_{n,p+r}):$$

One can easily read on the model the morphisms $u^{(r)}$, $v^{(r)}$, $\partial^{(r)}$, the exactness of C^r, the relations $C^0 = C$, $C^{r+1} = (C^r)^1$ and the usual properties of the spectral sequence (see [Mas1]).

7.5.6 The biuniversal homological model

We now construct the biuniversal ex3-model. As in 7.5.4, we use the semi-topological space $\Gamma \times \Gamma$ in the ordered plane $\mathbb{R} \times \mathbb{R}$ and its subspaces $\hat{D}_{np}$, $\hat{E}_{np}$.

Let $\mathsf{S} = \mathcal{J}_0^\sharp \langle \Sigma \rangle$, the *homological* category produced by the set Σ of all $\hat{D}_{np}$ and $\hat{E}_{np}$: by definition, it is the full subcategory of $\mathcal{J}_0^\sharp$ whose objects are the pairs (X, Y) such that there exists some semitopological space $S \in \Sigma$, for which $Y \subset X \subset S$ (as an inclusion of closed subspaces).

Also here the pair (X, Y) determines S, as soon as $X \neq \emptyset$. In fact, a non-empty closed subset of $\hat{D}_{np}$ (resp. $\hat{E}_{np}$) necessarily contains the point marked with a circle (resp. a dot), and all such points are different.

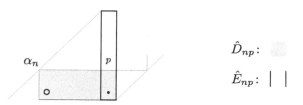

We define the model $\hat{C}$ in the homological category $\mathsf{S} = \mathcal{J}_0 < \Sigma >$, as above (in 7.5.4). Notice that the objects $\hat{D}_{np}$ are bifiltered by (7.52), and the lattice $\mathrm{Nsb}(\hat{D}_{np})$ is the free modular lattice generated by these two chains of normal subobjects.

Now, given a modular exact couple $C = (D, E, u, v, \partial) \colon \Delta \to \mathsf{E}$ with values in a homological category, let

$$R_{np} \colon \mathrm{Nsb}(\hat{D}_{np}) \to \mathrm{Nsb}(D_{np}),$$

be the (unique) homomorphism of lattices taking the two filtrations of $\hat{C}$ to the corresponding filtrations of C (this is legitimate, because C satisfies (e)). Similarly, we define

$$R'_{np} \colon \mathrm{Nsb}(\hat{E}_{np}) \to \mathrm{Nsb}(E_{np}),$$

the (unique) mapping of totally ordered sets that takes the filtration (7.53) of $\hat{C}$ to the corresponding one of C.

Finally, we have a functor $R \colon \mathsf{S} \to \mathsf{E}$,

$$R(X, Y) = RX/RY,$$
$$R((X, Y) \subset (X', Y')) = (RX/RY \to RX'/RY'). \tag{7.54}$$

From the characterisation of kernels and cokernels of induced morphisms (in 3.2.2) it follows easily that R is exact. Again, it is also easy to show that R satisfies the second universal property.

Also here, the universal model of the corresponding g-exact theory is given by the composition of S_* with the universal arrow $S \to \mathrm{Exq}\,S$. This yields again the exact universal model of the exact couple, as constructed in Part I.

7.5.7 The homotopy exact couple of a tower of fibrations

Finally, let us start from a tower of fibrations of pathwise connected, pointed spaces (cf. 6.3.2).

We already know that this gives an exact couple in the p-homological category Ngp. The additional modularity property (e) is always satisfied, as we have already remarked at the end of 7.5.1. Therefore, we can use the biuniversal homological model to study the associated spectral sequence.

In the non-connected case of Section 6.5, the application is less obvious, because part of the exactness conditions are not required.

Appendix A
Some points of category theory

In this book, category theory is used extensively, if at an elementary level. The notions of category, functor and natural transformation are used throughout, together with standard tools like limits, colimits and adjoint functors.

The review of this appendix, more reduced than Chapter A of Part I, is also meant to fix terminology and notation. More information, and the proofs of the results mentioned here, can be found in the texts of Mac Lane [M5], Borceux [Bo1], and Adámek, Herrlich, Strecker [AHS] or as specified below.

A1 Basic notions

A1.1 Smallness

As already said in the Introduction, we work within a set-theoretic setting widely used in category theory, which is based on standard set theory and the assumption of the existence of a *Grothendieck universe* (see [M5, Bo1]).

A basic universe $\mathcal{U}$ is fixed throughout; its elements are called *small* sets. A $\mathcal{U}$-*category* has objects and arrows belonging to this universe, and is said to be *small* if its set of morphisms belongs to $\mathcal{U}$, *large* if it does not. The concrete categories we consider are generally large $\mathcal{U}$-categories, e.g. the category Set of small sets, or Ab of small abelian groups, or Top of small topological spaces. In such cases, the term 'small' is generally understood. CAT denotes the 2-category of all $\mathcal{U}$-categories, with their functors and their natural transformations; it is a $\mathcal{V}$-category, for any universe $\mathcal{V}$ to which $\mathcal{U}$ belongs.

ObC and MorC denote the set of objects and morphisms of a category C.

We do *not* assume that a $\mathcal{U}$-category has small hom-sets $C(X, Y)$. Indeed, the hom-functors do not play a relevant role here, *while the subobject-functors do.* Hence we are more interested in other conditions of 'local smallness' of categories, for instance the fact of being 'normally' well-powered, i.e. to have small sets of normal subobjects with respect to an assigned ideal (see 1.5.1).

A set X can be viewed as a *discrete* category: its objects are the elements of X, and the only arrows are their (formal) identities. A preordered set X is also (implicitly) viewed as a category: the objects are its elements, and there is precisely one arrow $x \to y$ when $x \prec y$ in X, none otherwise; composition is uniquely determined. All diagrams in these categories commute.

A1.2 Categories and functors

We assume that the reader is familiar with the very basic concepts and notation of category theory (already reviewed in Section A1 of Part I), like:

- category; the identity morphism $\mathrm{id}X$ (or 1_X) of an object X in a category; isomorphism (or iso), monomorphism (or mono) and epimorphism (or epi) in a category (also reviewed here in 1.1.5);

- functor; the identity functor $\mathrm{id}\mathsf{C}$ (or 1_C) of a category C; forgetful functor between two categories of structured sets;

- a subcategory and its inclusion functor; a cartesian product of categories and its projection functors.

Given two morphisms $u\colon A \to X$ and $p\colon X \to A$ such that $pu = \mathrm{id}A$, we say that A is a *retract* of X, that u is a *split monomorphism* (or a *section*), and p is a *split epimorphism* (or a *retraction*).

A functor $F\colon \mathsf{C} \to \mathsf{D}$ is said to be *faithful* (resp. *full*) if all the mappings $\mathsf{C}(X,Y) \to \mathsf{D}(FX, FY)$ that define its action on the morphisms are injective (resp. surjective); it is *conservative* if it reflects isomorphisms: for every map u of C, if $F(u)$ is invertible in D so is u in C. A subcategory C' of C is *full* (resp. *conservative*) if so is the inclusion functor $\mathsf{C}' \to \mathsf{C}$.

Plainly, every functor preserves isomorphisms, split monos and split epis. A faithful functor reflects monos and epis, while a faithful and full functor also reflects isomorphisms. However, if C is *balanced* (i.e. epi and mono implies iso), every faithful functor $\mathsf{C} \to \mathsf{D}$ reflects isomorphisms.

A *congruence* $R = (R_{XY})$ in a category C consists of a family of equivalence relations R_{XY} in each set of morphisms $\mathsf{C}(X,Y)$; the family must be consistent with composition:

$$\text{if } f\, R_{XY}\, f' \text{ and } g\, R_{YZ}\, g', \text{ then } gf\, R_{XZ}\, g'f'. \qquad \text{(A.1)}$$

Then one defines the *quotient category* $D = C/R$: the objects are the ones of C, and $D(X, Y) = C(X, Y)/R_{XY}$; in other words, a morphism $[f] \colon X \to Y$ in D is an equivalence class of morphisms $X \to Y$ in C. The composition is induced by that of C, which is legitimate because of condition (A.1):

$$[g].[f] = [gf]. \tag{A.2}$$

The projection functor $P \colon C \to C/R$ sends each object to itself and each morphism f to $[f]$.

A1.3 Natural transformations

Given two functors $F, G \colon C \to D$ between the same categories, a *natural transformation* $\varphi \colon F \to G \colon C \to D$ consists of the following data:

- for each object X of C, a morphism $\varphi X \colon FX \to GX$ in D (called the *component* of φ on X, and also written as φ_X),

so that, for every arrow $f \colon X \to X'$ in C, we have a commutative square in D:

$$
\begin{array}{ccc}
FX & \xrightarrow{\ \varphi X\ } & GX \\
{\scriptstyle Ff}\downarrow & & \downarrow{\scriptstyle Gf} \\
FX' & \xrightarrow[\varphi X']{} & GX'
\end{array}
\qquad
\begin{array}{c}
\varphi X'.F(f) = G(f).\varphi X \\[2em]
(\textit{naturality condition}).
\end{array}
\tag{A.3}
$$

In particular, *the identity of a functor* $F \colon C \to D$ is the natural transformation $\mathrm{id}F \colon F \to F$, of components $(\mathrm{id}F)X = \mathrm{id}(FX)$.

Natural transformations have a *vertical composition*

$$
C \begin{array}{c} \xrightarrow{F} \\[-0.3em] {\scriptstyle \downarrow\varphi} \\[-0.3em] \xrightarrow{} \\[-0.3em] {\scriptstyle \downarrow\psi} \\[-0.3em] \xrightarrow[H]{} \end{array} D
\qquad \psi\varphi \colon F \to H, \tag{A.4}
$$

$$(\psi\varphi)(X) = \psi X.\varphi X \colon FX \to GX \to HX.$$

There is also a *whisker composition*, or *reduced horizontal composition*, of natural transformations with functors

$$
C' \xrightarrow{\ H\ } C \begin{array}{c} \xrightarrow{F} \\[-0.3em] {\scriptstyle \downarrow\varphi} \\[-0.3em] \xrightarrow[G]{} \end{array} D \xrightarrow{\ K\ } D' \tag{A.5}
$$

$$K\varphi H \colon KFH \to KGH \colon C' \to D',$$
$$(K\varphi H)(X') = K(\varphi(HX')).$$

An *isomorphism of functors* is a natural transformation $\varphi\colon F \to G\colon \mathsf{C} \to \mathsf{D}$ which is invertible, with respect to vertical composition. It is easy to see that this happens if and only if all the components φX are invertible in D.

This '2-dimensional structure' of categories, where natural transformations play the role of '2-dimensional arrows' between functors, will be further analysed below, in Section A4 (including the *full* horizontal composition of natural transformations).

Replacing the category C with a (directed) graph Δ, one can consider, as above, a *natural transformation* (resp. an *isomorphism*) $\varphi\colon F \to G\colon \Delta \to \mathsf{D}$ between two *morphisms of graphs* with values in a category.

A1.4 Products and equalisers

Many definitions in category theory are based on a universal property (at least since Mac Lane's paper [M1]).

For instance, in a category C, the *product* of a family $(X_i)_{i\in I}$ of objects (indexed on a *small set* I), is defined as an object X equipped with a family of morphisms $p_i\colon X \to X_i$ $(i \in I)$, called *projections*, which satisfy the following universal property:

$$
\begin{array}{ccc}
Y & \overset{f}{-\!\!\!-\!\!\!\rightarrow} & X \\
& {\scriptstyle f_i}\searrow & \downarrow{\scriptstyle p_i} \\
& & X_i
\end{array}
\qquad\qquad (\mathrm{A.6})
$$

(i) for every object Y and every family of morphisms $f_i\colon Y \to X_i$, there exists a unique morphism $f\colon Y \to X$ such that, for all $i \in I$, $p_i f = f$. (The map f can be written as (f_i), by its *components*.)

The product of a family need not exist. If it does, it is determined up to a unique *coherent* isomorphism, in the sense that if also Y is a product of the family $(X_i)_{i\in I}$ with projections $q_i\colon Y \to X_i$, then the unique morphism $f\colon X \to Y$ which commutes with all projections (i.e. $q_i f = p_i$, for all indices i) is an iso. Therefore, one speaks of *the* product of the family (X_i), denoted as $\prod_i X_i$.

We say that a category C *has products* (resp. *finite products*) if every family of objects indexed on a *small* set (resp. on a finite set) has a product in C.

In particular, the product of the empty family of objects $\emptyset \to \mathrm{Ob}\mathsf{C}$ means an object X (equipped with no projections) such that for every object Y (equipped with no maps) there is a unique morphism $f\colon Y \to X$ (satisfying no conditions). The solution is called the *terminal* object of C; again, it need not exist, but is determined up to a unique isomorphism. It can be written as $\top$.

In Set, Top and Ab all products exist, and are the usual cartesian ones; the terminal object is the singleton. In the category X associated to a preordered set (see A1.1), the categorical product of a family of points $x_i \in X$ amounts to their inf, while the terminal object amounts to the greatest element of X (when such things exist; notice that they are determined up to the equivalence relation associated to our preorder, and uniquely determined for an ordering).

It is easy to prove that a category has finite products if and only if it has binary products $X_1 \times X_2$ and a terminal object.

Products are a basic instance of a much more general concept recalled below (in A2.2), the limit of a functor. Another basic instance is the *equaliser* of a pair $f, g \colon X \to Y$ of 'parallel' maps of C; this is (an object E equipped with) a map $m \colon E \to X$ such that $fm = gm$ and the following universal property holds:

$$E \xrightarrow{\ m\ } X \underset{g}{\overset{f}{\rightrightarrows}} Y$$

$$\text{with } w, h, Z \tag{A.7}$$

(ii) every map $h \colon Z \to X$ such that $fh = gh$ factorises uniquely through m (i.e. there exists a unique map $w \colon Z \to E$ such that $mw = h$).

The equaliser morphism is necessarily a monomorphism, and - by definition - a *regular* monomorphism.

In Set (resp. Top, Ab), the equaliser of two parallel maps $f, g \colon X \to Y$ is the embedding in X of the whole subset (resp. subspace, subgroup) of X on which they coincide. Therefore, regular monomorphisms coincide with monomorphisms (or injective morphisms) in Set and Ab, but 'amount' to *inclusion of subspaces* in Top (while every injective continuous mapping is a monomorphism).

In the category X associated to a preordered set, two parallel maps $x \to x'$ always coincide, so that their equaliser is the identity 1_x.

The reader may also be interested to know that all subgroups are regular subobjects in Gp (a non-trivial fact, see [AHS], Exercise 7H), while a subsemigroup or a subring need not be a regular subobject ([AHS], Section 7.58). Embeddings in concrete categories may give a better notion of 'good subobject' (see A2.7).

A1.5 Duality, sums and coequalisers, biproducts

If C is a category, the *opposite* (or *dual*) category, written C* or C^{op}, has the same objects as C and 'reversed' arrows,

$$C^{op}(X, Y) = C(Y, X), \tag{A.8}$$

with 'reversed composition' $g * f = fg$.

Every notion, or statement, of category theory has a dual instance, which comes from the opposite category (or categories): thus, monomorphism and epimorphism are dual to each other, while isomorphism is a selfdual notion. A dual notion is often distinguished by the prefix 'co-'.

The *sum*, or *coproduct*, of a family $(X_i)_{i \in I}$ of objects of C is dual to their product. Explicitly, it is an object X equipped with a family of morphisms $u_i \colon X_i \to X$ ($i \in I$), called *injections*, which satisfy the following universal property:

$$
\begin{array}{ccc}
X & \overset{f}{\dashrightarrow} & Y \\
{\scriptstyle u_i} \big\uparrow & \nearrow {\scriptstyle f_i} & \\
X_i & &
\end{array}
\tag{A.9}
$$

(i*) for every object Y and every family of morphisms $f_i \colon X_i \to Y$, there exists a unique morphism $f \colon X \to Y$ such that, for all $i \in I$, $f u_i = f_i$. (The map f will be written as $[f_i]$, by its *co-components*.)

Again, if the sum of the family (X_i) exists, it is determined up to a unique coherent isomorphism, and denoted as $\sum_i X_i$, or $X_1 + \ldots + X_n$ in a finite case. The sum of the empty family is the *initial* object $\bot$: this means that, for every object X, there is precisely one map $\bot \to X$.

Sums in Set and Top are realised as disjoint unions, and the initial object is the empty set or space. In Ab, categorical sums are realised as direct sums and the initial object is the null group. In an ordered set, categorical sums amount to joins. A lattice is the same as an ordered set with finite (categorical) products and sums.

The *coequaliser* of a pair $f, g \colon X \to Y$ of parallel maps of C is a map $p \colon Y \to C$ such that $pf = pg$ and:

$$
\begin{array}{ccc}
X & \underset{g}{\overset{f}{\rightrightarrows}} Y & \overset{p}{\longrightarrow} C \\
& {\scriptstyle h}\big\downarrow & \swarrow {\scriptstyle w} \\
& Z &
\end{array}
\tag{A.10}
$$

(ii*) every map $h \colon Y \to Z$ such that $hf = hg$ factors uniquely through p (i.e. there exists a unique map $w \colon C \to Z$ such that $wp = h$).

A coequaliser morphism is necessarily an epimorphism, and - by definition - a *regular epimorphism*. A reader not familiar with these notions should begin by performing these constructions in Set and Top, where a regular epimorphism amounts to a projection on a quotient set or space.

Sums and coequalisers are particular instances of the *colimit* of a functor (see A2.2).

As already recalled in 1.1.6, a category is said to be *pointed* if it has a *zero object*; by definition, this object is both initial and terminal, and will generally be written as 0; the zero morphism $A \to 0 \to B$ between two given objects is written as 0_{AB}, or just 0.

In a pointed category C, the *biproduct* of a finite family (A_i) of objects is an object A, often denoted as $\bigoplus_i A_i$, equipped with a family of injections $u_i \colon A_i \to A$ and a family of projections $p_i \colon A \to A_i$ so that:

- $p_i u_i = \mathrm{id} A_i$, for all indices i,

- $p_i u_j = 0$ whenever $i \neq j$,

- $(A, (p_i))$ is the product of (A_i) and $(A, (u_i))$ is their sum.

Plainly, the biproduct of the empty family is the zero object. C is said to be *semiadditive* if it has finite biproducts. Then (see I.2.1) one can define the sum of two maps $f, g \colon A \to B$ as $f + g = \partial(f \oplus g)d$

$$A \xrightarrow{\ d\ } A \oplus A \xrightarrow{\ f \oplus g\ } B \oplus B \xrightarrow{\ \partial\ } B \tag{A.11}$$

where $d \colon A \to A \oplus A$ is the diagonal (defined by $pd = qd = \mathrm{id}A$), while $\partial \colon B \oplus B \to B$ is the codiagonal (defined by $\partial u' = \partial v' = idB$). All hom-sets of C are thus abelian monoids (possibly large) and composition distributes over the sum (cf. I.2.1.3).

Furthermore, C is said to be additive if its hom-sets are abelian groups (possibly large).

All categories of modules are additive. Ltc is just semiadditive. (See 1.2.7 and Section 2.4).

A1.6 Subobject and quotients

Let A be an object of the category C. A subobject of A cannot be based on the notion of subset, but is defined as an equivalence class of monomorphisms, or better as a *selected* representative of such a class.

More precisely, given two monos $m \colon M \rightarrowtail A$, $n \colon N \rightarrowtail A$ *with values in* A, we say that $m \prec n$ if there is a (uniquely determined) morphism u such that $m = nu$. We say that m, n are *equivalent*, or $m \sim n$, if $m \prec n \prec m$, i.e. if there is an isomorphism u such that $m = nu$.

In every class of equivalent monos (with codomain A), precisely one is *selected* and called a *subobject* of A; in the class of isomorphisms, we always choose the identity 1_A. In a category of structured sets there is often a natural choice, given by convenient subsets equipped with the induced structure.

The subobjects of A in C form the (possibly large) *ordered* set Sub$A$, with maximum 1_A; here, the induced order $m \prec n$ is also written as $m \leqslant n$. Equalisers (if exstant) are always assumed to be subobjects, letting their choice depend on the general one.

Epimorphisms with domain A are dealt with in a dual way. Their preorder and equivalence relation are also written as $p \prec q$ (meaning that p factorises through q) and $p \sim q$.

A *quotient* of A is a *selected* representative of an equivalence class of epimorphisms with domain A; they form the ordered set Quo$A$, with maximum 1_A; again the induced order is also written as $p \leqslant q$. Coequalisers are always chosen as quotients.

For the sake of simplicity, *we often follow the common abuses of notation* for subobjects (and quotients). Thus, a subobject $m\colon M \rightarrowtail A$ is denoted by means of its domain M; if $N \leqslant M$ in Sub$A$ (i.e. we have a subobject $n\colon N \rightarrowtail A$ with $n \leqslant m$), N can also denote the corresponding subobject of M, equivalent to the monomorphism $n'\colon N \rightarrowtail M$ such that $mn' = n$.

In a pointed category, the *kernel* (or cokernel) of a morphism $f\colon A \to B$ is defined as the equaliser (or coequaliser) of f and the zero-morphism $A \to B$ (see 1.1.7). They are thus, respectively, a (normal) subobject of A and a (normal) quotient of B.

In this book we extend these notions with respect to an assigned ideal of null maps (not necessarily produced by a zero object), and our interest is about normal monos and normal epis in this generalised sense (see Section 1.3).

A1.7 Isomorphism and equivalence of categories

An *isomorphism of categories* is a functor $F\colon \mathsf{C} \to \mathsf{D}$ which is invertible. This means that F admits an inverse $G\colon \mathsf{D} \to \mathsf{C}$, i.e. a functor such that $GF = \mathrm{id}\mathsf{C}$ and $FG = \mathrm{id}\mathsf{D}$. Being isomorphic categories is an equivalence relation, written as $\mathsf{C} \cong \mathsf{D}$.

For instance, the category Ab of abelian groups is (clearly) isomorphic to the category of $\mathbb{Z}$-modules (and $\mathbb{Z}$-homomorphisms).

More generally, an *equivalence of categories* is a functor $F\colon \mathsf{C} \to \mathsf{D}$ which is invertible up to isomorphism of functors (A1.3), i.e. there exists a functor $G\colon \mathsf{D} \to \mathsf{C}$ such that $GF \cong \mathrm{id}\mathsf{C}$ and $FG \cong \mathrm{id}\mathsf{D}$.

An *adjoint equivalence of categories* is a coherent version of this notion, namely a four-tuple $(F, G, \eta, \varepsilon)$ where:

- $F \colon \mathsf{C} \to \mathsf{D}$ and $G \colon \mathsf{D} \to \mathsf{C}$ are functors,

- $\eta \colon \mathrm{id}\mathsf{C} \to GF$ and $\varepsilon \colon FG \to \mathrm{id}\mathsf{D}$ are isomorphisms of functors satisfying the *coherence conditions*:

$$F\eta = (\varepsilon F)^{-1} \colon F \to FGF, \qquad \eta G = (G\varepsilon)^{-1} \colon G \to GFG.$$

The following conditions on a functor $F \colon \mathsf{C} \to \mathsf{D}$ are equivalent, forming a very useful *characterisation of the equivalence of categories*:

(i) F is an equivalence of categories,

(ii) F can be completed to an adjoint equivalence $(F, G, \eta, \varepsilon)$,

(iii) F is faithful, full and *essentially surjective on objects*.

The last condition means that: for every object Y of D there exists some object X in C such that $F(X)$ is isomorphic to Y in D. The proof of the equivalence of these three conditions is rather long and requires the axiom of choice [M5].

One says that two categories C, D are *equivalent*, written as $\mathsf{C} \simeq \mathsf{D}$, if there exists an equivalence of categories between them (or, equivalently, an adjoint equivalence of categories). This is indeed an equivalence relation, as follows easily from condition (iii), above.

For instance, the category of finite sets (and mappings between them) is equivalent to its full subcategory of finite cardinals, which is small (and therefore cannot be isomorphic to the former).

A category is said to be *skeletal* if it has no pair of distinct isomorphic objects. It is easy to show, by the previous characterisation of equivalences, that every category has a *skeleton*, i.e. an equivalent skeletal category. The latter can be obtained by *choosing* precisely one object in every class of isomorphic objects.

We have described above a skeleton of the category of finite sets, that can be constructed without using choice. A preordered set is a skeletal category if and only if it is ordered.

A1.8 Categories of functors and presheaves

Let S be a small category and $S = \mathrm{Ob}\mathsf{S}$ its set of objects. For any category C, one writes C^S for the category whose objects are the functors $F \colon \mathsf{S} \to \mathsf{C}$ and whose morphisms are the natural transformations $\varphi \colon F \to G \colon \mathsf{S} \to \mathsf{C}$, with vertical composition (see (A.4)).

In particular, the arrow category **2** (with two objects $0, 1$ and one non-identity arrow, $0 \to 1$) gives the *category of morphisms* $\mathsf{C}^{\mathbf{2}}$ of C, where a map $(u_0, u_1)\colon f \to g$ is a commutative square of C; these are composed by 'pasting' commutative squares, as in the right diagram below

$$
\begin{array}{ccc}
A_0 & \xrightarrow{u_0} & B_0 \\
f\downarrow & & \downarrow g \\
A_1 & \xrightarrow{u_1} & B_1
\end{array}
\qquad
\begin{array}{ccccc}
A_0 & \xrightarrow{u_0} & B_0 & \xrightarrow{v_0} & C_0 \\
f\downarrow & & \downarrow g & & \downarrow h \\
A_1 & \xrightarrow{u_1} & B_1 & \xrightarrow{v_1} & C_1
\end{array}
\qquad . \qquad \text{(A.12)}
$$

A natural transformation $\varphi\colon F \to G\colon A \to B$ can be viewed as a functor $A \times 2 \to B$ or, equivalently, as a functor $A \to B^2$.

A functor $F\colon C \to \mathsf{Set}$ is said to be *representable* if it is isomorphic to a functor $C(X_0, -)\colon C \to \mathsf{Set}$, for some object X_0 in C, which is said *to represent* F, and is determined by the latter, up to isomorphism. Then, the Yoneda lemma describes the natural transformations $F \to G$, for every functor $G\colon C \to \mathsf{Set}$ [M5].

A functor $S^{op} \to C$, defined on the opposite category S^{op}, is also called a *presheaf* of C on the (small) category S. They form the presheaf category $\mathsf{Psh}(S, C) = C^{S^{op}}$.

S is canonically embedded in the latter, by the *Yoneda embedding*

$$
y\colon S \to C^{S^{op}}, \qquad y(i) = S(-, i)\colon S^{op} \to \mathsf{Set}, \qquad \text{(A.13)}
$$

which sends every object i to the corresponding representable presheaf $y(i)$.

Taking as S the category Δ of finite positive ordinals (and increasing maps), one gets the category $C^{\Delta^{op}}$ of simplicial objects in C, and - in particular - the well-known category of simplicial sets. Here, the Yoneda embedding sends the ordinal n to the simplicial set Δ^n, freely generated by one simplex of dimension n.

A1.9 A digression on mathematical structures and categories

When studying a mathematical structure with the help of category theory, it is crucial to choose the 'right' kind of structure and the 'right' kind of morphisms, so that the result is sufficiently general and 'natural' to have good properties (with respect to the goals of our study) - even if we are interested in more particular situations.

For instance, the category Top of topological spaces and continuous mappings is a classical framework for studying topology. Among its good properties there is the fact that all (co)products and (co)equalisers exist, and are computed *as in* Set, then equipped with a suitable topology determined by the structural maps. (More generally, this is true of all limits and colimits, and is a consequence of the fact that the forgetful functor Top → Set has a left *and* a right adjoint, corresponding to discrete and chaotic topologies; see below).

Hausdorff spaces are certainly important, but it is often better to view them in Top, as their category is less well behaved: coequalisers exist, but are not computed as in Set, i.e. preserved by the forgetful functor to Set.

(Many category theorists would agree with Mac Lane [M5], saying that even Top is not sufficiently good, because it is not a cartesian closed category, and prefer - for instance - the category of compactly generated spaces; however, since our main interests are about homology and homotopy, we are essentially satisfied with the fact that the standard interval is exponentiable in Top. See A4.3.)

Similarly, if we are interested in ordered sets, it is generally better to view them in the category of *pre*ordered sets and (weakly) increasing mappings, where (co)products and (co)equalisers not only exist, but again are computed *as in* Set, with a suitable preorder determined by the structural maps.

Another point to be kept in mind is that the isomorphisms of the category (i.e. its invertible arrows) should indeed 'preserve' the structure we are interested in, or we risk of studying something different from our purpose.

As a trivial example, the category T of topological spaces and *all* mappings between them has practically nothing to do with topology: an isomorphism of T is any bijection between topological spaces. Indeed, T is *equivalent to the category of sets* (according to the previous definition, in A1.7), and is a 'deformed' way of looking at the latter.

Less trivially, the category M of metric spaces and continuous mappings misses crucial properties of metric spaces, since its invertible morphisms do not preserve completeness. In fact, M is equivalent to the category of *metrisable topological spaces* and continuous mappings, and should be viewed in this way. A 'reasonable' category of metric spaces should be based on *Lipschitz* maps, or - more particularly - on weak contractions.

Excluding particular cases is, generally speaking, a bad option. Ab and Gp are both important, while the category of *non-commutative groups* seems to be of no importance; certainly, it has practically none of the good categorical properties of Ab and Gp (like completeness and cocompleteness, see A2.4). Of course, one can always consider such groups within Gp, when useful.

A striking example of this kind is concerned with the category Sgr of semigroups and their homomorphisms. According to their theory *as developed in Universal Algebra*, a semigroup should be non-empty; but this exclusion would destroy much of their good properties, both from a categorical and a 'practical' point of view: for instance, if we follow it, subsemigroups are not closed under intersection and counterimages, and no longer form a lattice; equalisers need not exist; etc.

More generally, let us recall that a *variety of algebras* $\mathrm{Alg}(\Omega, \Sigma)$ is formed of all algebras of signature Ω that satisfy the equational laws of Σ, with the obvious homomorphisms (see [Bo1, AHS]). We do not follow the usual conventions of Universal Algebra that exclude a priori the empty algebra (as for instance in Cohen's text [Co]).

In fact, when there are no zero-ary operations in Ω, this exclusion has the effects mentioned above for semigroups: it destroys many of the good practical properties of the setting, and most of the categorical ones.

A2 Limits and colimits

A2.1 Universal arrows

There is a general way of formalising universal properties, based on a functor $U: \mathsf{A} \to \mathsf{C}$ and an object X of C.

A *universal arrow from the object X to the functor U* is a pair

$$(A, \eta: X \to UA)$$

consisting of an object A of A and arrow η of C which is universal, in the sense that every similar pair $(B, f: X \to UB)$ factorises uniquely through (A, η): namely, there exists a unique map $g: A \to B$ in A such that the following triangle commutes in C

$$
\begin{array}{ccc}
X & \xrightarrow{\ \eta\ } & UA \\
 & \searrow^{f} & \downarrow^{Ug} \\
 & & UB
\end{array}
\qquad Ug.\eta = f. \qquad (A.14)
$$

Dually, a *universal arrow from the functor U to the object X* is a pair $(A, \varepsilon: UA \to X)$ consisting of an object A of A and arrow ε of C such that every similar pair $(B, f: UB \to X)$ factorises uniquely through (A, ε): there exists a unique $g: B \to A$ in A such that

$$
\begin{array}{ccc}
UA & \xrightarrow{\ \epsilon\ } & X \\
{\scriptstyle Ug}\uparrow & \nearrow_{f} & \\
UB & &
\end{array}
\qquad \epsilon.Ug = f. \qquad (A.15)
$$

A reader which is not familiar with this notion might begin by constructing the universal arrow from a set X to the forgetful functor $\mathsf{Ab} \to \mathsf{Set}$, or from a group G to the inclusion functor $\mathsf{Ab} \to \mathsf{Gp}$. Then, one can describe (co)products and (co)equalisers in a category C as universal arrows for suitable functors (as we shall do in A2.5, in a more general way).

Universal arrows for '2-dimensional categories' are considered in A4.6.

A2.2 Limits

We now review the categorical notion of the *limit* of a functor, which contains, as particular cases, cartesian products, equalisers (already considered in A1.4), pullbacks (see A2.3) and the classical 'projective limits'.

Let S be a small category and $X: \mathsf{S} \to \mathsf{C}$ a functor, written in 'index notation' (with $S = \mathrm{ObS}$):

$$i \mapsto X_i, \quad a \mapsto (X_a: X_i \to X_j) \qquad (i \in S;\ a: i \to j \text{ in } \mathsf{S}). \qquad (A.16)$$

A *cone* for X is an object A of C equipped with a family of maps $(f_i: A \to X_i)_{i \in S}$ in C such that the following triangles commute

$$
\begin{array}{l}
A \xrightarrow{\ f_i\ } X_i \\
\quad \searrow_{f_j} \quad \downarrow{X_a} \\
\qquad\quad X_j
\end{array}
\qquad X_a.f_i = f_j \qquad (a: i \to j \text{ in } \mathsf{S}). \qquad (A.17)
$$

The *limit* of $X: \mathsf{S} \to \mathsf{C}$ is a universal cone $(L, (u_i: L \to X_i)_{i \in S})$. This means a cone of X such that every cone $(A, (f_i: A \to X_i)_{i \in S})$ 'factorises uniquely through the former'; in other words, there is a unique map $f: A \to L$ such that, for all $i \in S$, $u_i f = f_i$.

The solution need not exist. When it does, it is determined up to a unique coherent isomorphism, and the object L is denoted as $\mathrm{Lim}(X)$.

More generally, a *weak limit* of the functor $X: \mathsf{S} \to \mathsf{C}$ is any cone $(L, (u_i: L \to X_i)_{i \in S})$ such that every cone $(A, (f_i: A \to X_i)_{i \in S})$ factorises through the former; uniqueness is *not* assumed, and there can be non-isomorphic solutions. Weak limits often 'subsist' in homotopy categories, and intervene in the construction of ordinary limits in the epi-mono completion of a category (see Sections 2.6-2.8).

Dually, the *colimit* of the functor X is a universal *cocone*. It contains, as particular cases, sums, coequalisers (already considered in A1.5), pushouts (see A2.3) and the classical 'injective limits'. Again, the existence part of the universal property defines a *weak colimit*.

A2.3 Particular cases, pullbacks and pushouts

The product ΠX_i of a family $(X_i)_{i \in S}$ of objects of C is the limit of the corresponding functor $X: \mathsf{S} \to \mathsf{C}$, defined on the *discrete* category whose objects are the elements $i \in S$ (and whose morphisms reduce to formal identities of such objects).

The equaliser in C of a pair of parallel morphisms $f, g \colon X_0 \to X_1$ is the limit of the obvious functor defined on the category $0 \rightrightarrows 1$.

The *pullback* of a pair of morphisms $f \colon X_1 \to X_0 \leftarrow X_2 \colon g$ (with the same codomain) is the limit of the obvious functor defined on the category $1 \to 0 \leftarrow 2$. This amounts to the usual definition: an object A equipped with two maps $u_i \colon A \to X_i$ $(i = 1, 2)$ which form a commutative square with f and g, in a universal way:

$$
\begin{array}{ccc}
A & \xrightarrow{\;u_1\;} & X_1 \\
{\scriptstyle u_2}\downarrow & & \downarrow{\scriptstyle f} \\
X_2 & \xrightarrow[g]{} & X_0
\end{array}
\qquad (A.18)
$$

that is, $fu_1 = gu_2$, and for every triple (B, v_1, v_2) such that $fv_1 = gv_2$, there exists a unique map $w \colon B \to A$ such that $u_1 w = v_1$, $u_2 w = v_2$. In Set, the pullback-object can be realised as

$$
A = \{(x, x') \in X_1 \times X_2 \mid f(x) = g(x')\}.
$$

Generalising this construction, it is easy to prove that a category that has binary products and equalisers also has pullbacks: A is constructed as the equaliser of the maps $fp_1, gp_2 \colon X_1 \times X_2 \to X_0$.

Notice that, when $X_1 = X_2$, the pullback of the pair (f, g) is not the same as their equaliser. If $f = g$, the pullback $R \rightrightarrows X_1$ of the diagram $X_1 \to X_0 \leftarrow X_1$ is called the *kernel pair* of f. In Set, it can be realised as $R = \{(x, x') \in X_1 \times X_1 \mid f(x) = f(x')\}$, and amounts to the equivalence relation associated to f.

Dually, the *pushout* of a pair of morphisms with the same domain is the colimit of the obvious functor defined on the category $1 \leftarrow 0 \to 2$. The existence of binary sums and coequalisers implies the existence of pushouts.

A2.4 *Complete categories and the preservation of limits*

A category C is said to be *complete* (resp. *finitely complete*) if it has a limit for every functor $\mathsf{S} \to \mathsf{C}$ defined over a *small* category (resp. a *finite* category).

One says that a functor $F \colon \mathsf{C} \to \mathsf{D}$ *preserves the limit*

$$
(L, (u_i \colon L \to X_i)_{i \in S})
$$

of a functor $X \colon \mathsf{S} \to \mathsf{C}$ if the cone

$$
(FL, (Fu_i \colon FL \to FX_i)_{i \in S})
$$

is the limit of the composed functor $FX \colon \mathsf{S} \to \mathsf{D}$. One says that F *preserves limits* if it preserves those limits *which exist* in C. Analogously for the preservation of products, equalisers, etc.

One proves, by a constructive argument, that a category is complete (resp. finitely complete) if and only if it has equalisers and products (resp. finite products). Moreover, if C is complete, a functor $F: C \to D$ preserves all limits (resp. all finite limits) if and only if it preserves equalisers and products (resp. finite products).

Dual results hold for colimits; for instance, a category is *cocomplete* (resp. *finitely cocomplete*) if and only if it has coequalisers and sums (resp. finite sums).

The categories Set, Top, Ab are complete and cocomplete; the forgetful functor Top → Set preserves limits and colimits, while Ab → Set only preserves limits.

The category associated to a preordered set X is complete if and only if the latter has all inf; since this fact is (well known to be) equivalent to the existence of all sup, X is complete if and only if it is cocomplete. In the ordered case, this amounts to a complete lattice.

A representable functor preserves all (the existing) limits.

A2.5 Limits and colimits as universal arrows

Consider the category C^S of functors $S \to C$ and their natural transformations (A1.8). The *diagonal functor*

$$D: C \to C^S, \quad (DA)_i = A, \quad (DA)_a = \mathrm{id}A \quad (i \in S, \ a \text{ in } S), \qquad (A.19)$$

sends an object A to the constant functor at A; similarly, it sends a morphism $f: A \to B$ to the natural transformation $Df: DA \to DB: S \to C$ whose components are constant at f.

Then, the limit of a functor $X: S \to C$ in C is the same as a universal arrow $(L, \varepsilon: DL \to X)$ from the functor D to the object X of C^S. Dually, the colimit of X in C is the same as a universal arrow $(L, \eta: X \to DL)$ from the object X of C^S to the functor D.

A2.6 Barr-exact and Quillen-exact categories

A *regular* category [Ba, Bo1] is a category with finite limits, where the kernel pair $R \rightrightarrows A$ of every map $f: A \to B$ (cf. A2.3) has a coequaliser; moreover, regular epis must be stable under pullback along any arrow.

In a regular category, a pair of maps $R \rightrightarrows X$ is said to be an *equivalence relation* if the associated map $R \to X \times X$ is a monomorphism and satisfies the obvious 'translation' of the usual properties: reflexivity, symmetry and transitivity.

A *Barr-exact* category [Ba, Bo1] is a regular category where every equivalence relation is *effective*, meaning that it is the kernel pair of some map; equivalently, every equivalence relation has a coequaliser, and is the kernel pair of the latter.

The following categories are Barr-exact: all the abelian ones, every elementary topos, every category monadic over Set, and in particular the category of each variety of algebras [M5, Bo1].

Finally, let us mention the fact that a *Quillen-exact* category [Qu] is a full additive subcategory E of an abelian category, closed under extensions; this means that, for every short exact sequence $A \rightarrowtail C \twoheadrightarrow B$ of the abelian category, if A and B are in E, so is C. The notion can be defined intrinsically, without reference to an abelian environment, as an additive category equipped with a class of 'short exact sequences' satisfying some axioms.

A2.7 Concrete categories

Concrete categories are extensively studied in the text of Adámek, Herrlich and Strecker [AHS]. We are interested in the notion of 'embedding', that covers subalgebras and topological subspaces, and can be of use for our categories of pairs (as defined in 2.5.1).

A category C is made *concrete* over a category X by assigning a faithful functor $U: C \to X$, called the *forgetful functor*. For the sake of simplicity, we will only consider here categories concrete over Set (but the category of topological groups should rather be viewed as concrete over Gp.)

For given objects A, B of C one says that a set-theoretical mapping $u: U(A) \to U(B)$ *is* in C if it can be lifted to a C-map $A \to B$, obviously in a unique way (because U is faithful). This agrees with the standard use of saying that a mapping between topological spaces is continuous when it can be lifted with respect to the obvious forgetful functor Top $\to$ Set.

We are interested in the following notion: a morphism $m: M \to A$ of C is called an *embedding* ([AHS], Section 8.6) if:

- the underlying mapping $U(m)$ is injective (which implies that m is mono by A1.2),

- an arbitrary Set-mapping $u: U(B) \to U(M)$ is in C if (and only if) the composite $U(m).u$ is.

Embeddings are clearly closed under composition. An *initial subobject* is an embedding that is a subobject (i.e. a selected monomorphism).

In a variety of algebras $\mathrm{Alg}(\Omega, \Sigma)$ (see A1.9), all monomorphisms are embeddings (with respect to the obvious forgetful functor) and the initial subobjects amount to subalgebras.

In Top, with the standard forgetful functor, an initial subobject amounts to an ordinary subspace (equipped with the initial topological structure). More generally, for a 'topological construct' $U: \mathsf{C} \to \mathsf{Set}$ (as defined in [AHS], Section 21.7), an initial subobject is viewed as expressing the idea of a 'subspace'.

A2.8 Concrete categories of pairs

We end with some remarks about categories of pairs for concrete categories. The proofs should be obvious, for a reader interested in this domain.

(a) Let C be a (concrete) category, equipped with a faithful functor $U: \mathsf{C} \to \mathsf{Set}$. We also suppose that:

(i) C is complete and U preserves all limits,

(ii) U *has small fibres* (i.e. for every set X the objects A of C such that $U(A) = X$ form a small set).

Then the initial subobjects of every object form a small, complete lattice; C is a normal ds-category (see 2.5.1) with respect to them, and C_2 is a homological category.

(b) For a variety of algebras $\mathsf{C} = \mathrm{Alg}(\Omega, \Sigma)$ all monomorphisms are embeddings and the hypotheses (i), (ii) are satisfied. Thus, as already said in 2.5.1, C is a normal ds-category when equipped with all subalgebras as distinguished subobjects.

(c) Let $U: \mathsf{C} \to \mathsf{Set}$ be a topological construct (as recalled above) with small fibres. Then, again, C is a normal ds-category with respect to its initial subobjects.

A3 Adjoint functors

A3.1 Main definitions

An *adjunction* $F \dashv G$, with a functor $F: \mathsf{C} \to \mathsf{D}$ *left adjoint* to a functor $G: \mathsf{D} \to \mathsf{C}$, can be equivalently presented in four main forms. (An elegant, concise proof of the equivalence can be seen in [M5]; again, one needs the axiom of choice.)

(i) We assign two functors $F: \mathsf{C} \to \mathsf{D}$ and $G: \mathsf{D} \to \mathsf{C}$ together with a family of bijections

$$\varphi_{XY}: \mathsf{D}(FX, Y) \to \mathsf{C}(X, GY) \qquad (X \text{ in } \mathsf{C}, \ Y \text{ in } \mathsf{D}),$$

which is natural in X, Y. More formally, the family (φ_{XY}) is an invertible natural transformation

$$\varphi\colon \mathsf{D}(F(-),.) \to \mathsf{C}(-, G(.))\colon \mathsf{C}^{\mathrm{op}} \times \mathsf{D} \to \mathsf{Set}.$$

(ii) We assign a functor $G\colon \mathsf{D} \to \mathsf{C}$ and, for every object X in C, a *universal arrow*

$$(F_0 X, \eta X\colon X \to GF_0 X) \quad \text{from the object } X \text{ to the functor } G.$$

(ii*) We assign a functor $F\colon \mathsf{C} \to \mathsf{D}$ and, for every object Y in D, a *universal arrow*

$$(G_0 Y, \varepsilon Y\colon FG_0 Y \to Y)) \quad \text{from the functor } F \text{ to the object } Y.$$

(iii) We assign two functors $F\colon \mathsf{C} \to \mathsf{D}$ and $G\colon \mathsf{D} \to \mathsf{C}$, together with two natural transformations

$$\eta\colon \mathrm{id}\mathsf{C} \to GF) \quad \text{(the } unit\text{)}, \quad \varepsilon\colon FG \to \mathrm{id}\mathsf{D}) \quad \text{(the } counit\text{)},$$

which satisfy the *triangular identities*: $\varepsilon F.F\eta = \mathrm{id}F$, $G\varepsilon.\eta G = \mathrm{id}G$

$$F \xrightarrow{F\eta} FGF \xrightarrow{\varepsilon F} F \qquad G \xrightarrow{\eta G} GFG \xrightarrow{G\varepsilon} G \quad . \quad (A.20)$$
$$\underset{\mathrm{id}F}{} \qquad \qquad \underset{\mathrm{id}G}{}$$

A3.2 Remarks

The previous forms have different features.

Form (i) is the classical definition of an adjunction, and is at the origin of the name, by analogy with adjoint maps of Hilbert spaces. Form (ii) is used when one starts from a 'known' functor and wants to construct its left adjoint (possibly less easy to define). Form (ii*) is dual to the previous one, and used in a dual way. The 'algebraic' form (iii) is adequate to the formal theory of adjunctions (and makes sense in an abstract 2-category, cf. A4.5).

Duality of categories interchanges left and right adjoint.

An adjoint equivalence (defined in A2.5) amounts to an adjunction where the unit and counit are invertible.

A (covariant) Galois connection is an adjunction between ordered sets, viewed as categories; see 1.1.3.

A3.3 Main properties of adjunctions

(a) *Uniqueness and existence.* Given a functor, its left adjoint (if it exists) is uniquely determined up to isomorphism. A crucial theorem for proving the existence (under suitable hypothesis) is the Adjoint Functor Theorem of P. Freyd (see [M5, Bo1]).

(b) *Composing adjoint functors.* Given two consecutive adjunctions

$$F: \mathsf{C} \rightleftarrows \mathsf{D} : G, \quad \eta: 1 \to GF, \quad \varepsilon: FG \to 1,$$
$$H: \mathsf{D} \rightleftarrows \mathsf{E} : K, \quad \rho: 1 \to KH, \quad \sigma: HK \to 1, \tag{A.21}$$

there is a composed adjunction from the first to the third category:

$$HF: \mathsf{C} \rightleftarrows \mathsf{E} : GK,$$
$$G\rho F.\eta: 1 \to GF \to GK.HF, \tag{A.22}$$
$$\sigma.H\varepsilon K: HF.GK \to HK \to 1.$$

There is thus a category Adj of small categories and adjunctions, with morphisms $(F, G; \eta, \varepsilon)$.

(c) *Adjoints and limits.* A left adjoint preserves (the existing) colimits, a right adjoint preserves (the existing) limits.

(d) *Faithful and full adjoints.* Suppose we have an adjunction $F \dashv G$, with counit $\varepsilon: FG \to 1$. Then

(i) G is faithful if and only if all the components εY of the counit are epi;

(ii) G is full if and only if all the components εY are split monos;

(iii) G is full and faithful if and only if the counit is invertible.

A3.4 Reflective and coreflective subcategories

A subcategory $\mathsf{C}' \subset \mathsf{C}$ is said to be *reflective* (notice: not 'reflexive') if the inclusion functor $U: \mathsf{C}' \to \mathsf{C}$ has a left adjoint (called *reflector*); it is said to be *coreflective* if U has a right adjoint (called *coreflector*).

For instance, Ab is reflective in Gp, while the full subcategory of Ab formed by all torsion abelian groups is coreflective in Ab.

A4 Monoidal categories and two-dimensional categories

A4.1 Monoidal categories

A *monoidal category* $(\mathsf{C}, \otimes, E)$ is a category equipped with a functor in two variables, often called a *tensor product*

$$\mathsf{C} \times \mathsf{C} \to \mathsf{C}, \quad (A, B) \mapsto A \otimes B. \tag{A.23}$$

Without entering into details, this operation is assumed to be associative up to a natural isomorphism $(A \otimes B) \otimes C \cong A \otimes (B \otimes C)$, and the object E is assumed to be an identity, up to natural isomorphisms $E \otimes A \cong A \cong A \otimes E$.

All these isomorphisms must form a 'coherent' system, which allows one to forget them and write $(A \otimes B) \otimes C = A \otimes (B \otimes C)$, $E \otimes A = A = A \otimes E$. See [M4, Ke1, EK, Ke2].

A *symmetric* monoidal category is further equipped with a symmetry isomorphism, coherent with the other ones:

$$s(X, Y) \colon X \otimes Y \to Y \otimes X. \qquad (A.24)$$

The latter cannot be omitted: notice that $s(X, X) \colon X \otimes X \to X \otimes X$ is not the identity, in general.

A category C with finite products has a symmetric monoidal structure given by the categorical product; this structure is called cartesian. The category Ab of abelian groups also has a different (more important) symmetric monoidal structure, the usual tensor product.

A4.2 Exponentiable objects and internal homs

In a *symmetric* monoidal category C, an object A is said to be *exponentiable* if the functor $- \otimes A \colon \mathsf{C} \to \mathsf{C}$ has a right adjoint, often written as $(-)^A \colon \mathsf{C} \to \mathsf{C}$ or $\mathrm{Hom}(A, -)$, and called an *internal hom*.

There is thus a family of bijections, natural in the variables X, Y

$$\varphi^A_{XY} \colon \mathsf{C}(X \otimes A, Y) \to \mathsf{C}(X, \mathrm{Hom}(A, Y)) \qquad (X, Y \text{ in } \mathsf{C}). \qquad (A.25)$$

Since adjunctions compose, all the tensor powers $A^{\otimes n}$ are also exponentiable, with

$$\mathrm{Hom}(A^{\otimes n}, -) = (\mathrm{Hom}(A, -))^n. \qquad (A.26)$$

A symmetric monoidal category is said to be *closed* if all its objects are exponentiable. The category Ab of abelian groups is symmetric monoidal closed, with respect to the usual tensor product and Hom functor. The same holds for every category of modules.

In the non-symmetric case, one should consider a left and a right hom functor, as is the case with cubical sets.

A4.3 Cartesian closed categories

A category C with finite products is said to be *cartesian closed* if all the objects are exponentiable for this structure.

Set is cartesian closed, with obvious 'internal' hom

$$\mathsf{Hom}(A, Y) = \mathsf{Set}(A, Y).$$

The category Cat of *small* categories is cartesian closed, with the internal hom $\mathsf{Cat}(\mathsf{S}, \mathsf{C}) = \mathsf{C}^{\mathsf{S}}$ described in A1.8. Every category of presheaves of sets is cartesian closed.

Ab is not cartesian closed: for every non-trivial abelian group A, the product $- \times A$ does not preserves sums and cannot have a right adjoint.

Top is not cartesian closed: for a fixed Hausdorff space A, the product $- \times A$ preserves coequalisers (if and) only if A is locally compact ([Mi], Theorem 2.1 and footnote (5)).

But, as a crucial fact for homotopy, the standard interval $[0, 1]$ is exponentiable, with all its powers. More generally, it is well known (and not difficult to prove) that every locally compact Hausdorff space A is exponentiable: $\mathsf{Hom}(A, Y)$ is then the set of maps $\mathsf{Top}(A, Y)$ endowed with the compact-open topology (for an arbitrary space Y).

A4.4 Sesquicategories

A *sesquicategory* [Sr] is a category C equipped with:

(a) for each pair of parallel morphisms $f, g\colon X \to Y$, a set of *2-cells*, or *homotopies*, $C_2(f, g)$ whose elements are written as $\varphi\colon f \to g\colon X \to Y$ (or $\varphi\colon f \to g$), so that each map f has an *identity endocell* $\mathrm{id} f\colon f \to f$;

(b) a *whisker composition*, or *reduced horizontal composition*, for maps and homotopies, written as $k{\circ}\varphi{\circ}h$ or $k\varphi h$

$$X' \xrightarrow{h} X \underset{g}{\overset{f}{\underset{\Downarrow\varphi}{\rightrightarrows}}} Y \xrightarrow{k} Y' \qquad k{\circ}\varphi{\circ}h\colon kfh \to kgh\colon X' \to Y'; \qquad (\mathrm{A}.27)$$

(c) a *concatenation*, or *vertical composition* of 2-cells $\psi.\varphi$

$$X \underset{\underset{h}{\downarrow\,\psi}}{\overset{\overset{f}{\downarrow\,\varphi}}{\longrightarrow}} Y \qquad \psi.\varphi\colon f \to h\colon X \to Y. \qquad (\mathrm{A}.28)$$

These data must satisfy the following axioms (of *associativities, identities* and *distributivity of the vertical composition*):

$$k'{\circ}(k{\circ}\varphi{\circ}h){\circ}h' = (k'k){\circ}\varphi{\circ}(hh'), \quad \chi.(\psi.\varphi) = (\chi.\psi).\varphi,$$

$$1_Y{\circ}\varphi{\circ}1_X = \varphi, \quad k{\circ}\mathrm{id}(f){\circ}h = \mathrm{id}(kfh), \quad \varphi.\mathrm{id}(f) = \varphi = \mathrm{id}(g).\varphi, \qquad (\mathrm{A}.29)$$

$$k{\circ}(\psi.\varphi){\circ}h = (k{\circ}\psi{\circ}h).(k{\circ}\varphi{\circ}h).$$

A4.5 Two-categories

A *2-category* can be defined as a sesquicategory which satisfies the following *reduced interchange property*:

$$X \; \underset{g}{\overset{f}{\rightrightarrows}} \downarrow\varphi \; Y \; \underset{k}{\overset{h}{\rightrightarrows}} \downarrow\psi \; Z \qquad (\psi\circ g).(h\circ\varphi) = (k\circ\varphi).(\psi\circ f). \qquad \text{(A.30)}$$

To recover the usual definition [Be1, KeS], one defines the *horizontal composition* of 2-cells φ, ψ which are *horizontally consecutive*, as in diagram (A.30), using the previous identity:

$$\psi\circ\varphi = (\psi\circ g).(h\circ\varphi) = (k\circ\varphi).(\psi\circ f)\colon hf \to kg\colon X \to Z. \qquad \text{(A.31)}$$

Then, one proves that the horizontal composition of 2-cells is associative, has identities (namely, the identity cells of identity arrows) and satisfies the *middle-four interchange property* with vertical composition (an extension of the previous reduced interchange property):

$$X \; \underset{\downarrow\psi}{\overset{\downarrow\varphi}{\rightrightarrows}} \; Y \; \underset{\downarrow\tau}{\overset{\downarrow\sigma}{\rightrightarrows}} \; Z \qquad (\tau.\sigma)\circ(\psi.\varphi) = (\tau\circ\psi).(\sigma\circ\varphi). \qquad \text{(A.32)}$$

As a prime example of such a structure, Cat (resp. CAT) will also denote the *2-category* of small (resp. all) $\mathcal{U}$-categories, their functors *and* their natural transformations.

We already noticed that the usual definition of a 2-category is based on the complete horizontal composition, rather than the reduced one. But in practice one usually works with the reduced horizontal composition.

Furthermore, there are important sesquicategories where the reduced interchange property does not hold (and one does not define a full horizontal composition): for instance, the sesquicategory of chain complexes, chain morphisms and homotopies, over an additive category.

Two-dimensional limits are studied in [Ke4, Sr, Gry].

Adjunctions, equivalences and adjoint equivalences can be defined *inside* every 2-category. (For an adjunction one should use the 'algebraic' form (iii), in definition A3.1.)

A more general notion of *bicategory* was introduced by Bénabou [Be1]; it is a laxified version of a 2-category, where the horizontal composition is associative and has units up to (coherent) vertical isomorphisms.

A4.6 Two-dimensional functors and universal arrows

A 2-functor $U\colon A \to X$ between 2-categories sends objects to objects, arrows to arrows and cells to cells, strictly preserving the whole structure: (co)domains, units and compositions. (Lax versions can be found in [Be1, KeS], but are not used here.)

Universal arrows of functors (cf. A2.1) have a strict and a weak extension to the 2-dimensional case, and we need both.

A *2-universal arrow* from an object X of X to the 2-functor $U \colon A \to X$ is a pair $(A_0, h \colon X \to UA_0)$ which gives an isomorphism of categories (of arrows and cells, with vertical composition):

$$A(A_0, A) \to X(X, UA), \qquad g \mapsto Ug.h. \qquad (A.33)$$

This amounts to saying that the functor (A.33) is bijective on objects, full and faithful, that is:

(i) for every A in A and every $f \colon X \to UA$ in X there exists precisely one $g \colon A_0 \to A$ in A such that $f = Ug.h$;

(ii) for every pair of arrows $g, g' \colon A_0 \to A$ in A and every cell $\varphi \colon Ug.h \to Ug'.h$ in X, there is precisely one cell $\psi \colon g \to g'$ in A such that $\varphi = U\psi.h$.

(Equivalently, one can use a global universal property: for every cell $\varphi \colon f \to f' \colon X \to UA$ in X, there is precisely one cell $\psi \colon g \to g'$ in A such that $\varphi = U\psi.h$. This implies that $f = Ug.h$ and $f' = Ug'.h$.)

More generally, a *biuniversal arrow* from X to $U \colon A \to X$ is a pair $(A_0, h \colon X \to UA_0)$ so that the functor (A.33) is an *equivalence* of categories.

This can be rephrased saying that the functor (A.33) is essentially surjective on objects, full and faithful (cf. A1.7). In other words, we replace (i) with a weaker version (and keep (ii) as it is)

(i′) for every A in A and every $f \colon X \to UA$ in X there exists *some* $g \colon A_0 \to A$ in A such that $f \cong Ug.h$ (*isomorphic objects* in the category $X(X, UA)$).

The property of surjectivity on objects, which is intermediate between (i) and (i′), is often satisfied:

(i″) there exists *some* $g \colon A_0 \to A$ such that $f = Ug.h$.

We will use this intermediate property whenever possible. For instance, it is always the case for biuniversal models of theories, as proved in Lemma 7.2.4.

Of course, the solution of a 2-universal problem is *determined up to isomorphism*, while the solution of a biuniversal one is *determined up to equivalence* (in a 2-category).

If B is 2-subcategory of a 2-category A (in the obvious sense), B is said to be bireflective (resp. bi-coreflective) in A if, for every object A in A there exists a biuniversal arrow $A \to UR^-(A)$ from A to the inclusion 2-functor

$U \colon \mathsf{B} \to \mathsf{A}$ (resp. a biuniversal arrow $UR^+(A) \to A$ from U to the object A).

A4.7 Double categories

Double categories were introduced by C. Ehresmann [Eh1, Eh3]. The weak notion, corresponding to a bicategory and called a *pseudo double category*, is studied in a series of four papers [GP1] - [GP4], by R. Paré and the author.

Here we just need the basic terminology, for the strict case. A double category $\mathbb{A}$ has horizontal morphisms $f \colon A \to B$ (with composition gf), vertical morphisms $u \colon A \to A'$ (with composition $u' \bullet u$, or $u \otimes u'$), and (double) cells α as in the left diagram below

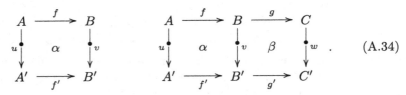

$$(A.34)$$

Cells have a horizontal composition $(\alpha | \beta)$, as in the right diagram above, that agrees with the horizontal composition of maps; notice that the vertical map v is the *horizontal* codomain of α and the horizontal domain of β (for horizontal composition). Similarly, there is a vertical composition $\alpha \otimes \gamma$, when the *vertical* codomain f' of α coincides with the vertical domain of γ.

The *boundary* of the cell α in diagram (A.34) is written as $\alpha \colon (u \, {}^{f}_{f'} \, v)$, or also as $\alpha \colon u \to v$.

The axioms essentially say that both laws are 'categorical', and satisfy the interchange law.

The notions of *double subcategory* and *double functor* are obvious. A 2-category amounts to a double category whose vertical arrows (for instance) are identities.

Here we are mostly interested in the elementary case of *flat* double categories, where each cell is determined by the four maps of its boundary.

This is the case of the double category $\mathbb{L}\mathsf{thc}$ of lattices, lattice homomorphisms and Galois connections (see 1.7.1), or its double subcategory $\mathbb{M}\mathsf{lhc}$ of modular lattices, homomorphisms and modular connections (again in 1.7.1), or the double category $\mathbb{I}\mathsf{nd}\mathsf{E}$ of inductive squares of the homological category E (see 3.2.7), or the double category $\mathbb{R}\mathsf{o}(A)$ of RO-squares over a RO-category (see I.2.4.9).

References

[AHRT] J. Adámek, H. Herrlich, J. Rosický and W. Tholen, Weak factorization systems and topological functors, Appl. Categorical Structures 10 (2002), 237-249.

[AHS] J. Adámek, H. Herrlich and G. Strecker, Abstract and concrete categories, Wiley Interscience Publ., New York 1990.

[Ba] M. Barr, Exact categories, in Lecture Notes in Math. Vol. 236, Springer, Berlin 1971, pp. 1-120.

[Bau] H.J. Baues, Algebraic homotopy, Cambridge Univ. Press, Cambridge 1989.

[Be1] J. Bénabou, Introduction to bicategories, in: Reports of the Midwest Category Seminar, Lecture Notes in Math. Vol. 47, Springer, Berlin 1967, pp. 1-77.

[Be2] J. Bénabou, Some remarks on 2-categorical algebra (Part I), Bull. Soc. Math. Belgique 41 (1989), 127-194.

[BeG] R. Betti and M. Grandis, Complete theories in 2-categories, Cah. Top. Géom. Diff. 29 (1988), 9-57.

[BKPS] G.J. Bird, G.M. Kelly, A.J. Power and R.H. Street, Flexible limits for 2-categories, J. Pure Appl. Algebra 61 (1989), 1-27.

[Bi] G. Birkhoff, Lattice theory, 3rd ed., Amer. Math. Soc. Coll. Publ. 25, Providence 1973.

[Bo1] F. Borceux, Handbook of categorical algebra 1-3, Cambridge University Press, Cambridge 1994.

[Bo2] F. Borceux, A survey of semi-abelian categories, in: Galois theory, Hopf algebras, and semiabelian categories, Fields Inst. Commun. Vol. 43, Amer. Math. Soc., Providence, RI, 2004, pp. 27-60.

[BoB] F. Borceux and D. Bourn, Mal'cev, protomodular, homological and semi-abelian categories, Kluwer Academic Publishers, Dordrecht, 2004.

[BoC] F. Borceux and M.M. Clementino, Topological protomodular algebras, Topology Appl. 153 (2006), 3085-3100.

[BoG] F. Borceux and M. Grandis, Jordan-Hölder, modularity and distributivity in non-commutative algebra, J. Pure Appl. Algebra 208 (2007), 665-689.

[Bou1] D. Bourn, Normalization equivalence, kernel equivalence and affine categories, in: Category theory (Como, 1990), Lecture Notes in Math. Vol. 1488, Springer, Berlin 1991, pp. 43-62.

[Bou2] D. Bourn, Moore normalization and Dold-Kan theorem for semi-abelian categories, in: Categories in algebra, geometry and mathematical physics, Amer. Math. Soc., Providence, RI, 2007, pp. 105-124.

[BouK] A.K. Bousfield and D.M. Kan, Homotopy limits, completions and localizations, Lecture Notes in Math. Vol. 304, Springer, Berlin 1972.

[BP] H.B. Brinkmann and D. Puppe, Abelsche und exacte Kategorien, Korrespondenzen, Lecture Notes in Math. Vol. 96, Springer, Berlin 1969.

[Br] R. Brown, Elements of modern topology, Mc Graw-Hill, New York 1968.

[Bur] M.S. Burgin, Categories with involution and correspondences in γ-categories, Trans. Moscow Math. Soc. 22 (1970), 181-257.

[Ca1] A. Carboni, Categories of affine spaces, J. Pure Appl. Algebra 61 (1989), 243-250.

[Ca2] A. Carboni, Some free constructions in realizability and proof theory, J. Pure Appl. Algebra 103 (1995), 117-148.

[CaG] A. Carboni - M. Grandis, Categories of projective spaces, J. Pure Appl. Algebra 110 (1996) 241-258.

[CaLW] A. Carboni, S. Lack and R.F.C. Walters, Introduction to extensive and distributive categories, J. Pure Appl. Algebra 84 (1993), 145-148.

[CaV] A. Carboni and E. Vitale, Regular and exact completions, J. Pure Appl. Algebra 125 (1998), 79-116.

[CE] H. Cartan and S. Eilenberg, Homological algebra, Princeton Univ. Press, Princeton 1956.

[Co] P.M. Cohen, Universal algbra, D. Reidel Publ. Co., Dordrecht 1981.

[Cp] L. Coppey, Algèbres de décompositions et précategories, Diagrammes 3 (Suppl.), 1980.

[DT] A. Dold and R. Thom, Quasifaserungen und unendliche symmetrische Produkte, Ann. of Math. 67 (1958), 239-281.

[EcH] B. Eckmann and P.J. Hilton, Unions and intersections in homotopy theory, Comm. Math. Helv. 38 (1963-64), 293-307.

[Eh1] C. Ehresmann, Catégories structurées, Ann. Sci. Ecole Norm. Sup. 80 (1963), 349-425.

[Eh2] C. Ehresmann, Cohomologie à valeurs dans une catégorie dominée, Extraits du Colloque de Topologie, Bruxelles 1964, in: C. Ehresmann, Oeuvres complètes et commentées, Partie III-2, 531-590, Amiens 1980.

[Eh3] C. Ehresmann, Catégories et structures, Dunod, Paris 1965.

[EK] S. Eilenberg and G.M. Kelly, Closed categories, in Proc. Conf. Categorical Algebra, La Jolla 1965, Springer-Verlag 1966, pp. 421-562.

[EM] S. Eilenberg and J.C. Moore, Limits and spectral sequences, Topology 1 (1961), 1-23.

[ES] S. Eilenberg and N. Steenrod, Foundations of algebraic topology, Princeton Univ. Press, Princeton 1952.

[F1] P. Freyd, Representations in abelian categories, in: Proc. Conf. Categ. Algebra, La Jolla, 1965, Springer, Berlin 1966, pp. 95-120.

[F2] P. Freyd, Stable homotopy, in: Proc. Conf. Categ. Algebra, La Jolla, 1965, Springer, Berlin 1966, pp. 121-176.

[F3] P. Freyd, Stable homotopy II, in: Proc. Symp. Pure Maths. XVII, Amer. Math. Soc., Providence RI 1970, pp. 161-183.

[F4] P. Freyd, On the concreteness of certain categories, in: Symposia Mathematica, Vol. IV (INDAM, Rome, 1968/69), Academic Press, London 1970, pp. 431-456.

[F5] P. Freyd, Homotopy is not concrete, in: The Steenrod algebra and its applications, Battelle Memorial Inst., Columbus, OH, 1970, Lecture Notes in Math., Vol. 168, Springer, Berlin, 1970, pp. 25-34.

[FK] P. Freyd and G.M. Kelly, Categories of continuous functors, J. Pure Appl. Algebra 2 (1972), 1-18.

[FS] P.J. Freyd and A. Scedrov, Categories, allegories, North-Holland Publishing Co., Amsterdam 1990.

[FuM] T. Fujiwara and K. Murata, On the Jordan-Hölder-Schreier theorem, Proc. Japan Acad. 29 (1953), 151-153.

[Ga] P. Gabriel, Des catégories abéliennes, Bull. Soc. Math. France 90 (1962), 323-348.

[GaZ] P. Gabriel and M. Zisman, Calculus of fractions and homotopy theory, Springer, Berlin 1967.

[G1] M. Grandis, Symétrisations de catégories et catégories quaternaires, Atti Accad. Naz. Lincei Mem. Cl. Sci. Fis. Mat. Natur. 14 sez. 1 (1977), 133-207.

[G2] M. Grandis, Exact categories and distributive lattices (Orthodox symmetrizations, 2), Ann. Mat. Pura Appl. 118 (1978), 325-341.

[G3] M. Grandis, Concrete representations for inverse and distributive exact categories, Rend. Accad. Naz. Sci. XL Mem. Mat. 8 (1984), 99-120.

[G4] M. Grandis, Transfer functors and projective spaces, Math. Nachr. 118 (1984), 147-165.

[G5] M. Grandis, On distributive homological algebra, I. RE-categories, Cahiers Top. Géom. Diff. 25 (1984), 259-301.

[G6] M. Grandis, On distributive homological algebra, II. Theories and models, Cahiers Top. Géom. Diff. 25 (1984), 353-379.

[G7] M. Grandis, On distributive homological algebra, III. Homological theories, Cahiers Top. Géom. Diff. 26 (1985), 169-213.

[G8] M. Grandis, Convergence in homological algebra, Rend. Accad. Naz. Sci. XL Mem. Mat. 11 (1987), 205-236.

[G9] M. Grandis, Distributive homological algebra: the universal model for the exact system and the exact couple, Boll. Un. Mat. Ital. 2-B (1988), 613-640.

[G10] M. Grandis, On homological algebra, I. Semiexact and homological categories, Dip. Mat. Univ. Genova, Preprint 186 (1991). Available in a slightly revised version:
http://www.dima.unige.it/~grandis/HA1.Pr1991(Rev.09).ps.

[G11] M. Grandis, On homological algebra, II. Homology and satellites, Dip. Mat. Univ. Genova, Preprint 187 (1991). Available in a slightly revised version:
http://www.dima.unige.it/~grandis/HA2.Pr1991(Rev.09).ps.

[G12] M. Grandis, Fractions for exact categories, Rend. Accad. Naz. Sci. XL Mem. Mat. 16 (1992), 21-38.

[G13] M. Grandis, A categorical approach to exactness in algebraic topology, in Atti del V Convegno Nazionale di Topologia, Lecce-Otranto 1990, Rend. Circ. Mat. Palermo 29 (1992), 179-213.

[G14] M. Grandis, On the categorical foundations of homological and homotopical algebra, Cahiers Top. Géom. Diff. Catég. 33 (1992), 135-175.

[G15] M. Grandis, Weak subobjects and weak limits in categories and homotopy categories, Cah. Topol. Géom. Diff. Catég. 38 (1997), 301-326.

[G16] M. Grandis, Weak subobjects and the epi-monic completion of a category, J. Pure Appl. Algebra 154 (2000), 193-212.

[G17] M. Grandis, On the monad of proper factorisation systems in categories, J. Pure Appl. Algebra 171 (2002) 17-26.

[G18] M. Grandis, Directed Algebraic Topology, Models of non-reversible worlds, Cambridge Univ. Press, Cambridge 2009. Available online:
http://www.dima.unige.it/~grandis/BkDAT_page.html.

[G19] M. Grandis, Homotopy spectral sequences, J. Homotopy Relat. Struct. 5 (2010), 213-252.

[G20] M. Grandis, Homological algebra: The interplay of homology with distributive lattices and orthodox semigroups, World Scientific Publishing Co., Singapore 2012.
[This book is referred to as Part I.]

[GP1] M. Grandis and R. Paré, Limits in double categories, Cah. Topol. Géom. Differ. Catég. 40 (1999), 162-220.

[GP2] M. Grandis and R. Paré, Adjoint for double categories, Cah. Topol. Géom. Differ. Catég. 45 (2004), 193-240.

[GP3] M. Grandis and R. Paré, Kan extensions in double categories (On weak double categories, Part III), Theory Appl. Categ. 20 (2008), No. 8, 152-185.

[GP4] M. Grandis and R. Paré, Lax Kan extensions for double categories (On weak double categories, Part IV), Cah. Topol. Géom. Differ. Catég. 48 (2007), 163-199.

[GrT] M. Grandis and W. Tholen, Natural weak factorisation systems, Archivum Mathematicum (Brno) 42 (2006), 397-408.

[GrV1] A.R. Grandjeán and L. Valcárcel, Homología en categorías exactas, Dep. de Algebra y Fundamentos, Santiago de Compostela 1970.

[GrV2] A.R. Grandjeán and L. Valcárcel, Pares exactos y sucesiones espectrales, Dep. de Algebra y Fundamentos, Santiago de Compostela 1974.

[Grz] G. Grätzer, General lattice theory, Academic Press, New York 1978.

[Gry] J.W. Gray, The existence and construction of lax limits, Cah. Topol. Géom. Differ. 21 (1980), 277-304.

[Gt] A. Grothendieck, Sur quelques points d'algèbre homologique, Tôhoku Math. J. 9 (1957), 119-221.

[HaK] K.A. Hardie and K.H. Kamps, Exact sequence interlocking and free homotopy theory, Cahiers Top. Géom. Diff. 26 (1985), 3-31.

[Har] R. Hartshorne, Residues and duality, Lecture Notes in Math. Vol. 20, Springer, Berlin 1966.

[HeS] H. Herrlich and G.E. Strecker, Category theory, an introduction, Allyn and Bacon, Boston 1973.

[Hg] P.J. Higgins, Notes on categories and groupoids, van Nostrand Reinhold, London 1971.

[Hi] P. Hilton, Correspondences and exact squares, in: Proc. of the Conf. on Categorical Algebra, La Jolla 1965, Springer, Berlin 1966, pp. 255-271.

[HiL] P.J. Hilton and W. Ledermann, On the Jordan-Hölder theorem in homological monoids, Proc. London Math. Soc. 10 (1960), 321-334.

[HiW] P.J. Hilton and S. Wylie, Homology theory, Cambridge Univ. Press, Cambridge 1962.

[JMT] G. Janelidze, L. Márki and W. Tholen, Semi-abelian categories, in: Category theory 1999 (Coimbra), J. Pure Appl. Algebra 168 (2002), pp. 367-386.

[JT] G. Janelidze and W. Tholen, Functorial factorisation, well-pointedness and separability, J. Pure Appl. Algebra 142 (1999), 99-130.

[Ke1] G.M. Kelly, Single-space axioms for homology theories, Proc. Cambridge Philos. Soc. 55 (1959), 10-22.

[Ke2] G.M. Kelly, On Mac Lane's conditions for coherence of natural associativities, commutativities, etc., J. Algebra 1 (1964), 397-402.

[Ke3] G.M. Kelly, Basic concepts of enriched category theory, Cambridge University Press, Cambridge 1982.

[Ke4] G.M. Kelly, Elementary observations on 2-categorical limits, Bull. Austral. Math. Soc. 39 (1989), 301-317.

[KeS] G.M. Kelly and R. Street, Review of the elements of 2-categories, in: Category Seminar, Sydney 1972/73, Lecture Notes in Math. Vol. 420, Springer, Berlin 1974, pp. 75-103.

[KT] M. Korostenski and W. Tholen, Factorisation systems as Eilenberg-Moore algebras, J. Pure Appl. Algebra 85 (1993), 57-72.

[Ku] A.G. Kurosh, The theory of groups, Chelsea Publ. Co., New York, 1960.

[La] R. Lavendhomme, Un plongement pleinement fidèle de la catégorie des groupes, Bull. Soc. Math. Belgique 17 (1965), 153-185.

[LaV] R. Lavendhomme and G. Van Den Bossche, Homologisation de catégories pre-régulières, Inst. de Math. Pure Appl., Univ. Catholique de Louvain 41 (1973), 1-24.

[Law] F.W. Lawvere, Adjoints in and among bicategories, in: 'Logic and Algebra', Pontignano, 1994, Marcel Dekker, New York, 1996, pp. 181-189.

[LiM] Y.T. Lisitsa and S. Mardešić, Steenrod homology, in: Geometric and algebraic topology, Banach Center Publ. 18, Warsaw 1986, p. 330-343.

[M1] S. Mac Lane, Duality for groups, Bull. Am. Math. Soc. 56 (1950), 485-516.

[M2] S. Mac Lane, An algebra of additive relations, Proc. Nat. Acad. Sci. USA 47 (1961), 1043-1051.

[M3] S. Mac Lane, Homology, Springer, Berlin 1963.

[M4] S. Mac Lane, Natural associativity and commutativity, Rice Univ. Studies 49 (1963), 28-46.

[M5] S. Mac Lane, Categories for the working mathematician, Springer, Berlin 1971.

[MaM] S. Mardešić and Z. Miminoshvili, The relative homeomorphism and wedge axiom for strong homology, Glasnik Mat. 25 (1990), 387-416.

[Mas1] W.S. Massey, Exact couples in algebraic topology, Ann. Math. 56 (1952), 363-396.

[Mas2] W.S. Massey, Some problems in algebraic topology and the theory of fibre bundles, Ann. of Math. 62 (1955), 327-359.

[Mat] M. Mather, Pull-backs in homotopy theory, Can. J. Math. 28 (1976) 225-263.

[Mi] E. Michael, Local compactness and Cartesian products of quotient maps and k-spaces, Ann. Inst. Fourier (Grenoble) 18 (1968), 281-286.

[Mi1] J. Milnor, Construction of universal bundles, I, Ann. of Math. 63 (1956), 272-284.

[Mi2] J. Milnor, On the Steenrod homology theory, Mimeographed notes, Univ. of California, Berkeley 1960.

[Mt] B. Mitchell, Theory of categories, Academic Press, New York 1965.

[Mta] K. Morita, Čech cohomology and covering dimension for topological spaces, Fund. Math. 87 (1975), 31-52.

[Ne] A. Neeman, New axioms for triangulated categories, J. of Algebra 139 (1991) 221-255.

[P1] D. Puppe, Homotopiemengen und ihre induzierten Abbildungen, I, Math. Z. 69 (1958), 299-344.

[P2] D. Puppe, Korrespondenzen in abelschen Kategorien, Math. Ann. 148 (1962), 1-30.

[P3] D. Puppe, On the formal structure of stable homotopy theory, in: Colloquium on Algebraic Topology, Mat. Inst., Aarhus Univ., 1962, pp. 65-71.

[P4] D. Puppe, Stabile Homotopietheorie I, Math. Ann. 169 (1967), 243-274.

[Qu] D. Quillen, Higher algebraic K-theory, I. in Lecture Notes in Math. Vol. 341, Springer, Berlin 1973, pp. 85-147.

[Ra] D.A. Raĭkov, Semiabelian categories, Dokl. Akad. Nauk SSSR 188 (1969), 1006-1009.

[Ri] L. Ribes, On a cohomology theory for pairs of groups, Proc. Amer. Math. Soc. 21 (1969), 230-234.

[RT] J. Rosický and W. Tholen, Lax factorization algebras, J. Pure Appl. Algebra 175 (2002), 355-382.

[Se] Z. Semadeni, Banach spaces of continuous functions, Polish Sci. Publ., Warszawa 1971.

[Ser] J.P. Serre, Classes de groupes abéliens et groupes d'homotopie, Ann. of Math. 58 (1953), 258-294.

[Sc] R. Schmidt, Subgroup lattices of groups, de Gruyter, Berlin 1994.

[Si] I. Singer, Bases in Banach spaces, II, Springer, Berlin 1984.

[St] N.E. Steenrod, Regular cycles of compact metric spaces, Ann. of Math. 41 (1940), 833-851.

[Sr] R. Street, Limits indexed by category-valued 2-functors, J. Pure Appl. Alg. 8 (1976), 149-181.

[Su] M. Suzuki, Group theory I, Springer, Berlin 1982.

[Sw] R.M. Switzer, Algebraic topology - homotopy and homology, Springer, Berlin 1975.

[Ta] S. Takasu, Relative homology and relative cohomology theory of groups, J. Fac. Sci. Univ. Tokyo, sect. I, 8 (1959-60), 75-110.

[T1] M.S. Tsalenko, Correspondences over a quasi exact category, Dokl. Akad. Nauk SSSR 155 (1964), 292-294.

[T2] M.S. Tsalenko, Correspondences over a quasi exact category, Mat. Sbornik 73 (1967), 564-584.

[Vb] G. Van Den Bossche, Catégories pre-régulières: suite longue d'homologie associée à une suite exacte courte de complexes, Inst. de Math. Pure Appl., Univ. Catholique de Louvain 68 (1977), 1-33.

[Ve] J.L. Verdier, Catégories dérivées, in: Séminaire de Géomtrie algébrique du Bois Marie SGA 41/2, Cohomologie étale, Lecture Notes in Math. Vol. 569, Springer, Berlin 1977, pp. 262-311.

[Wa] T. Watanabe, Čech homology, Steenrod homology and strong homology, I, Glasnik Mat. 22 (1987), 187-238.

[Ze] E.C. Zeeman, On the filtered differential group, Ann. Math. 66 (1957), 557-585.

Index